OPTICS

OPTICS

LECTURES ON THEORETICAL PHYSICS, VOL. IV

BY ARNOLD SOMMERFELD
UNIVERSITY OF MUNICH

TRANSLATED BY
OTTO LAPORTE AND PETER A. MOLDAUER
UNIVERSITY OF MICHIGAN

ACADEMIC PRESS
NEW YORK AND LONDON

ACADEMIC PRESS INC.
111 Fifth Avenue, New York, New York 10003

United Kingdom Edition published by
ACADEMIC PRESS INC. (LONDON) LTD.
Berkeley Square House, London, W.1

FIRST PAPERBACK EDITION, JANUARY 1964
SECOND PRINTING, AUGUST 1964

PRINTED IN THE UNITED STATES OF AMERICA

PREFACE

This volume is closely connected with "Electrodynamics," Vol. III of my lectures. Not only the formalism of Maxwell's equations but also their intrinsic character, the invariance with respect to the group of Lorentz transformations, is adopted from Vol. III and is assumed to be known.

Chapter I is entitled "Reflection and Refraction of Light." Only the (never realizable) ideal case of the monochromatic plane wave which is necessarily completely (and, in general, elliptically) polarized is treated in this chapter. Reflection and refraction are regarded throughout as boundary value problems associated with a single boundary surface or (in the case of the plate) with two boundary surfaces. It is surprising how much material falls in this category: It extends from the classical Fresnel formulae to the very timely problem of the tunnel effect and covers non-reflecting lenses, the Perot-Fabry etalon, and the (no longer timely) problem of the "black submarine." The fundamental question of the "coherence or non-coherence of light" is touched upon briefly only in fig. 2 of this chapter. Not until the last chapter, Sec. 49, will we return to the problem of characterizing white light.

Chapter II deals at once with the optics of moving media. Indeed, these questions seem to me to be basically simpler and more fundamental than the contents of the later chapters because one is dealing here with the universal character of the velocity of light and with its physical and astronomical consequences. The first doubts about the classical wave nature of light appear at the end of this chapter in connection with the Doppler and photoelectric effects, and the equivalent corpuscular nature of light makes its first appearance.

Chapter III deals with the theory of dispersion from Drude's semi-phenomenological point of view, which is based on the classically formulated resonance oscillations of electrons bound to atoms. However, it seemed to me unavoidable to add to this chapter a section in which the theory of dispersion is treated *wave-mechanically*, that is, where the characteristic oscillations are replaced by transitions between two different energy levels.

Chapter IV is dedicated to crystal optics, the favorite subject of physics in the last century. Here again the treatment is phenomenological even in the problem of the rotation of the plane of polarization in acentric crystals, which turns out to be particularly simple thanks to our use of the complex notation.

Chapter V and most of Chapter VI are devoted to the problem of diffraction. Diffraction by gratings (including three-dimensional ones) is treated first. Then follows Huygens' principle for scalar diffraction problems, which is applied to the question of "light and shadow" with its manifold paradoxical contradictions of geometrical optics. Chapter V closes with a presentation of the rigorously solvable boundary value problem of the perfectly reflecting straight edge.

Chapter VI begins with the problem of the narrow slit, which Lord Rayleigh solved in the first approximation more than fifty years ago. The problem leads to an integral equation from which higher approximations can be derived if proper use is made of the insight gained in the problem of the straight edge by regarding the behavior of the branched solutions at the edge of the screen. In the succeeding paragraphs a more or less new comprehensive point of view is applied to the question of the resolving powers of spectral apparatus (including Michelson's mirrors for the measurement of the diameters of fixed stars). Thomas Young's theory of diffraction in the formulation given to it by Rubinowicz, and Debye's formulation of focal point diffraction are presented next. Finally, the difference between the scalar and the vector diffraction problems is emphasized and the vectorial generalization of Huygens' principle is discussed. This latter discussion follows the most recent and particularly lucid treatment of the problem by W. Franz.

The presentation of the Cerenkov electron in Sec. 47 reaches beyond the limits of the conventional conception of optics and enters, so to speak, the realm of velocities greater than that of light. Section 48 deals with the (so far almost entirely neglected) *geometrical optics*. The introduction of the eikonal (and the unit vector associated with it) enables us to give a very brief presentation of several of the fundamental problems of geometrical optics. The very large field of *physiological optics*, on the other hand, could only be touched on in the introduction even though it is of primary importance with regard to our actual experience.

The last section is concerned with the nature of *white light* which possesses not a trace of periodicity and attains its wave character only upon passing through a spectral apparatus. The wave representation which appears here only as a secondary attribute of light is missing entirely in geometrical optics and is replaced by a corpuscular conception in Fermat's principle. The corpuscular concept points the way to the modern *theory of photons* and the *complementarity* of wave and corpuscle which was already stated at the end of the second chapter. Finally, it is impressed upon the reader that our presentation, which is essentially based upon the classical wave concept, forms only a part of the entire field of optics; in particular it does not encompass the primary processes in the retina because these are photoelectric in nature and therefore their discussion must be based on the theory of photons and not on the wave theory.

The text of this volume is based upon a careful record of my lectures on optics made by L. Waldmann in 1934. However, the last few subjects discussed go considerably beyond the contents of my lectures at that time.

As in the case of Vol. III, I enjoyed the invaluable cooperation of Mr. J. Jaumann in the preparation of this volume. In our many discussions he not only communicated to me his rich experience in experimental optics, but in many instances he prepared the first drafts for the manuscript. I mention, in particular, Sections 3 C, 6 C, 7 C, 30 C, 41, and 42. His part in the writing of this book should not be underestimated. My colleague, Dr. O. Buhl, subjected the entire manuscript to his critical inspection and has helped me with many useful remarks. Dr. P. Mann has kindly checked the exercises.

Munich, end of 1949.

Arnold Sommerfeld

Translators' Note

The translators of this volume have endeavored to adhere to the spirit of the original as much as possible and to keep changes to a minimum. In addition to certain changes in notation, some modifications of the text have proved to be inevitable. These are especially contained in Sections 27 and 28 which were kindly contributed by Professor P. P. Ewald. Furthermore, Sec. 47 should be read in the light of a recent paper by H. Motz and L. I. Schiff, Am. J. Phys. **21**, 258, 1953. A completely new author and subject index was prepared.

O. L.

P. A. M.

CONTENTS

INTRODUCTION

1. Geometrical, Physical, and Physiological Optics. Historical Chart

The eye is our noblest sense organ. It is therefore not surprising that even the natural philosophers of antiquity were concerned with the science of light. Leonardo da Vinci called optics "the paradise of mathematicians". Of course, by optics he meant only *geometrical* or *ray optics*, the theory of perspective and the distribution of light and shadow. How much more justified would his assertion have been had he known the *wave optics* with its marvelous color phenomena arising from diffracted light or the polarized light of crystals. It is in particular these latter phenomena which one has in mind when one speaks of *physical optics*. Physical optics is related to ray optics in the same way in which wave mechanics is related to classical mechanics. This fact was recognized by Schrödinger on the basis of the profound work of Hamilton.

There is, however, still a third branch of optics which is called *physiological optics* after the title of Helmholtz's principal work. Also in this field fundamental laws hold which, however, are based on the operation of the sense organs and the mind. But these laws are not encompassed by our physical theory. It was the tragedy in the life of Goethe that he would not recognize the distinction between physical and physiological optics; this was the reason for his fruitless fight against Newton. Today we understand without difficulty that the sensation yellow which is caused by the D-lines of sodium is a phenomenon which is entirely different from the wavelengths $\lambda = 5890$ Å and $\lambda = 5896$ Å by which we must describe these lines physically. For, we know that the psychological response to an event is something entirely different from the physical event itself; the two are different in nature and incommensurable.

In this volume we shall be able to deal only briefly with ray optics and unfortunately not at all with physiological optics. Wave optics, which we shall develop directly from the results of Vol. III and which, through spectroscopy, opens the way to modern atomic physics, will give us enough to do. We shall not, for instance, enter upon the interesting field of the theory of color which was formulated in a classical manner by Thomas Young and Helmholtz, was further developed particularly by Grassmann, Maxwell and

Schrödinger, and is even today not a closed subject. We shall here only demonstrate very briefly that, quite aside from the quality of colors and their contrast effects, there exists a profound difference between subjective perception and objective fact even in regard to the quantitative determination of intensity. The phenomenon in question is that of the so-called "half-shadow".

This phenomenon played a role in the earliest attempts to determine the wavelength of X-rays. On X-ray plates there appear half-shadow regions between the complete shadow and the region of full illumination. These are due to secondary X-rays which originate, for instance, at the edges of a slit. To the eye these half-shadow regions appear as bright and dark fringes which were at first interpreted as interference lines. However, Haga and Wind were able to show that these fringes were subjective in origin and they called attention to a phenomenon which had been investigated by E. Mach[1] and had also been recognized by H. Seeliger in his studies of eclipses of the moon. We shall describe it here as our sole example of physiological-optical phenomena.

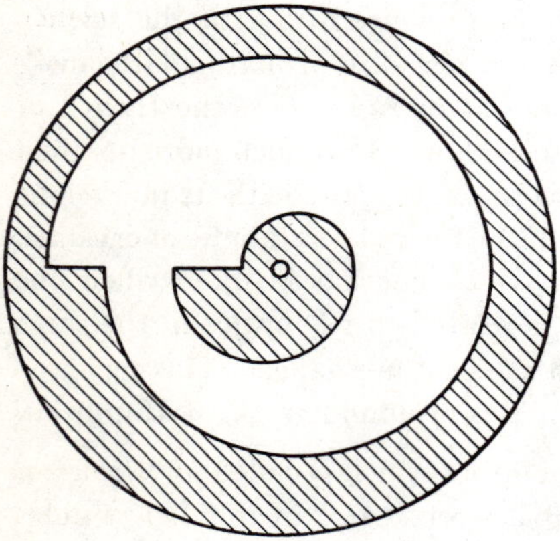

Fig. 1.

Rotating disc for the demonstration of a physiological optical illusion.

Consider a white circular cardboard disc which is partially blackened as shown in fig. 1. The boundary between the black and white fields consists of two spirals of Archimedes and portions of a radius of the disc. Let us consider the *average* brightness (or blackness) along each circle concentric with the edge of the disc; this quantity determines, in accordance with a law due to Talbot, the perception of brightness when the disc is rotated sufficiently fast. The center of the disc is then perfectly black and so is its edge. Between the center and the edge there is a zone of maximum brightness. The transition between darkness and brightness consists of two half-shadow regions. Since the radius vectors of the spirals of Archimedes increase (or decrease) linearly with the central angle, the intensity in the half-shadow region also increases (or decreases) linearly with the distance from the center of the disc. If the disc is set into rapid rotation on the axis of a motor, then

[1] See, for instance, his book Prinzipien der Physikalischen Optik, p. 158, J. A. Barth, publ. 1921.

the intensity distribution presented to the eye is that represented by the dotted line in fig. 1a. But what does the eye see? Instead of the linearly varying half-shadows the eye perceives a uniform *average* brightness; where the half-shadow borders on the completely black regions it perceives *dark* fringes which are considerably blacker than the regions of complete blackness; at the limits of full brightness it sees *bright* fringes which appear much brighter than the region of full brightness. The eye (or the mind?) is, as it were, startled by the transition from the half-shadow to full illumination; it exaggerates the contrast. The same exaggeration takes place at the transition from the half-shadow to complete blackness. The eye (or the mind) judges only contrasts and not objective intensity values; it is affected more by the derivatives of the intensity curve than by the absolute values of its ordinate. The bright and dark fringes (which on the rotating disc are, of course, circles about the center) are so definitely pronounced that a naive observer would swear to their genuineness.

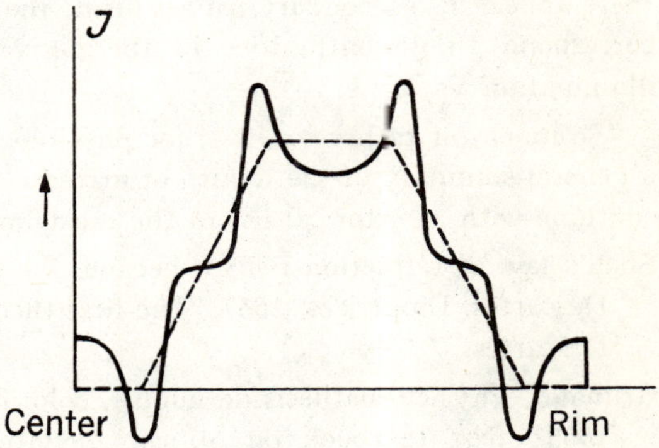

Fig. 1a.

The subjective intensity distribution perceived by the eye (full line) and the objective distribution of intensity (dotted line) when the disc is rotated.

Similar fringes are seen wherever extended light sources produce half-shadows according to geometrical optics, as for instance, behind a pencil which is illuminated by a Welsbach mantle. Also the bright border which one sees about ones own shadow on the road when the sun is behind ones back and which has the effect of a sort of halo about the head and limbs is at least partly due to this optical illusion. Such fringes also played a part in certain occasional arguments between the author and a group of Munich painters which revolved around the old controversy "Goethe vs. Newton". The opponents in these discussions understandably enough considered these subjective phenomena as objective and offered them as proof of the falseness of the physical theories.

It might be thought that this illusion could not be photographed and would thereby betray its subjective character. This is not so. Even though the number of blackened grains on the photographic plate corresponds to the correct intensity, the eye interprets the photographic image in the same way as the original object and is deceived by its subjective contrast perceptions. This is illustrated by the following experiment[1]: a micrometer slit illuminated from the rear with parallel light is photographed. At the beginning of the exposure the slit may have a width $2b$. It is then slowly and uniformly opened to a width $2a$, whereupon the exposure is terminated. Thus the center portion $2b$ of the photographic plate is continuously illuminated during the exposure; the adjoining portions $a - b$ are illuminated a shorter time which decreases linearly to zero. On the photograph one sees again bright and dark fringes at the limits of the half-shadows (if the slit is opened non-uniformly, there appear also secondary fringes inside the half-shadow regions $a - b$ which correspond to discontinuities in the derivative of the curve depicting illumination vs. time).

So much (or rather so little) for physiological optics. In order to provide a general summary of the wealth of material to be covered in this volume we continue with a historical list of the most important optical discoveries.

Snell's law of refraction (which became known only through Huygens) and Descartes, Dioptrices, 1637. The first theory of the rainbow is also due to Descartes.

Grimaldi, Physico-mathesis de lumine, coloribus et iride, Bononiae (Bologna) 1655; first textbook on optics; deviations from rectilinear ray paths; diffraction.

Olaf Roemer, 1675; determination of the velocity of light from the eclipses of the satellites of Jupiter.

Christian Huygens, Traité de la Lumière, Leiden 1690; wave theory without closer investigation of the nature of the oscillations (whether longitudinal or transverse). Huygens' principle; wave surfaces. Double refraction in calcite.

Newton, Opticks 1706, English 1675. Colors of thin plates. Spectral colors and their composition into white light. Theory of emission with lateral "fits".

Bradley, 1728; aberration of light.

Thomas Young, Lectures on Natural Philosophy, 1807. Interference of light; diffraction; theory of color; the color triangle; Young also deciphered hieroglyphics.

Malus, Sur une propriété des forces répulsives qui agissent sur la lumière, 1809. Polarization by reflection.

Biot, Brewster, Arago, crystal physics, Arago, 1811, rotatory power of quartz.

[1] J. Drecker, Physikal. ZS. **2**, 145, 1900.

Fraunhofer, 1787 – 1826, Fresnel, 1788 – 1827, the two classics of wave optics, both of whom died young after lives filled with work, success and fame. Fraunhofer was the greatest glass technician and telescope constructor of his time; he made the first diffraction gratings and was the father of spectroscopy and astrophysics because of his discovery of the Fraunhofer lines in the spectrum of the sun and the planets. Fresnel developed the wave theory; his *dragging coefficient* was a forerunner of the theory of relativity; he was an untiring experimenter in crystal optics. Fraunhofer and Fresnel diffraction.

Bessel, in 1838, measured the first fixed star parallaxes in the constellation Cygnus by means of a Fraunhofer telescope.

Christian Doppler, 1842, "Über das farbige Licht der Doppelsterne und einiger anderer Gestirne des Himmels". (On the Colored Light of Double Stars and Several Other Stars.)

Terrestial determination of the velocity of light, Fizeau 1849 by means of a toothed wheel; Foucault 1850 by means of a rotating mirror; also Michelson, beginning in 1926.

Faraday, 1845, "On the Magnetization of Light, and the Illumination of Magnetic Lines of Force".

Maxwell, 1861, discovery of the electro-magnetic theory of light; Treatise, 1873.

The experiment planned by Maxwell to determine by interferometry a possible dependence of the velocity of light on the azimuth of the earth in its orbit about the sun was carried out by Michelson in 1881, improved by Michelson and Morley in 1887, and repeated with greatest accuracy by Joos in 1930 at the Zeiss works, Jena.

The theory of dispersion based on an elastic theory was developed by Ketteler and Sellmeier. The electromagnetic theory of dispersion was started by Helmholtz and was completed by Drude on the basis of the theory of electrons; Drude, Lehrbuch der Optik, 1900. 1926, wave-mechanical theory of dispersion by Schrödinger.

Abbe, 1840 – 1905, diffraction theory of optical images; also simultaneous work by Helmholtz and Lord Rayleigh.

Standing light waves, O. Wiener, 1890, on their basis Lippmann's color photography.

Lord Rayleigh, 1842 – 1919, explanation of the blue color of the sky; introduction of group velocity into optics; resolving power of the prism. Conception of natural white light as a completely random, non-periodic process.

Zeeman effect, 1896; explanation of the normal Zeeman effect by H. A. Lorentz.

Einstein in 1905 deduced from the quantum theory the notion of photons (light quanta).

CHAPTER I

REFLECTION AND REFRACTION OF LIGHT

2. Review of Electrodynamics. Basic Principles of Ideal and Natural Light

In Vol. II, Sec. 45 we showed that at the interface between two optically different media the elastic theory of light provides more boundary conditions than are consistent with the known facts of polarization, i. e. the transverse character, of light. Hence we turn our attention to the electromagnetic theory of light, which, in contrast to the elastic theory, specifies only two boundary conditions for the electric field strength $\mathbf{E}$ and two for the associated magnetic "disturbance" $\mathbf{H}$, namely, the equality of the components *tangential* to the boundary surface.

We shall assume the light to be monochromatic and will hence perform our calculations using a *single frequency* ω. This assumption involves a far-reaching idealization of true conditions and has the following meaning: we consider the light to be spectrally decomposed and use only an infinitesimally small portion of the spectrum as a light source. The spectroscopic apparatus used for this purpose is called a *monochromator*. Furthermore, we do not consider any arbitrary bundle of light rays but, again by a far-reaching idealization of true conditions, we consider the mathematically much simpler case of a plane wave with a well-defined direction of propagation. This means that we use a *collimator* (a tube with a convex lens in whose focal plane is a slit) by means of which we can obtain a system of parallel rays of a certain width from an originally diverging bundle of light rays.

Natural light does not possess either of these two properties. This holds even for sunlight which is completely irregular as regards frequency, and also lacks sufficient parallelism because of the finite size of the sun's disc.

We will first consider ideal light which has passed through an ideal monochromator and collimator. Later, we will discuss the characteristics of natural light.

We choose the x-axis in the direction of propagation of our plane wave, denote its electromagnetic field by the two vectors $\mathbf{E}$ and $\mathbf{H}$, and represent these by the real parts of the following expressions in which we suppress[1] the symbol Re which is to be thought of as attached to the right-hand sides:

$$\mathbf{E} = \mathbf{A}\, e^{i(kx-\omega t)}, \qquad \mathbf{H} = \mathbf{A}'\, e^{i(kx-\omega t)} \tag{1}$$

k is called the wave number of the light. $\mathbf{A}$ is a constant which is independent of x and t but which has different and generally complex values for the different components of $\mathbf{E}$. $\mathbf{A}'$ is determined by $\mathbf{A}$. Instead of $\mathbf{H}$ we could, of course, have used $\mathbf{B}$ to represent the wave. We prefer $\mathbf{H}$ mainly because the conditions at the boundary between two optically different media require the continuity of $\mathbf{H}$ as well as of $\mathbf{E}$; and partly also because the radiation vector $\mathbf{S}$ which is fundamental in optics is given by

$$\mathbf{S} = \mathbf{E} \times \mathbf{H} \tag{1 a}$$

(this contains no additional factors, thanks to our M K S Q system of units which was introduced in Vol. III and upon which also this volume will be based). Furthermore, we are thus in better agreement with the usual literature in which $\mathbf{H}$ (given the designation "field strength", however) is almost always used in conjunction with $\mathbf{E}$.

In an isotropic medium which is free of charges electromagnetism requires the transverse character of light because of the condition div $\mathbf{E} = 0$ which, see Vol. III, Sec. 6, leads to $E_x = 0$ for the case of our plane wave given by (1). Hence, there exist only the two components E_y and E_z. The same is true of $\mathbf{H}$. k is connected with ω by the equation (see Vol. III, Sec. 6)

$$k = \sqrt{\varepsilon\mu}\,\omega \qquad \begin{cases} \varepsilon = \text{dielectric constant} \\ \mu = \text{permeability.} \end{cases} \tag{2}$$

Dimensionally, $\sqrt{\varepsilon\mu}$ is an inverse velocity. We shall call it $1/u$ where u is the phase velocity of propagation in the particular medium. This follows from the exponents in expression (1), which, on differentiation with respect to t and setting equal to zero, give:

$$k\frac{dx}{dt} - \omega = 0.$$

In a vacuum with $\varepsilon = \varepsilon_0$, $\mu = \mu_0$ we have

$$u = c, \qquad c = \frac{1}{\sqrt{\varepsilon_0\mu_0}} \sim 3\,.\,10^8 \text{ meters/second.} \tag{3}$$

---

[1] In exercise I. 1. we will illustrate by means of a very simple example the advantage of computing wave problems using complex exponential functions rather than using their real parts.

The connection between H and E or, what is the same, between the constants A' and A occurring in (1) follows from Maxwell's equations for nonconducting media

$$\mu \frac{\partial \mathsf{H}}{\partial t} = - \text{ curl } \mathsf{E}, \qquad \varepsilon \frac{\partial \mathsf{E}}{\partial t} = \text{curl } \mathsf{H}. \tag{4}$$

When specialized to the case of our plane wave, the first of these expressions gives, since $E_x = 0$, $H_x = 0$,

$$-i \omega \mu A_y' = i k A_z, \qquad -i \omega \mu A_z' = -i k A_y$$

and hence

$$A_y' = -\frac{k}{\mu \omega} A_z = -\sqrt{\frac{\varepsilon}{\mu}} A_z, \qquad A_z' = \frac{k}{\mu \omega} A_y = \sqrt{\frac{\varepsilon}{\mu}} A_y. \tag{5}$$

These expressions also follow, of course, from the second eq. (4).

Depending on the choice of the constants A_y, A_z, there results as a *necessary consequence of Maxwell's equations* a uniquely determined state of oscillation or, as we can also say, a uniquely determined *polarization* of our monochromatic plane wave.

In exercise I.2. we shall observe that in general eq. (1) represents *elliptic polarization*. This means that if the electric vector E is drawn so as to originate from the point $y = 0, z = 0$, its tip describes during the time $\tau = \dfrac{2\pi}{\omega}$ an ellipse with its principal axes in the yz-plane. The same is true of the vector H. It need hardly be said that, according to our conception of the Maxwell theory, nothing material oscillates in this process, no motion of the "light ether" takes place. We will see later how this ideal case of elliptic polarization can be realized practically to a good degree of approximation. (Total reflection, reflection on metals, crystal optics.)

An important special case of elliptic polarization is *circular polarization* which results when

$$|A_y| = |A_z|, \qquad \frac{A_z}{A_y} = \pm i. \tag{6}$$

Linear polarization is characterized by

$$\frac{A_z}{A_y} \text{ real (positive or negative)}, \tag{6 a}$$

in particular, of course, by the special cases $A_z = 0$ or $A_y = 0$.

In the following discussion we shall have occasion to speak not only of a monochromator and collimator but also of a polarizer (Nicol prism, quarter wave plate, etc.). According to the above discussion, such a polarizer does not actually produce polarized light but rather serves to *transform* the light from one type of polarization to another. The presence of general elliptic polarization is already guaranteed by the monochromator and collimator and by Maxwell's equations, paradoxical as this may sound to the experimentalist in practical optics. Theoretically it is correct to say that even after using an ideal monochromator and collimator, the four parameters contained in the arbitrary constants A_y and A_z (two amplitudes $|A_y|$ and $|A_z|$ and two phase constants α_y and α_z) remain undetermined; only through the use of a polarizer are the values of these parameters restricted.

Of course, an actual monochromator or collimator never functions ideally. Consequently, we are, in reality, never faced with ideal light but rather with a continuous superposition of ideal cases in which, at best, a certain region of frequency or a certain direction of propagation is strongly favored. Even light which has passed through the monochromator is, thus, not strictly monochromatic but has a certain spectral width even when it is reduced to a single spectral line. By the same token, the light emerging from the collimator is actually characterized by an integral over a region of spatial directions, though one of these directions contributes most strongly. Likewise, the various polarization devices also are endowed with a certain indefiniteness.

For *natural* light the region of integration is in every respect unlimited; it comprises all frequencies $0 < \omega < \infty$ and in the diffuse case all possible directions of incidence. Moreover, for natural light no one direction of polarization takes precedence over all others. Planck had to analyze the consequences of this situation in order to set up his thermodynamic law of radiation and this law, in turn, led to the discovery of the quantum theory. He represented the radiation of a black body not by eq. (1), but for every ever-so-small frequency range $\Delta\,\omega$ by an infinite sum of terms of the form (1) in which the constants A, even for neighboring ω's change their absolute values and phases arbitrarily. This formulation is due to the fact that the elementary processes of black-body radiation originate in single atoms which emit independently of one another. Only the quantity $\sqrt{A_y{}^2 + A_z{}^2}$ has a given value which is determined by the mean energy of all elementary processes. The A_y, A_z taken singly, however, remain completely undetermined especially as regards their phases. All this is also true for what we call "natural light". To be sure, even the naked eye exercises a certain amount of selection among the infinitely numerous components of the natural light. By fixating a certain point the eye acts as a collimator since, like the latter, it is equipped with a lens. Furthermore, by its spectrally selective sensitivity and its perception of color, the eye limits the frequency region.

The more the preference for a certain frequency and direction is emphasized in any single case, the larger is the *region of space and time in which the plane wave represents a sufficiently good approximation of the natural light field.* Outside of this region the phases are subject to statistically distributed variations in space and time. Fig. 2. attempts to illustrate this by means of an example.

This figure represents the superposition of six plane monochromatic waves; that is, the superposition of the six real parts of exponential functions of the form (1); their frequencies ω are in the ratios

$$95 : 97 : 99 : 101 : 103 : 105;$$

their directions of propagation k deviate in pairs by $\pm$ 1/20 radian from the direction of the middle pair; they are denoted successively by + 1/20, + 1/20, 0, 0, − 1/20, − 1/20.

The system of solid lines which are mostly straight shows the instantaneous nodes of the resulting state of oscillation. Between these nodes lie alternately wave crests and troughs whose amplitudes are denoted by the broken lines in a manner similar to the contour lines on a map. Successive contour lines have a difference in amplitude of 1. Except for the numbers 0, only the height of the tallest amplitude crest is given in each case. One sees that the regular sequence of waves is interrupted only at points where the amplitude vanishes; at such points a wave of new period infiltrates, so to speak. From this it follows that behind such a point the wavefronts overtake, or fall behind, the progress of the undisturbed waves corresponding to a locally closer succession of waves, or, effectively, to an increase in the frequency by 1. One must imagine that this whole wave picture propagates with the velocity of light in the direction of the arrows while its shape changes gradually.

Visual inspection shows that there are finite regions, several wavelengths in extent, which have, to a sufficient degree of approximation, the character of a homogeneous plane wave and also retain this character as the wave propagates. Hence, these regions indeed satisfy the above postulated conditions for "regions of good approximation". Only the zero points seem to be exceptions. However, just because the amplitude vanishes there, they do not produce any stronger effect than other points of varying intensity.

In order to verify experimentally any results calculated by means of plane monochromatic waves, one must make certain that the entire object under observation is contained in such a region of good approximation during the entire time of observation. If this condition is satisfied, one speaks of the production of *coherent light*. To be more precise, one should speak of the

Fig. 2. Instantaneous picture of a "wave-packet" composed of 6 simple waves: — lines of constant phase, --- lines of constant amplitude.

production of a sufficiently large *region of coherence* in space-time because even a natural light field always has small regions of coherence.

The size of the region that is sufficient in each particular case, and hence the demands which are to be made upon the monochromatic quality and the directional uniformity of the rays, depend upon the relative sizes of the object and the wavelength. Anticipating a later discussion, we shall give a few examples.

No optical instruments are required in order to see the interplay of colors resulting from colloidal particles or in a lunar corona.

For the observation of the colors of *thin plates* a parallel direction of the incident light is of almost no importance. Also, in this case, the spectral selectivity of the eye suffices to limit the frequency range.

From *thick plates,* however, one obtains interference fringes only if sharp spectral lines and well-defined rays are used; a telescope is necessary to separate the fringes.

One cannot observe diffraction at a slit in sunlight unless the slit is extremely narrow. A collimator or auxiliary slit is necessary. Only rays originating from one and the same region of coherence can interfere with one another. For such rays the *light vectors* add. Radiation fields from more distant regions combine energetically in a time average sense. For these the intensities add.

Before we turn to the actual subject of this chapter, namely reflection and refraction, we shall review briefly the units used above as well as some of the quantities from Vol. III which we will use later on.

Using the four units M (meter), K (kilogram of mass), S (second), Q (electric charge)[1], one obtains

Unit of force: 1 newton $= \text{MKS}^{-2} = 10^5$ dyne
Unit of energy: 1 joule $= 10^7$ erg
Unit of power: 1 joule $\text{S}^{-1} = 1$ watt

Electric field strength:

$$\frac{\text{Force}}{\text{Charge}} = \frac{\text{newton}}{\text{Q}} = \frac{\text{volt}}{\text{M}}, \quad 1\text{V} = 1\,\frac{\text{joule}}{\text{Q}}$$

Current strength: $\text{QS}^{-1} = \text{amp.}$

Current density:

$$J = \frac{\text{amp.}}{\text{M}^2} = \text{QM}^{-2}\,\text{S}^{-1}$$

[1] For numerical calculations one uses the coulomb as a unit for Q, i. e. the charge one ampere second.

The "displacement current" $\dot{\mathbf{D}}$ has the same dimension. Hence the dimension of Maxwell's "electric displacement" $\mathbf{D}$ which is our "electric excitation" becomes QM^{-2}. Because $\mathbf{D} = \varepsilon\, \mathbf{E}$, the dimension of the dielectric constant becomes $\frac{Q^2M^{-2}}{\text{newton}} = M^{-1}\, S\Omega^{-1}$, $1\,\Omega = 1$ joule $S/Q^2 = 1$ ohm.

Magnetic moment = Pole strength × Lever arm = Pl Following Wilhelm Weber this is defined as Current × Area. Hence,

Pole strength $P = QMS^{-1}$ = "moving charge" in the sense of Ampere.

Magnetic Induction (really "field strength") $\mathbf{B}$ = newton/P = newton $SQ^{-1}M^{-1}$ = volt S/M^2

Magnetic Excitation $\mathbf{H} = P/M^2 = QM^{-1}S^{-1}$ = amp/M

Permeability $\mu = \frac{B}{H}$ = newton $S^2Q^{-2} = M^{-1}S\Omega$

$\varepsilon\mu = M^{-2}S^2 = (\text{velocity})^{-2}$

$\mu/\varepsilon = \Omega^2$, $\sqrt{\mu_0/\varepsilon_0}$ = wave impedance of the vacuum

Convention for the purpose of eliminating 4π from the formulae: $\mu_0 = 4\pi \cdot 10^{-7}\, M^{-1}S\Omega$. From this it follows that: $4\pi c^2 \varepsilon_0 = 10^7\, MS^{-1}\Omega^{-1}$

Radiation vector $\mathbf{S} = \mathbf{E} \times \mathbf{H}$ = joule $M^{-2}S^{-1}$

We shall not introduce here the magnetic pole strength P as a fifth unit (see Vol. III 8 B).

3. Fresnel's Formulae. Transitions from Rarer to Denser Media

We can consider the boundary between two optically different media to be plane if we limit our consideration to a small portion of that boundary (for example, a piece several hundred wavelengths in size). We call this boundary the plane $y = 0$ of a right-handed cartesian coordinate system. Let a linearly polarized plane wave coming from the half-space $y > 0$ impinge upon this boundary. The plane of incidence shall be the plane of the paper in figures 3 a, b, that is, the yx-plane. Let the direction of propagation of the wave form the angle α with the negative y-axis. For the solution of the "boundary value problem" under consideration we will introduce, besides the refracted wave in the second medium (angle β with the negative y-axis), a reflected plane wave whose angle with respect to the positive y-axis shall, for the present, be termed α'. We denote the electric amplitudes of the three waves by A, B, C, where A belongs to the incident ray $\mathbf{S}_i$, B to the refracted or "transmitted" ray $\mathbf{S}_d$, C to the reflected ray $\mathbf{S}_r$. We first consider the case:

A. ELECTRIC VECTOR PERPENDICULAR TO PLANE OF INCIDENCE.

Besides omitting the symbol for the real part in the relations (2.1) we shall also omit the time factor $\exp(-i\,\omega\,t)$, which is the same for all three waves. **E** is everywhere in the direction of the z-axis. We now introduce, temporarily, polar coordinates r, φ in the xy-plane. For the direction of propagation occurring in (2.1) and there denoted by x, we substitute $x = r\cos\varphi$ or, more generally, we take account of the differences between the three directions of propagation by writing $x = r\cos(\varphi-\gamma)$. According to fig. 3 a we must then use the following values for γ:

for the incident wave $-\frac{1}{2}\pi+\alpha$, $\qquad r\cos(\varphi-\gamma) = x\sin\alpha - y\cos\alpha$,

for the refracted wave $-\frac{1}{2}\pi+\beta$, $\qquad r\cos(\varphi-\gamma) = x\sin\beta - y\cos\beta$,

for the reflected wave $+\frac{1}{2}\pi-\alpha'$, $\qquad r\cos(\varphi-\gamma) = x\sin\alpha' + y\cos\alpha'$,

where the x, y in the last column indicate, in contrast to the relation (2.1), the coordinates defined in the figure. The resulting superposition of the incident and reflected wave in medium 1 and the resulting refracted wave in medium 2 are given by

$$E_z = A\,e^{i k_1 (x\sin\alpha - y\cos\alpha)} + C\,e^{i k_1 (x\sin\alpha' + y\cos\alpha')} \tag{1}$$

and

$$E_z = B\,e^{i k_2 (x\sin\beta - y\cos\beta)} \tag{1 a}$$

respectively. The boundary condition at $y = 0$ demands

$$A\,e^{i k_1 x\sin\alpha} + C\,e^{i k_1 x\sin\alpha'} = B\,e^{i k_2 x\sin\beta}. \tag{2}$$

Because of its dependence on x, this condition can be fulfilled by a choice of the constants $A : B : C$ only if the exponential factors cancel; hence if

$$\alpha = \alpha', \qquad k_1\sin\alpha = k_2\sin\beta. \tag{3}$$

The first of these equations is the *law of reflection*; the second is the *law of refraction* which, recalling (2.2), becomes

$$\frac{\sin\alpha}{\sin\beta} = \frac{k_2}{k_1} = \sqrt{\frac{\varepsilon_2\mu_2}{\varepsilon_1\mu_1}}. \tag{3 a}$$

The right-hand side of this double equation defines the (relative) index of refraction of the media 1 and 2

$$n_{12} = \sqrt{\frac{\varepsilon_2\mu_2}{\varepsilon_1\mu_1}}. \tag{3 b}$$

If we assume medium 1 to be air, whose constants ε_1 and μ_1 are almost the same as those of vacuum, and if we denote the constants of medium 2 merely by ε and μ, then we obtain the definition of the *index of refraction with respect to air*:

$$n = \sqrt{\frac{\varepsilon\mu}{\varepsilon_0\mu_0}} = \frac{c}{u}. \tag{4}$$

Here, as in (2.3), u denotes the phase velocity in medium 2. After putting $\mu = \mu_0$ and $\varepsilon/\varepsilon_0 = \varepsilon_{rel}$ (dielectric constant relative to vacuum), this is usually written:

$$n = \sqrt{\varepsilon_{rel}}, \tag{4 a}$$

which is Maxwell's relation. As Boltzmann has shown, this relation is fairly well satisfied for homopolar gases and vapors but not at all for solid and liquid media, especially not for those with infrared resonance vibrations. For water, for instance, one gets $\sqrt{\varepsilon_{rel}} \sim 9$ as against $n \sim 4/3$. Maxwell's relation does not explain dispersion (dependence of n on frequency) at all.

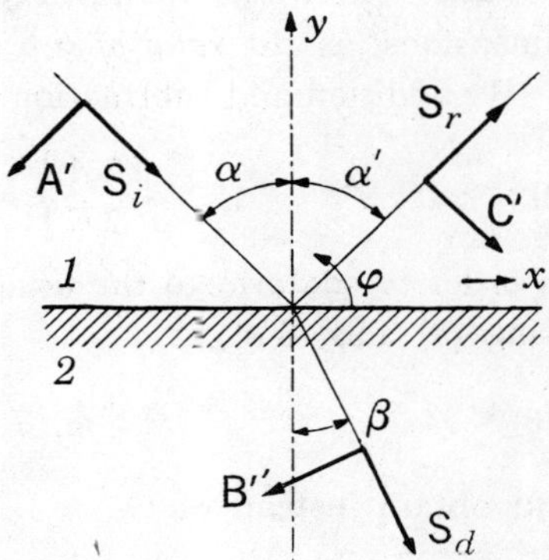

Fig. 3 a.

Illustration for the derivation of Fresnel's formulae for transition from a rarer to a denser medium. Electric vector perpendicular to the plane of the drawing.

Because of (3), (2) reduces simply to

$$A + C = B. \tag{5}$$

We obtain a second equation for the ratio of the constants $A : B : C$ by writing the boundary condition for the tangential component H_x. According to (2.5) the amplitude factor of **H** follows from that of **E** by multiplication by $\pm\sqrt{\varepsilon/\mu}$ where the sign is to be decided according to the right-handed screw rule: **E**, **H** and **S** form for all three waves a right-handed coordinate system. At an instant when **E** is positive, that is, directed out of the paper in fig. 3 a, **H** is generated by a clockwise rotation from the **S** direction. Hence, according to the ray directions as drawn, **H** must point to the *left* for the incident and refracted wave and to the *right* for the reflected wave. Accordingly, H_x is found by multiplying the corresponding amplitudes of the electric vector by $-\cos\alpha$ and $-\cos\beta$ in the case of the incident and refracted wave, respectively, and by $+\cos\alpha$ in the case of the reflected wave. One obtains then instead of (1)

(6) $$H_x = \sqrt{\frac{\varepsilon_1}{\mu_1}} \cos\alpha \, e^{ik_1 x \sin\alpha} \{-A\, e^{-ik_1 y \cos\alpha} + C\, e^{ik_1 y \cos\alpha}\}$$

and instead of (1 a)

(6 a) $$H_x = -\sqrt{\frac{\varepsilon_2}{\mu_2}} \cos\beta \, e^{ik_2 x \sin\beta}\, B\, e^{-ik_2 y \cos\beta}.$$

Upon simplification by means of the law of refraction, the boundary condition takes the form

(7) $$\sqrt{\frac{\varepsilon_1}{\mu_1}} \cos\alpha \{-A + C\} = -\sqrt{\frac{\varepsilon_2}{\mu_2}} \cos\beta\, B,$$

which we shall write in the form

(8) $$A - C = m_{12} \frac{\cos\beta}{\cos\alpha} B. \qquad m_{12} = \sqrt{\frac{\varepsilon_2}{\mu_2}\frac{\mu_1}{\varepsilon_1}}.$$

m is generally (for $\mu_2 \neq \mu_1$) different from n. While n indicates the *ratio of two wave velocities*, we have to look upon m, according to our table of dimensions, as the *ratio of two wave impedances*.

By addition and subtraction of the two equations (5) and (8), one obtains

(9) $$\left.\begin{matrix} 2A \\ 2C \end{matrix}\right\} = \left(1 \pm m_{12} \frac{\cos\beta}{\cos\alpha}\right) B.$$

In order to conform to the usual way of writing Fresnel's formulae, we put, using (8) and (3 b),

(10) $$m_{12} = n_{12} \frac{\mu_1}{\mu_2} = \frac{\mu_1}{\mu_2} \frac{\sin\alpha}{\sin\beta}$$

and obtain instead of (9)

(11) $$\left.\begin{matrix} 2\mathrm{A} \\ 2\mathrm{C} \end{matrix}\right\} = \left(1 \pm \frac{\mu_1}{\mu_2} \frac{\sin\alpha\cos\beta}{\sin\beta\cos\alpha}\right) \mathrm{B}.$$

In the case $\mu_2 \sim \mu_1 = \mu_0$ which usually obtains, we can write the *first Fresnel formula* more simply as:

(12) $$A : B : C = \sin(\beta + \alpha) : [\sin(\beta + \alpha) + \sin(\beta - \alpha)] : \sin(\beta - \alpha).$$

B. Magnetic vector perpendicular to plane of incidence.

We start now with the magnetic vector H in the direction of the z-axis. As in (2.1), we denote the amplitude factor of the incident wave by A' and the amplitudes of the refracted and reflected waves by B' and C', respectively. Because of the continuity of H_z at $y = 0$, we have instead of (2) the boundary condition:

(13) $$A'\, e^{ik_1 x \sin\alpha} + C'\, e^{ik_1 x \sin\alpha'} = B'\, e^{ik_2 x \sin\beta}$$

The laws of reflection and refraction follow from this just as in (3), (3 a) and (13) reduces to

$$A' + C' = B'. \tag{14}$$

For the E-waves (factor $\sqrt{\varepsilon_1/\mu_1}$ with A' and C', factor $\sqrt{\varepsilon_2/\mu_2}$ with B') this gives

$$A + C = m_{12}\, B. \tag{14 a}$$

where the meaning of m_{12} is given by (8).

A second condition follows from the continuity of E_x at $y = 0$. In order to determine the signs we look at fig. 3 b. and consider again an instant when H, represented as an arrow perpendicular to the plane of the paper, is directed toward the reader. From the fact that E, H and S, in that order, must form a right-handed system, the E arrows follow as drawn in fig. 3 b. Projecting these onto the x-axis one obtains, instead of (8), as a second boundary condition

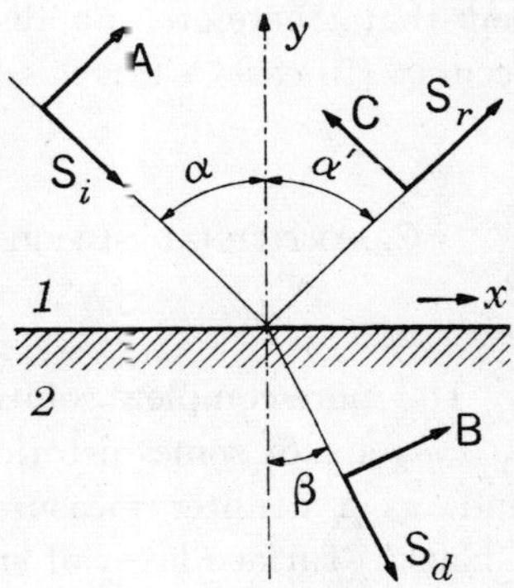

FIG. 3 b.

Illustration for the derivation of Fresnel's formulae for transitions from a rarer to a denser medium. Electric vector in the plane of the drawing.

$$\cos\alpha\,(A - C) = \cos\beta\, B,$$

and combining this with (14 a)

$$\left.\begin{matrix} 2A \\ 2C \end{matrix}\right\} = \left(m_{12} \pm \frac{\cos\beta}{\cos\alpha}\right) B = \left(\frac{\mu_1}{\mu_2}\frac{\sin\alpha}{\sin\beta} \pm \frac{\cos\beta}{\cos\alpha}\right) B. \tag{15}$$

In the usual case $\mu_1 \sim \mu_2$ one obtains the simplification

$$\frac{4A}{B} = \frac{\sin 2\alpha + \sin 2\beta}{\sin\beta\cos\alpha} = \frac{2\sin(\alpha+\beta)\cos(\alpha-\beta)}{\sin\beta\cos\alpha}, \tag{15 a}$$

$$\frac{4C}{B} = \frac{\sin 2\alpha - \sin 2\beta}{\sin\beta\cos\alpha} = \frac{2\cos(\alpha+\beta)\sin(\alpha-\beta)}{\sin\beta\cos\alpha}. \tag{15 b}$$

From this it follows immediately that

$$A : C = \tan(\alpha+\beta) : \tan(\alpha-\beta).$$

On the other hand, from (15 a) one can easily calculate

$$A : B = \tan(\alpha+\beta) : \frac{2\sin\beta\cos\alpha}{\cos(\alpha+\beta)\cos(\alpha-\beta)} = \tan(\alpha+\beta) :$$

$$: \frac{\sin(\alpha+\beta) - \sin(\alpha-\beta)}{\cos(\alpha+\beta)\cos(\alpha-\beta)} = \tan(\alpha+\beta) : \left(\frac{\tan(\alpha+\beta)}{\cos(\alpha-\beta)} - \frac{\tan(\alpha-\beta)}{\cos(\alpha+\beta)}\right).$$

Collecting these two ratios, one obtains the *second Fresnel formula*

$$(16) \qquad A : B : C = \tan(\alpha + \beta) : \left(\frac{\tan(\alpha + \beta)}{\cos(\alpha - \beta)} - \frac{\tan(\alpha - \beta)}{\cos(\alpha + \beta)}\right) : \tan(\alpha - \beta).$$

In order to complete the above calculation, we will convince ourselves in exercise I.3 of the fact that no electric charge is induced on the surface $y = 0$ and that, therefore, no discontinuity in the electric excitation $D_y = \varepsilon E_y$ occurs. (In case A this is self-evident because of $E_y = 0$.)

C. Artificial suppression of reflection for perpendicular incidence.

The more complete solutions (9) and (15) of our problem in which we have $\mu_1 \neq \mu_2$ are of some historic interest. During the war the problem arose to find, as a counter measure against allied radar, a largely non-reflecting ("black") surface layer of small thickness. This layer was to be particularly non-reflecting for perpendicular or almost perpendicular incidence of the radar wave. In this case α, and because of the law of refraction also β, are almost equal to zero. The problem is solved according to (9) and (15) by making

$$(17) \qquad m_{12} = 1.$$

The criterion is, thus, not the index of refraction n but the ratio of wave resistances m. In order to "camouflage" an object against radar waves, one must cover it with a layer for which this ratio of wave resistances has the value 1 in the region of centimeter waves. According to (8) this means that if we call the constants of the desired material ε and μ and those of air ε_0 and μ_0, then

$$(18) \qquad \frac{\varepsilon}{\varepsilon_0} = \frac{\mu}{\mu_0}.$$

Hence, the problem concerns not only the dielectric constant but also the relationship between the dielectric constant and the permeability. A substance must be formed whose *relative permeability* μ/μ_0 is of the same magnitude as its relative dielectic constant $\varepsilon/\varepsilon_0$.

But thereby the problem is not yet solved. For at its back surface the layer borders on the object (metal) which is to be camouflaged, and this second surface still reflects strongly. Hence, the further condition must be

imposed that the layer should absorb sufficiently strongly. This requires a *complex* rather than a real dielectric constant and because of the requirement (18) a corresponding *complex permeability.* The material must, therefore, be *ferromagnetic* and must possess a *strong hysteresis* or a structural relaxation that acts correspondingly. Thus, a difficult technological problem was posed which, though not unsolvable, required extensive preparatory work.

Because of the urgent war situation the solution which had to be used resulted from the following considerations. According to our presentation reflection appears as a consequence of the discontinuity of the material constants between the two media 1 and 2. The question of whether a completely *continuous* transition also causes reflection is an old one and was disputed for a long time. It has been answered only lately and in a completely affirmative sense, owing to the special interest which wave mechanics[1] as well as ionosphere research has recently brought to this question.

It turns out, however, that reflection becomes *extremely small* when the change of the material constants is spread over a distance equal to or larger than one wavelength, while an increase within a distance which is less than 1/4 wavelength acts appreciably like a discontinuous increase. By material constants we mean the complex dielectric constant (it must be complex because of the necessary absorption; the permeability can be left out of this consideration).

In practice it was necessary to approximate the required continuous rise of ε by a series of steps, that is, by the application of several layers whose dielectric constants (especially their imaginary parts, which are the more important) increase with depth from one layer to the next. *In this manner the reflected intensity could be reduced to* 1% *of the value given by Fresnel's formula* for all wavelengths less than an upper limit whose value depends on the thickness of the layer. This could be accomplished without exceeding the admissible additional weight of the layers.

We will treat another method for diminishing reflection (extinction by interference) in Sec. 7.

---

[1] In wave mechanics one is concerned with the penetration of an electron into a region of increasing repulsive potential which according to the energy theorem of classical mechanics, would be inaccessible to the electron with its given kinetic energy. See also the later remarks on the tunnel effect, Sec. 5 c. In particular S. Epstein found a special profile for the increase in refractive index for which the reflection can be calculated rigorously (by means of hypergeometric functions). For further details see, for instance, "Atombau und Spektrallinien" Vol. II p. 29. A general discussion of the various methods of computation is given in Kofink and Menzer, Ann d. Phys. (Lpz) **39**, 388, 1941; Kofink, ibid. **1**, 119, 1947.

4. Graphical Discussion of Fresnel's Formulae. Brewster's Law.

Let medium 1 be the optically *rarer* medium, for instance air, and let medium 2 be the optically *denser* medium, for instance water or glass. Since these media are non-magnetic ($\mu = \mu_0$), m has the same value as n. The designations "denser" and "rarer" have their origins in the elastic (or rather quasielastic) theory of Fresnel.

In our fig. 4 we use for the abscissa the angle of incidence α, where $0 < \alpha < \pi/2$. In the direction of the ordinate we plot the amplitude ratios for the transmitted and reflected rays

$$D = \frac{B}{A}, \qquad R = -\frac{C}{A}. \tag{1}$$

The negative sign of R is desirable because throughout the greater part of our figure the sign of the reflected amplitude C is opposite from that of the incident amplitude A. A change in sign during reflection obviously means a phase difference of π, that is, the addition of a phase factor

$$e^{i\pi} = -1.$$

We shall now distinguish the two cases A and B of the preceding paragraph by means of the indices p and s. Their meaning is: "plane of polarization parallel or perpendicular to the plane of incidence", and they contain a definition of the otherwise arbitrary term "plane of polarization". The significance of this definition is merely historical; see the beginning of Sec. 8.

A. Plane of polarization parallel to plane of incidence.

We begin with R_p which is given according to (1) and (3.12) by

$$R_p = \frac{\sin(\alpha - \beta)}{\sin(\alpha + \beta)}. \tag{2}$$

For small α we have, according to the law of refraction,

$$\beta = \frac{\alpha}{n}, \quad \text{hence} \quad R_p = \frac{n-1}{n+1}; \tag{3}$$

For $n = 4/3$ (water) and $n = 3/2$ (mean value in the spectrum of light crown glass) this gives the results:

$$R = 1/7 \text{ and } 1/5, \text{ respectively.}$$

Correspondingly, the ratios of reflected to incident intensities are

$$R_p^2 = 2\% \text{ and } 4\%, \text{ respectively.}$$

Neither water nor glass can serve as *mirrors* for *perpendicular* incidence. If we look perpendicularly at water, we see our own mirror image less clearly than the bottom or the water's own color in the case of deep water. Ordinary mirrors are not glass mirrors but metal mirrors. The glass serves only for the protection of the silver on the reverse side[1].

We wish to improve the approximation for small α by one order. Hence, we set

$$\sin(\alpha \mp \beta) = (\alpha \mp \beta)\left\{1 - \frac{1}{6}(\alpha \mp \beta)^2\right\}$$

and obtain instead of (3), see exercise I.4,

$$R_p = \frac{n-1}{n+1}\left(1 + \frac{\alpha^2}{n}\right). \tag{4}$$

Hence our representation of R_p in fig. 4 begins at a distance $\frac{n-1}{n+1}$ from the abscissa with a *horizontal tangent* and increases parabolically.

We now move from perpendicular to grazing incidence, $\alpha = \pi/2$. Here we get, according to the law of refraction,

$$\sin\beta = \frac{1}{n}, \qquad \sin(\alpha \mp \beta) = \cos\beta = \frac{\sqrt{n^2-1}}{n}, \tag{5}$$

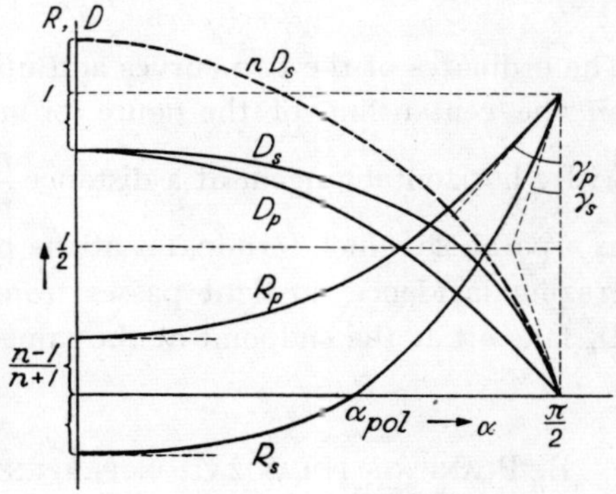

FIG 4.

The amplitude ratios R, for the reflected ray, and D for the transmitted ray, as functions of the angle of incidence α.

hence $R_p = 1$. For grazing incidence reflection is *complete* This is the reason for the beautiful mirror image of the opposite shore in the waters of a mountain lake, as well as for the mirror image of the setting sun in a smooth sea; this image approaches the sun itself in intensity.

We also wish to determine at what angle our R_p curve approaches its end point $R_p = 1$ at $\alpha = \pi/2$. For this purpose we compute $dR_p/d\alpha$ at that point. We note that (again because of the law of refraction)

$$\cos\alpha\, d\alpha = n\cos\beta\, d\beta, \quad \text{hence} \quad \frac{d\beta}{d\alpha} = 0 \text{ as we approach } \cos\alpha = 0. \tag{5}$$

[1] To be sure, the reflection from the front surface of the glass, weak as it is, makes such back-silvered mirrors unsuitable for optical purposes. For these the front surface of the glass must be covered with metal, preferably with rhodium.

Hence we need to differentiate (2) only partially with respect to α and by so doing obtain for $\alpha = \pi/2$:

$$\frac{dR_p}{d\alpha} = 2 \tan \beta = \frac{2}{\sqrt{n^2 - 1}} \quad \text{because} \quad \sin \beta = \frac{1}{n}. \tag{6 a}$$

It follows that the angle denoted in the figure by γ_p is given by:

$$\tan \gamma_p = \frac{\sqrt{n^2 - 1}}{2}. \tag{7}$$

It is now very simple to draw also the curve for D_p. In our present[1] notation (1) we have, according to (3.5),

$$D_p = 1 - R_p. \tag{7 a}$$

The ordinates of the two curves add up to 1. We obtain D_p by reflecting R_p on the center line of the figure (ordinate 1/2). Hence the curve D_p starts with a horizontal tangent at a distance $\frac{n - 1}{n + 1}$ below the ordinate 1 and proceeds as a parabola which terminates at the point $\alpha = \pi/2$, $D_p = 0$. In the case of grazing incidence no light passes from the rarer into the denser medium. D_p falls off at the endpoint at the same angle γ_p as determined by (7).

B. Plane of polarization perpendicular to plane of incidence.

According to (3.16) C/A is *positive* for angles of incidence that are not too large; hence, following our definition (1), R_s becomes *negative*:

$$R_s = -\frac{\tan (\alpha - \beta)}{\tan (\alpha + \beta)}. \tag{8}$$

If we let $\alpha \to 0$ then, except for the sign, equation (3) holds true also here and its next degree of approximation becomes, see exercise I.4, instead of eq. (4)

$$R_s = -\frac{n - 1}{n + 1}\left(1 - \frac{\alpha^2}{n}\right). \tag{9}$$

Hence the curve for R_s is a quadratic parabola which starts with a horizontal tangent at a distance $\frac{n - 1}{n + 1}$ *below* the abscissa. It ends according to (8) at $\alpha = \pi/2$ with the *positive* ordinate

[1] Our quantities R and D are not to be confused with the quantities r and d which will be defined on the basis of energy in section E.

(10) $$R_s = -\frac{\tan\left(\frac{\pi}{2} - \beta\right)}{\tan\left(\frac{\pi}{2} + \beta\right)} = +1.$$

Its slope is steeper than that of R_p; the angle γ_s, computed in a manner similar to (7), is given by

(11) $$\tan \gamma_s = \frac{\sqrt{n^2 - 1}}{2\,n^2} < \tan \gamma_p.$$

Between its negative beginning (9) and its positive end (10) the R_s curve crosses the abscissa. We call this point

(12) $$\alpha = \alpha_{pol} = \textit{angle of polarization.}$$

From eq. (8) we conclude that at this point the denominator must suddenly change its value from $+\infty$ to $-\infty$ and that hence

(13) $$\alpha_{pol} + \beta = \frac{\pi}{2}, \qquad \beta = \frac{\pi}{2} - \alpha_{pol}, \qquad \sin \beta = \cos \alpha_{pol}.$$

On the other hand, according to the law of refraction,

(13 a) $$\sin \beta = \frac{1}{n} \sin \alpha_{pol}.$$

By comparing (13) and (13 a) it follows that

(14) $$\tan \alpha_{pol} = n.$$

For glass $\left(n = \frac{3}{2}\right)$ and water $\left(n = \frac{4}{3}\right)$

$$\alpha_{pol} = 57° \text{ and } 53°, \text{ respectively.}$$

Since R_s vanishes at this angle, *the reflected light is polarized completely parallel to the plane of incidence.* This was the discovery of Malus. Our eq. (13) also contains "Brewster's law"; see the later fig. 5: *the reflected ray* S_r *is perpendicular to the refracted ray* S_d.

Next we complete fig. 4 by drawing the curve for D_s. This can be derived from the relation (3.14 a) which, in our present notation, reads

(15) $$1 - R_s = n\, D_s.$$

If we now make the same construction as in the case of D_p, that is, if we mirror the curve R_s about the center line of the figure, then we are carried beyond the ordinate 1, as indicated by the broken curve denoted by $n\, D_s$ in the figure. The starting point of this curve has the ordinate

$$1 + \frac{n - 1}{n + 1} = \frac{2\,n}{n + 1}.$$

To obtain D_s itself we must reduce this curve by a factor of $1/n$. Then we obtain the same starting point ordinate as in the plot for D_p, namely

$$1 - \frac{n-1}{n+1} = \frac{2}{n+1}.$$

C. Practical production of polarized light.

While the *reflected* component provides *complete* polarization at the angle of polarization, it provides only *low* intensity. Though the refracted component is only incompletely polarized, its intensity is *greater*. Indeed, according to (13) and (14), for $\alpha = \alpha_{pol}$

$$\sin(\alpha + \beta) = 1$$
$$\sin(\alpha - \beta) = \sin^2\alpha - \cos^2\alpha = \frac{n^2-1}{n^2+1}$$

and hence, according to (2),

$$R_p = \frac{n^2-1}{n^2+1} = \frac{5}{13} \quad \text{for } n = 3/2 \text{ (glass)}.$$

The efficiency of this "polarizer" (ratio of reflected p-intensity to the entire incident $(p+s)$ – intensity) amounts, therefore, to only

$$\frac{1}{2} R_p{}^2 = 7.4\%.$$

For other angles of incidence $\alpha \neq \alpha_{pol}$ the reflected light is also polarized parallel to the plane of incidence but only partially so.

On the other hand, a glance at fig. 4 shows that $D_s > D_p$ for every α. The refracted light is always *partially* polarized *perpendicularly* to the plane of incidence. For instance, for the special case $\alpha = \alpha_{pol}$ we get according to (15) and (7 a)

$$(16) \qquad D_s = \frac{1}{n}, \quad D_p = 1 - \frac{n^2-1}{n^2+1} = \frac{2}{n^2+1}, \quad \frac{D_s}{D_p} = \frac{n^2+1}{2n} > 1.$$

If we consider the passage of light through a plate at the back surface of which a second transition with index of refraction $1/n$ takes place, then at this point the amplitude ratios are again the same[1] because

$$\frac{\frac{1}{n^2}+1}{2/n} = \frac{1+n^2}{2n}.$$

[1] For the generality of this relationship, see exercise I.2.

Thus for a glass plate the resulting ratio of refracted amplitudes $D_s : D_p$ is

$$\left(\frac{13}{12}\right)^2 = 1.17$$

and the ratio of intensities is

$$(1.17)^2 = 1.37.$$

Hence, by means of a *stack of glass plates* the polarization can be successively increased without *decreasing the intensity* (if the material is completely transparent and the surfaces are clean). In this way half of the intensity of the incident natural light (namely the *s*-polarized half) is completely utilized. The efficiency of an ideal stack of glass plates would thus be 50%. To be sure, complete polarization is approached only asymptotically as the number of glass plates is increased to infinity.

While the production of polarization by means of crystal structure is readily understandable, its origin in an isotropic material is, because of the complete lack of structural elements, somewhat paradoxical. This situation will be clarified in the next section.

D. Brewster's law from the point of view of electron theory.

We now leave, temporarily, the phenomenological viewpoint of Maxwell's theory and interpret the process of refraction as scattering of the light by the atoms of the second medium (the first medium can be thought of as a vacuum). From this physically more profound viewpoint refraction takes place only because the electric field acting in the second medium sets the atomic electrons into oscillations, these oscillations being in the direction of the field. Thus, we are concerned with real material oscillations and not merely with alternating fields as before.

Fig. 5 represents the case of "plane of polarization *perpendicular* to the *plane of incidence*" in which the electric vector oscillates *in the plane of incidence*. Its direction of oscillation in the second medium is, of course, perpendicular to the direction of the refracted ray. The electrons oscillate in this same direction. They act like Hertzian oscillators and like them radiate *no* light in the direction of their oscillations (the same is well-known to be true for the antennas of radio transmitters). Regular reflection in the first medium can occur only if the electrons of the second medium deliver radiant energy in the direction of reflection (as determined by the law of reflection). This is not the case when this direction is parallel to the oscillations of the electrons, hence perpendicular to the refracted ray, *in agreement with Brewster's law.* In other directions of reflection the electrons yield part of their radiation, which explains the variation of the strength of reflection with varying angles of incidence.

We also see immediately that this consideration does not affect the other case: "plane of polarization *parallel* to plane of incidence". In this case the electric vector and hence also the electron oscillations are *perpendicular* to the plane of incidence, hence also perpendicular to every position of the reflected ray. Every one of these directions is a direction of maximum radiation from the electrons. Thus there is no reason for a forbidden direction of reflection such as Brewster's law demands.

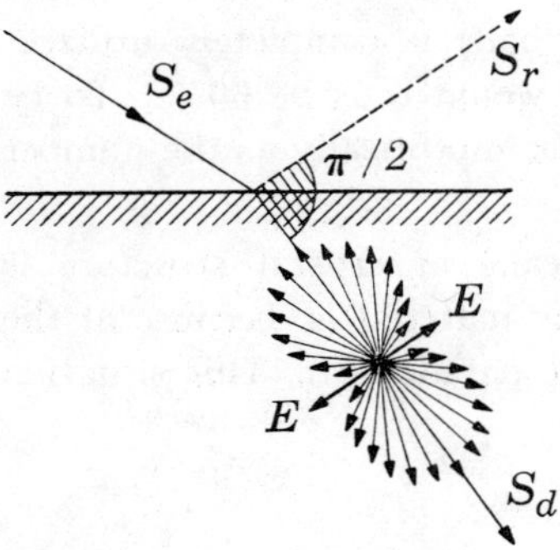

Fig. 5.

Brewster's law from the point of view of electron theory. At the angle of polarization the reflected and refracted rays are mutually perpendicular.

We do not claim that the reflecting power can be calculated in this simple manner; for that purpose our method is still too primitive. Furthermore, it must be observed that only the layer next to the surface can be taken into consideration because at greater depths the contributions of the single atoms cancel by interference. But nevertheless, the null effect of Brewster's law is incontestably illustrated by our method.

These considerations also show that polarization depends, even in the case of isotropic substances, on a structural property of the material. This structure is, however, not prescribed crystallographically but is brought about by the electromagnetic field itself in creating a directed dipole structure in the otherwise unordered atoms.

E. Energy considerations. Reflecting power r and transmissivity d.

Clearly these phenomena conserve energy in every case. In order to see this one may consider the flow of energy through an arbitrary cross section q of the incident wave. The corresponding cross section in the reflected wave is again q. But on the refracting plane q subtends the larger area $\frac{q}{\cos\alpha}$ and for the transmitted light the corresponding cross section is given by

(17) $$q' = q \frac{\cos\beta}{\cos\alpha}.$$

We write for the time average of the energy flow through these three cross-sections

$$S_i = q\,\overline{\mathsf{S}}_i, \qquad S_r = q\,\overline{\mathsf{S}}_r, \qquad S_d = q'\,\overline{\mathsf{S}}_d, \qquad \overline{\mathsf{S}} = \overline{E \times \mathsf{H}} = \sqrt{\frac{\varepsilon}{\mu}}\,\overline{\mathsf{E}^2}$$

and define the reflecting power and the transmissivity by

(18) $$r = \frac{S_r}{S_i} = \left|\frac{C}{A}\right|^2, \qquad d = \frac{S_d}{S_i} = \sqrt{\frac{\varepsilon_2\,\mu_1}{\mu_2\,\varepsilon_1}}\frac{q'}{q}\left|\frac{B}{A}\right|^2 = m\frac{\cos\beta}{\cos\alpha}\left|\frac{B}{A}\right|^2$$

where the meaning of m is the same as in (3.8). We will convince ourselves in exercise I.5 that, in compliance with the energy law, for every case

(19) $$r + d = 1,$$

and also that r and d are the same for both transitions: rarer ⇆ denser medium.

Our energy equation (19) is to be well distinguished from the amplitude equations (7 a) and (15)

$$R_p + D_p = 1 \qquad \text{and} \qquad R_s + n\,D_s = 1.$$

5. Total Reflection

In principle our formulae of Secs. 3 and 4 remain unchanged if we transfer the incident wave into the denser medium and investigate its reflection into the same medium and its refraction in the rarer medium. In particular, no changes need be made in the derivation of the law of refraction in (3.3). But because we want to preserve the former meaning of $n > 1$, we will substitute $1/n$ for n and must hence write

(1) $$\frac{\sin\alpha}{\sin\beta} = \frac{1}{n}.$$

From this it follows that $\beta > \alpha$ for small α but that β is *imaginary* for $n \sin\alpha > 1$. In this latter case the coefficients $A : B : C$ in Fresnel's formulae also become complex.

In the older literature this situation was rejected as being unphysical, but it is entirely consistent with our viewpoint for we consider the problem of reflection and refraction as a *boundary value problem.* Any procedure which leads to a self-consistent solution of this problem is justified. Calculations with complex magnitudes are just as admissible and advisable in optics as they are, for instance, in two-dimensional potential theory.

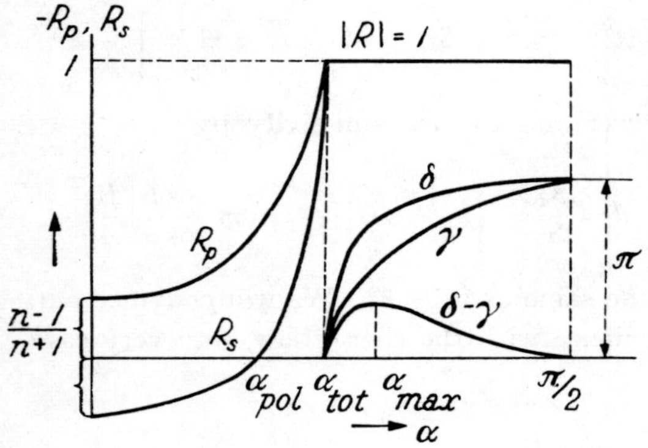

FIG. 6.
Reflection ratios R_p and R_s before and after total reflection occurs.

A. Discussion of Fresnel's formulae.

We shall adopt immediately the graphical method of Sec. 4 and again neglect the distinction which, in principle, exists between n and m. The abscissa of fig. 6 again represents the angle of incidence $0 < \alpha < \pi/2$. On it we mark the point $n \sin \alpha = 1$ at which the angle of refraction β attains its largest real value $\beta = \pi/2$. We call this point

(2) $$\alpha_{tot} = \text{Limiting angle of total reflection.}$$

For glass against air

$$\sin \alpha_{tot} \sim \frac{2}{3}, \qquad \alpha_{tot} \sim 42°.$$

Limiting ourselves for the time being to reflected light, we shall plot the following quantities along the ordinate:

(3) $$R = \frac{C}{A}, \qquad R_p = \frac{\sin(\beta - \alpha)}{\sin(\beta + \alpha)}, \qquad R_s = -\frac{\tan(\beta - \alpha)}{\tan(\beta + \alpha)}.$$

Consequently, we now choose the sign of R oppositely from that in (4.1). Thereby we attain the advantage that in spite of the opposite state of affairs ($\beta < \alpha$ in fig. 4., $\beta > \alpha$ in fig. 6) the curves for R_p and R_s are similar to those in fig. 4 for small α. They both start with horizontal tangents at the same ordinates $\pm\dfrac{n-1}{n+1}$ as before. But they reach the ordinate 1 not at $\alpha = \pi/2$ but already at $\alpha = \alpha_{tot}$. Indeed, since at that point $\beta = \pi/2$,

$$(4) \qquad R_p = \frac{\sin\left(\dfrac{\pi}{2} - \alpha\right)}{\sin\left(\dfrac{\pi}{2} + \alpha\right)} = 1, \qquad R_s = -\frac{\tan\left(\dfrac{\pi}{2} - \alpha\right)}{\tan\left(\dfrac{\pi}{2} + \alpha\right)} = 1.$$

Before reaching that point the curve for R_s will have crossed the abscissa at the angle of polarization

$$\alpha_{pol} + \beta = \frac{\pi}{2}, \qquad \tan\alpha_{pol} = \frac{1}{n}$$

which corresponds to equations (4.12) and (4.14).

We are interested in the slopes at which the R_p and R_s curves approach the ordinate 1 at α_{tot}. To this end we note that in contrast to (4.6) we now have

$$\frac{d\alpha}{d\beta} = 0 \quad \text{because} \quad \sin\beta = 1, \quad \cos\beta = 0,$$

and, hence, that we need to differentiate the expressions (3) only with respect to β in order to be able to compute the angle in question. Thus by setting $\beta = \pi/2$ after differentiating, we first obtain the result,

$$\frac{dR_p}{d\beta} = 2\,\frac{\sin\alpha}{\cos\alpha}.$$

Then recalling (4.6) with n replaced by $1/n$, we obtain in the neighborhood of the critical point

$$(5) \qquad \frac{dR_p}{d\alpha} = 2\,\frac{\sin\alpha}{\cos\alpha}\frac{d\beta}{d\alpha} = \frac{2\,n\sin\alpha}{\cos\beta} = \frac{2}{\cos\beta}.$$

The limit as $\beta \to \pi/2$, becomes ∞; the R_p curve has a *vertical tangent* at the point in question.

Correspondingly, one obtains

$$(5\text{ a}) \qquad \frac{dR_s}{d\alpha} = \frac{2}{\sin\alpha\cos\alpha}\frac{d\beta}{d\alpha} = \frac{2\,n}{\sin\alpha\cos\beta} = \frac{2\,n^2}{\cos\beta}.$$

The R_s curve runs even steeper than the R_p curve in the vicinity of the critical point. It also has a *vertical* tangent at that point.

This circumstance is of special experimental importance; it led Abbe and F. Kohlrausch to the construction of total reflection meters (total refraction meters, respectively). Because of the abrupt increase of the reflected light (the abrupt disappearance of the refracted light, respectively), the limit of total reflection can be determined very closely and thereby the index of refraction can be computed, according to the formula $n = \dfrac{1}{\sin \alpha_{tot}}$, with great accuracy.

We complete fig. 6 by drawing through the critical point $R = 1$ a line parallel to the abscissa and denoting it by $|R| = 1$. This means the following for both components R_p and R_s: *reflected intensity equals incident intensity*, hence indeed *total reflection*. To justify this assertion we follow the course of the β-point as given by the law of refraction in a complex β-plane. Its path follows the real axis from 0 to $\pi/2$, as α goes from 0 to α_{tot}. At this point the path of β splits into two, mathematically equally justifiable, branches $\beta = \pi/2 \pm i\,\beta'$ which run parallel to the imaginary axis. For both branches

$$\sin \beta = \sin\left(\frac{\pi}{2} \pm i\,\beta'\right) = \cos\,(\pm\, i\,\beta') = \cosh \beta' > 1 \tag{6}$$

as is required by the law of refraction $\sin \beta = n \cdot \sin \alpha > 1$. Using (6) one obtains from (3)

$$R_p = \frac{\sin\left(\dfrac{\pi}{2} \pm i\,\beta' - \alpha\right)}{\sin\left(\dfrac{\pi}{2} \pm i\,\beta' + \alpha\right)} = \frac{\cos\,(\alpha \mp i\,\beta')}{\cos\,(\alpha \pm i\,\beta')} = 1 \cdot e^{i\gamma} \tag{7}$$

$$R_s = -\frac{\tan\left(\dfrac{\pi}{2} \pm i\,\beta' - \alpha\right)}{\tan\left(\dfrac{\pi}{2} \pm i\,\beta' + \alpha\right)} = \frac{\cot\,(\alpha \mp i\,\beta')}{\cot\,(\alpha \pm i\,\beta')} = 1 \cdot e^{i\delta} \tag{7 a}$$

Since the numerators and denominators of R_p and R_s contain mutually conjugate quantities, the absolute values of these quotients are equal to 1. γ and δ are real phase angles whose experimental utilization we will consider in section *D*. In preparation we have already sketched their curves in the right-hand part of fig. 6. $|R| = 1$ immediately explains the excellent action of prismatic field glasses whose operation is based on the principle of total reflection.

B. Light penetrating into the rarer medium.

The general formulae for the refracted wave in Sec. 3 give a field in the rarer medium not only for $\alpha < \alpha_{tot}$ but also for $\alpha > \alpha_{tot}$.

In the case of p-polarization we start with eq. (3.1 a) by substituting

$$\beta = \frac{\pi}{2} \pm i\,\beta', \qquad \sin\beta = \cosh\beta', \qquad \cos\beta = \mp i \sinh\beta'$$

and obtain for $k_2 = k$ (vacuum)

$$E_z = B\, e^{i k (x \cosh\beta' \pm i y \sinh\beta')}$$

We see that only the lower sign in front of i is physically admissible (E_z must remain finite as $y \to -\infty$) so that we must set

(8) $$E_z = B\, e^{k y \sinh\beta'}\, e^{i k x \cosh\beta'}.$$

This wave has an entirely different structure from that of the usual "homogeneous" plane wave. It is called "inhomogeneous". Though this wave propagates without attenuation along the boundary surface, *its strength decreases perpendicularly to it.* Since $k = 2\,\pi/\lambda$, the wave is noticeable only within a distance of a few wavelengths from the boundary surface.

We compute, next, by means of Maxwell's equation $\mu_0\, \dot{\mathsf{H}} = -\operatorname{curl} \mathsf{E}$, the magnetic excitation H belonging to (8). We consider the right-hand side of (8) to be provided with the time factor $\exp(-i\omega t)$ and take into account the relationship $\omega/k = c = (\varepsilon_0 \mu_0)^{-1/2}$. We thus obtain besides $H_z = 0$:

(9) $$\left.\begin{aligned} H_x &= -i \sqrt{\frac{\varepsilon_0}{\mu_0}} \sinh\beta' \\ H_y &= -\sqrt{\frac{\varepsilon_0}{\mu_0}} \cosh\beta' \end{aligned}\right\} \cdot B\, e^{k y \sinh\beta'} e^{i k x \cosh\beta'}.$$

Both components of H have the same inhomogeneous structure as (8).

We now turn to the radiation vector $\mathsf{S} = \mathsf{E} \times \mathsf{H}$. We must not, of course, multiply the complex expressions (8) and (9) but only their real parts taking account thereby of both the time factor and the complex nature of (8) and (9). We write

(10) $$\left.\begin{aligned} S_x &= -E_z H_y \\ S_y &= +E_z H_x \end{aligned}\right\} = \sqrt{\frac{\varepsilon_0}{\mu_0}}\, |B|^2\, e^{+2 k y \sinh\beta'} \begin{cases} \cosh\beta' \cos^2\tau \\ \sinh\beta' \sin\tau \cos\tau \end{cases}$$

where $\tau = \omega t - k x \cosh\beta'$. We see that the x-component of S, which is the component parallel to the boundary surface, is always positive. On the other hand, the flow of energy in the direction perpendicular to the boundary surface changes its sign periodically. Its time average vanishes while an actual energy flow takes place parallel to the boundary surface.

This seems to contradict both the name "total reflection" and our oft-repeated statement that no energy is lost in this process. We must, however, consider the fact that we have always performed our calculations for the ideal case of an infinitely wide wave front. For the actual, laterally restricted waves energy can very well pass from the denser into the rarer medium or, respectively, flow back from the rarer into the denser medium at the lateral boundaries[1] of the wave. This is the energy which is transported parallel to the boundary surface or, so to say, meanders about it.

If one may be permitted to use a military analogy, we can describe the situation in the following manner: an army marching in closed ranks comes in its advance upon difficult territory which forces it to change its direction of march. The wing of the army detaches a weak patrol with orders to penetrate the difficult region and to secure the flank. This patrol needs to be only a few men strong in depth. After carrying out its orders the patrol returns to the army.

Just as the lack of such a precaution would violate all rules of military caution, so would a sudden discontinuity of our totally reflected wave violate all rules of electrodynamics.

C. THE TUNNEL EFFECT OF WAVE MECHANICS.

The experimental proof of the existence of the inhomogeneous wave in the rarer medium has posed a difficult problem. Quincke tried his hand at it for many decades. He placed two precisely cut glass plates side by side at a distance of a few wavelengths, and allowed light to be reflected totally in the first plate. He believed he was then able to observe traces of transmitted light in the second plate. He considered this as an indication that the air gap between the plates was bridged by the light field. Woldemar Voigt repeated similar experiments with an improved set-up.

The experiment becomes very simple with Hertz waves. In the *Bose-Institute*[2] in Calcutta the following set-up is demonstrated: two asphalt

[1] These boundaries also have to do with the "lateral displacement" of the totally reflected ray as studied recently by F. Goos and H. Hänchen experimentally, and by K. Artmann theoretically, Ann. d. Phys. (Lpz) **1**, 333, 1948 respectively **2**, 87, 1948. See also C. v. Fragstein ibid. **4**, 271, 1948.

[2] The botanist Sir Jagadis Bose in his younger years imitated experiments of classical optics with short Hertz waves, e.g. $\lambda = 20$ cm. See Collected Physical Papers, especially No. VI of the year 1897, Longmans, Green and Co., 1927.

prisms 1 and 2, fig. 7, are placed opposite each other at a distance of several centimeters. The waves are incident perpendicularly to 1 and are "totally reflected" on the back face of 1. Still, one obtains distinct signals in a receiver placed behind 2 and these increase in strength as the distance between the prism faces is decreased.

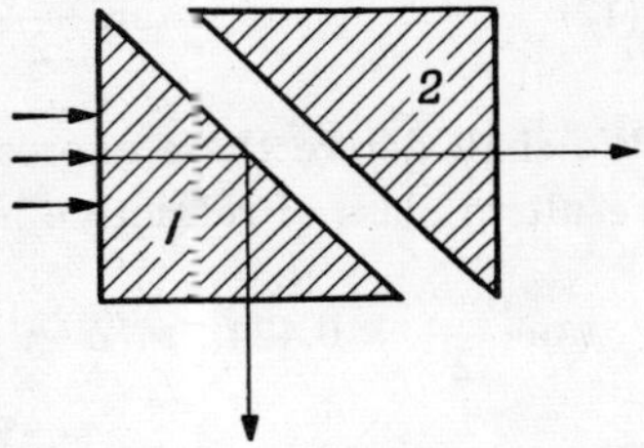

Fig. 7.
Experiment to prove that Hertz waves enter the rarer medium. The distance between the two prisms is a fraction of the wavelength.

In wave mechanics quite analogous situations occur under the name of "tunnel effects" (Condon and Gurney, 1928). By assigning a wave to a particle (electron, ion), according to L. de Broglie, and making the former obey the Schrödinger equation, one shows that the particle can, as a wave, pass through a potential barrier which, considering its kinetic energy, the particle could not surmount according to classical mechanics. This happens with a specific probability which depends on the thickness of the wall and the original energy of the particle. In the wave mechanical formulation the potential barrier plays exactly the same role as the air space in the experiments on total reflection. The parallelism between classical and quantum mechanics on the one hand and ray optics and wave optics on the other is hereby well illustrated. The wave mechanical tunnel effects proved to be fundamental in the theories of chemical binding, of "cold" electron emission of metals, of radioactivity, and also of the process of uranium fission.

D. Production of Elliptically and Circularly Polarized Light

Starting from equations (7), (7 a) we assume that the incident light is linearly polarized at an angle of 45° to the plane of incidence, a situation which is attainable by means of a Nicol prism. Then the amplitude factors A of the incident p- and s-waves will be equal and, according to the above-mentioned equations, the amplitudes of both totally reflected components will be the same but their phases γ and δ will differ; for by dividing (7) by (7 a) one obtains

$$e^{i(\gamma-\delta)} = \frac{\sin(\alpha \mp i\beta')}{\sin(\alpha \pm i\beta')}. \tag{11}$$

For $\alpha = \alpha_{tot}$ the right-hand side of this equation becomes 1, hence the phase difference becomes zero because there $\beta' = 0$. The same is also true for $\alpha = \pi/2$ because there $\sin(\pi/2 - i\,\beta') = \sin(\pi/2 + i\,\beta') = \cosh\beta'$. Hence between these two limits lies a point of maximum phase difference. The magnitude of this difference and the associated angle of incidence α_{max} are given by

$$\tan\frac{\delta-\gamma}{2} = \frac{n^2-1}{2n}, \qquad \sin^2\alpha_{max} = \frac{2}{n^2+1}. \tag{12}$$

We shall derive these expressions in exercise I.6. Here we shall apply the result to glass of refractive index $n = 1.51$ and find

$$\tan\frac{\delta-\gamma}{2} = 0.424, \qquad \delta-\gamma = 45°36', \qquad \sin\alpha_{max} = 0.781, \qquad \alpha_{max} = 51°20'.$$

We cannot attain the special case of *circular* polarization by a *single* total reflection since $\gamma - \delta \leqq 45°36'$ for all angles of incidence; but by two such reflections this case can be produced. For this purpose Fresnel constructed a glass prism with a parallelogram for a base. If linearly polarized light is incident perpendicularly upon the shorter prism face then, after being totally reflected twice at the longer prism faces, it emerges at the opposite short face as circularly polarized light.

6. Metallic Reflection

In Maxwell's theory metals are characterized by the conductivity σ. However, actual conduction of electricity has to be thought of as a phenomenon consisting of the interactions between free electrons and metallic ions at fixed positions and as being brought about by an averaging of many elementary processes. Only in stationary or slowly varying fields does this averaging lead to a constant which is independent of frequençy. One cannot expect that the phenomenological Maxwell theory will still suffice in the visible region of the spectrum. We have already encountered such a failure of the theory in the optics of transparent media (the failure of water, for instance, to satisfy the Maxwell relation $n^2 = \varepsilon_{rel}$. See p. 15). Again, the phenomenological description of metallic reflection proves to be insufficient in the visible domain though it agrees well with experiment in the infrared spectral range (see section B). Hence, the Maxwell theory of metallic reflection has general significance only in the sense of being a limiting theory.

As is well known, Maxwell's equations for conductors differ from the equations (2.4) for non-conductors only in that the Ohmic current $\sigma\,\mathbf{E}$ is added to the displacement current $\varepsilon\,\dot{\mathbf{E}}$. In the periodic case this means that $-\varepsilon\,i\,\omega$ is to be replaced by $-\varepsilon\,i\,\omega+\sigma$; hence ε is to be replaced by the complex dielectric constant

$$\varepsilon'=\varepsilon+i\,\frac{\sigma}{\omega}. \tag{1}$$

We accept this formulation in optics because it embodies, for a given ω, the most general linear relationship between $\mathbf{D}$ and $\mathbf{E}$. But according to the above discussion we must not expect that ε and σ will retain their electrodynamic, ω-independent meanings.

Together with ε the index of refraction of eq. (3.4) also becomes complex:

$$n'=\sqrt{\frac{\varepsilon'\,\mu}{\varepsilon_0\,\mu_0}}=n\,(1+i\,\varkappa). \tag{2}$$

The significance of the real quantities n and $\varkappa$ which are here introduced follows by squaring (2) and equating real and imaginary parts on both sides of the equation:

$$\frac{\varepsilon}{\varepsilon_0}\,\frac{\mu}{\mu_0}=n^2\,(1-\varkappa^2),\qquad \frac{\sigma}{\varepsilon_0\,\omega}\,\frac{\mu}{\mu_0}=2\,n^2\,\varkappa. \tag{2 a}$$

The metallic index of refraction n and the absorption coefficient $\varkappa$ thus defined are the optical constants of the metal. In connection with this usual designation "absorption coefficient" it should be remarked that the "ideal" conductor $\sigma\to\infty$ of electrodynamics is characterized, not by $\varkappa\to\infty$, but by

$$\varkappa\to 1,\qquad n\to\infty. \tag{2 b}$$

Indeed by dividing the two equations (2 a) one obtains

$$\frac{\varepsilon}{\sigma}=\frac{1-\varkappa^2}{2\,\omega\,\varkappa}\quad\text{and hence}\quad \varkappa^2\to 1 \text{ as } \sigma\to\infty$$

from which it also follows, by referring back to (2 a), that $n\to\infty$.

Together with n the resistance ratio m and the wave number k become complex. Corresponding to (3.8) and (2.2) we let

$$m'=n\,\frac{\mu_0}{\mu}\,(1+i\,\varkappa),\qquad k'=k\,n\,(1+i\,\varkappa),\qquad k=\frac{\omega}{c}. \tag{3}$$

We shall first concern ourselves with the structure of a monochromatic, linearly polarized, plane wave which propagates, for instance along the x-axis, in the metal. This wave is no longer homogeneous as in non-conductors. Its inhomogeneity is, however, of an entirely different nature from that which we encountered in the case of total reflection. We write, as in (2.1), omitting the time factors,

$$E_y = A\, e^{ik'x} = A\, e^{-\varkappa k n x}\, e^{iknx}$$

(4)
$$H_z = A'\, e^{ik'x} = A'\, e^{-\varkappa k n x}\, e^{iknx}, \qquad A' = m' A = n \frac{\mu_0}{\mu} (1 + i\varkappa)\, A.$$

It follows from this that the phase velocity is

$$\frac{dx}{dt} = \frac{\omega}{k\,n} = \frac{c}{n} \quad \text{and the wavelength} \quad \lambda = \frac{2\,\pi}{k\,n}.$$

Furthermore, we see from (4) that the wave is damped *longitudinally* in its propagation in the x-direction, not *transversely* as in total reflection. The decrease in amplitude per wavelength amounts to $\exp(-2\pi\varkappa)$. In addition, the complex nature of A' shows that there exists a constant phase difference between the magnetic and electric components. The nodes and the maxima of the two wave components no longer coincide as in the non-conductor but are displaced from each other by an amount depending on $\varkappa$.

A. Fresnel's formulae

Formally, we can take over Fresnel's formulae without change from Secs. 3 and 4, as well as the laws of reflection and refraction from (3.3). The former says again that angle of reflection = angle of incidence = real magnitude. The latter becomes because of (2)

(5)
$$\frac{\sin\alpha}{\sin\beta} = n\,(1 + i\varkappa)$$

which shows that the angle of refraction β is complex for all α, not merely for $\alpha > \alpha_{tot}$ as in total reflection.

Since the wave refracted into the interior of the metal is well-nigh unobservable because of its strong absorption, we must deduce the optical properties of a metal exclusively from the reflected light. Hence, we shall only have to discuss formulae (4.2) and (4.8) for R_p and R_s:

(6)
$$R_p = \frac{\sin(\alpha - \beta)}{\sin(\alpha + \beta)} = |R_p|\, e^{i\gamma},$$
$$R_s = -\frac{\tan(\alpha - \beta)}{\tan(\alpha + \beta)} = |R_s|\, e^{i\delta}.$$

Because of the complex value of β, γ and δ differ from zero and from each other.

We consider first reflection in the case of almost *perpendicular* incidence. Then α and $|\beta|$ are small and from (5)

$$\frac{\alpha}{\beta} = n\,(1 + i\,\varkappa).$$

Hence, according to (6)

$$R_p = \frac{n - 1 + i\,n\,\varkappa}{n + 1 + i\,n\,\varkappa} = -R_s.$$

Therefore, we obtain for both components the *reflecting power*

$$(7) \qquad r = |R|^2 = \frac{(n-1)^2 + n^2\varkappa^2}{(n+1)^2 + n^2\varkappa^2} = 1 - \frac{4\,n}{(n+1)^2 + n^2\varkappa^2}.$$

Assuming that we have a good conductor ($n \to \infty$ according to eq. (2 b)), we get $r \sim 1$. In contrast to the glass or water mirror (see p. 21) the metallic mirror is a *complete reflector*.

Turning now to *oblique* angles of incidence, we assume, as in eq. (5.11), that both components of the incident light, p and s, are equal in amplitude and phase. Then the quantities given by (6), namely the amplitude ratio $|R_p/R_s|$ and the phase difference $\gamma - \delta$, determine directly the nature of the reflected light. Generally, it is *elliptically* polarized. It will be *circularly* polarized only when $\gamma - \delta = \pi/2$ which we assume to be true for the special angle of incidence $\alpha = \alpha_h$, the so-called "principal angle of incidence". One observes this angle and converts the circular polarization, for instance by means of a $\lambda/4$ plate, to *linear* polarization. The azimuth α_p of the plane of polarization associated with the latter is called the "azimuth of the restored polarization". From α_p and α_h the metal constants n and $\varkappa$ can then be computed. The latter are to be looked upon as phenomenological substitutes for the real metal properties in the spectral range of visible light.

B. Experiments by Hagen and Rubens

We come now to the experiments which demonstrate for infrared rays the validity of eq. (7) which was derived from Maxwell's theory.

Hagen and Rubens, Ann. d. Phys. (Lpz) 1903, used for their experiments so-called residual rays which were left over from a larger spectral range after repeated reflections from crystals of alkaline earth halides (CaF_2, $CaCl_2$). These crystals possess pronounced resonances in the region from $\lambda = 10$ to $25.5\ \mu$ and hence have a highly selective reflecting power for such wavelengths. Then, according to (2 b), $\varkappa \sim 1$ and, according to (7),

$$(8) \qquad 1 - r = \frac{4\,n}{2\,n^2 + \ldots} = \frac{2}{n} \quad \text{is proportional to } \lambda^{-1/2}.$$

This proportionality to $\lambda^{-1/2}$ follows from the second eq. (2 a) (n^2 proportional to ω^{-1}). $1-r$ is the loss in reflection and $100\,(1-r)$ is the loss in reflection in %. Hagen and Rubens observed r and computed from it the values of $100\,(1-r)$ given in the following table:

	Ag	Au	Cu	Pt
$\lambda_1 = 12\,\mu$	9.05	13.8	12.1	10.6
$\lambda_2 = 25\,\mu$	7.07	8.10	6.67	6.88
Ratio	1.2	1.7	1.8	1.5

The last line gives the ratio of the two numbers above. According to (8) this should be constant and equal to

$$\sqrt{\frac{\lambda_2}{\lambda_1}} = \sqrt{\frac{25}{12}} = 1.46.$$

This is almost exactly the arithmetic mean of the four numbers in the last line of our table. Hagen and Rubens were also able to confirm the dependence of their observations on temperature which is to be expected according to (2 a) because σ is proportional to $1/T$ (T = absolute temperature) as well as the agreement of the conductivity σ thus obtained with the electromagnetic value. At very low temperatures these simple laws no longer hold even for infrared light[1] because then the mean time between two collisions of an electron with metallic ions becomes comparable with the period of oscillation of the light and, hence, the averaging procedure mentioned at the beginning of this paragraph fails.

C. Some remarks on the color of metals, glasses and pigments.

The dependence on wavelength contained in equations (7) and (8) would in itself result in a kind of coloring of the reflected light. The real color of metals, however, is caused by the characteristic oscillations of the electrons or ions which will be treated in Chapter III for the case of transparent bodies. Gold appears yellow when viewed directly. Very thin layers of gold allow green light to shine through. Except for metals real surface colors appear only where the optical constants n and $\varkappa$ are of a different *order of magnitude* from those of the bordering air and depend strongly upon wavelength. Dried *red* ink (solution of fuchsin) shines *yellow-green* in the incident light. Its red color on white paper is due to the light passing through the ink.

[1] K. Weiss, Ann. d. Phys. (Lpz) 2, p. 1., 1948 or E. Vogt, ibid. 3, p. 82., 1948 (Planck volume).

All other materials reflect practically unselectively; after reflection the incident white light remains almost *white*. This is the origin of the "glare" well-known to painters. The latter can also be observed, for instance, on the beautiful blue copper sulfate crystals or on ruby glass. One must however be careful to prevent any internally reflected light which has passed through the material and is therefore colored from entering the eye in addition to the glare. For otherwise the glare light itself will appear complementarily colored owing to the purely physiological "simultaneous contrast".

Colored glass obtains its color only from the transmitted light. Since the lengths of the light paths through the glass always amount to many hundreds of wavelengths even a *very weak selective absorptivity* suffices to color the glass intensively. That the glass shows the same color when looked at as when looked through is due to the fact that the light seen by looking at it has actually passed through the glass from the other side. If the back surface of the glass is painted with black lacquer, the color of the glass becomes invisible and only the colorless glare of the front surface remains. If, however, the glass is laid on a white surface, it looks colored because the light reflected by the white surface and leaving through the front surface has passed through the glass twice.

If a white textile is soaked in a dye solution, the crystal-clear fiber substance becomes selectively absorbing. The light reflected by the rear surface of the fibers, or by the fiber surfaces lying further back, has passed through the fibers several times. If the dyed fabric is soaked in water — or better, in a mixture of alcohol and benzene, then it appears dark and colorless because the reflection at the inner fiber surfaces has been stopped as a result of the equalization of the index of refraction.

Many inorganic "pigments" are pulverized melts. In their compact state they are dark; after pulverization and mixture with a binding substance reflecting surfaces are formed in the interior; the color becomes visible.

The green of leaves consists of transparent green grains. In order for foliage to appear light green, it must have in its interior sufficiently many inhomogeneities at which the light is reflected. If these are absent (coniferae, box-wood trees), a black-green color appears. Nevertheless, the chlorophyll of the coniferae is the same as that of other trees and plants.

When mixed, pigments act *subtractively* like color filters placed behind each other. Every component of the mixture extinguishes by absorption its own region of the spectrum. The colors of pigments placed *alongside each other*, on the other hand, add when mixed, as for instance, in the sectorially colored disc of a top. Also the illumination coming through colored church windows and the picture of a Lumière plate is put together *additively* from its single colors.

Very beautiful colors are produced by diffraction of light by larger aggregates of atoms, so-called *colloidal particles.* Lapis lazuli, for instance, owes its deep blue color to colloidal sulfur particles. The blue of the sky is, according to Einstein, brought about by the density variations of the air molecules which are to be expected statistically; Lord Rayleigh originally explained this phenomenon in a somewhat more special way by the diffraction on the (irregularly distributed) air molecules themselves.

Nature attains her most beautiful color ornament through interference colors, see Secs. 7 and 8 c, as witnessed in the wings of butterflies, the plumage of the tropical humming bird, the opal and mother-of-pearl. What opportunities would present themselves to painting if it were possible to develop a convenient interference color technique!

7. Colors of Thin Membranes and Thick Plates

In this section we shall discuss on one hand the age-old observations of Newton which motivated him to assume a kind of spatial structure of light and which might almost have led him to interference and the wave theory. On the other hand we shall describe the most modern experimental arrangements which serve in the most exact analysis of spectra. The mathematical problem which is the basis of both is that of the transparent plate with parallel surfaces, that is to say, the *problem of reflection and refraction at two boundary surfaces.* Until now we have actually solved this problem for only *one* boundary surface. Ordinarily, the two-surface problem is reduced to that of one boundary surface by dealing with repeated reflections and refractions. In contrast to this we shall treat the problem of the plate directly as a *boundary value problem*[1]. We seek, therefore, the extension of Fresnel's

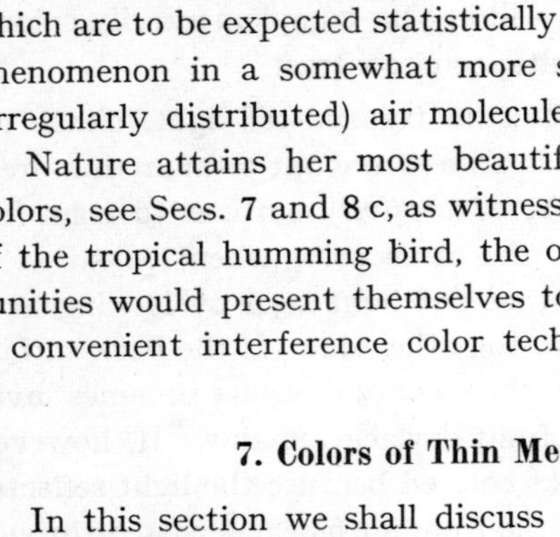

Fig. 8.
Reflection and refraction in a plate with plane parallel surfaces, considered as a boundary value problem.

[1] Of course this method has occasionally been used before; however only for special cases. See, e. g., M. Born, Optik, Berlin, Springer 1933, p. 125.

formulae corresponding to the generalized problem as described above; we thereby avoid the trouble of performing the summations over an infinite number of single processes which are necessitated by the other method. It is obvious, and will be shown in section E, that both methods must lead to the same result. However, we emphasize here already that in the case of the plate the above-mentioned single process is no longer a fundamental process and disappears in our extended boundary value problem.

A. The general case

While in the case of *one* boundary condition *two* amplitude ratios $A:B:C$ were sufficient, we now require *four* of these ratios:

$$A:B:C:D:E \tag{1}$$

The meaning of the five amplitude factors $A \ldots E$ is evident from fig. 8. The angle of incidence α is also the angle of reflection and would be encountered again as the angle of emergence of the transmitted light D if we allowed the bottom surface of the plate to border on the same material (air) as the top surface. We prefer, however, to let the medium behind the plate be arbitrary and to denote the angle of emergence by γ. The index of refraction of the plate with respect to that medium shall be n_1 and that with respect to air n. The thickness of the plate shall be $2\,h$; at the top and bottom surfaces we let $y = \pm h$. The z-axis is to be thought of as pointing out of the paper. The fact that the upward-reflected wave is represented by two arrows, bracketed together, is due only to the nature of the drawing. In reality all arrows in the figure represent, as before, not rays but unbounded plane waves.

We consider, for instance, the case of p-polarization, that is E parallel to the z-axis, and obtain above the plate

$$E_z = A\, e^{i k_1 (x \sin \alpha - y \cos \alpha)} + C\, e^{i k_1 (x \sin \alpha + y \cos \alpha)} \tag{I}$$

as in (3.1). Inside the plate eq. (3.1 a) holds true but is to be completed by adding to the right-hand side the second particular solution in this region ($+\,i\,y$ instead of $-\,i\,y$) multiplied by the arbitrary factor E:

$$E_z = B\, e^{i k_2 (x \sin \beta - y \cos \beta)} + E\, e^{i k_2 (x \sin \beta + y \cos \beta)}. \tag{II}$$

Only below the plate does the field consist of a single wave because we must include among the conditions of the problem the fact that the plate is not irradiated from below:

$$E_z = D\, e^{i k_3 (x \sin \gamma - y \cos \gamma)}. \tag{III}$$

The two laws of refraction at $y = \pm h$

$$\frac{\sin\alpha}{\sin\beta} = \frac{k_2}{k_1} = n, \qquad \frac{\sin\gamma}{\sin\beta} = \frac{k_2}{k_3} = n_1 \tag{2}$$

are necessary conditions to enable us to cancel the x-dependent factors from all terms in the following expressions. If, furthermore, we write k for k_1, we obtain for $y = +h$, instead of (3.5)

$$A\,e^{-ikh\cos\alpha} + C\,e^{+ikh\cos\alpha} = B\,e^{-iknh\cos\beta} + E\,e^{+iknh\cos\beta} \tag{3}$$

and for $y = -h$

$$B\,e^{iknh\cos\beta} + E\,e^{-iknh\cos\beta} = D\,e^{ik\left(\frac{n}{n_1}\right)h\cos\gamma} \tag{4}$$

Next we must write the expressions for H_x in I, II, and III in analogy to (3.5) and must require their continuity at $y = \pm h$. If we assume the plate to be non-magnetic, i. e. we set $m = n$, we obtain for $y = +h$ instead of (3.8):

$$A\,e^{-ikh\cos\alpha} - C\,e^{+ikh\cos\alpha} = n\,\frac{\cos\beta}{\cos\alpha}\left(B\,e^{-iknh\cos\beta} - E\,e^{+iknh\cos\beta}\right) \tag{5}$$

and for $y = -h$

$$B\,e^{iknh\cos\beta} - E\,e^{-iknh\cos\beta} = \frac{\cos\gamma}{n_1\cos\beta}\,D\,e^{ik\left(\frac{n}{n_1}\right)h\cos\gamma}. \tag{6}$$

We have, then, four linear homogeneous equations for the 5 unknowns $A \ldots E$, which we can collect in the form

$$a_i A + b_i B + c_i C + d_i D + e_i E = 0, \qquad i = 1, 2, 3, 4. \tag{7}$$

From this, the values of $A:B\ldots:E$ are computed as the ratios between the five corresponding four-rowed determinants of the scheme of coefficients $a\,b\,c\,d\,e$. Hence the Fresnel formulae of our plate problem assume the same form as before for the case of the single boundary surface, except that in place of the three-fold proportion, a five-fold proportion appears.

Since, however, the computation of this scheme of determinants becomes too cumbersome for the general case, we will in the following special examples do better to use the unsolved equations (3) to (6) as a starting point.

B. The oil spot on wet asphalt.

Everyone has seen on the pavement the beautiful interference colors on a thin layer of oil. Medium I is air, medium II a layer of oil which is to be assumed as bounded by parallel planes. If the oil were to lie on dry asphalt, its lower boundary would not be plane but would be optically rough.

Hence, as medium III we must add a layer of water wetting the asphalt. The asphalt as a black material serves to absorb the transmitted wave D and thereby prevents further reflection processes.

We look at the oil spot perpendicularly from above and assume for convenience that the (actually diffuse) illumination also comes perpendicularly from above. Then $\alpha = \beta = \gamma = 0$. Furthermore, to abbreviate, we let

$$\eta = e^{ikh}. \tag{8}$$

From (4) and (6) we obtain by eliminating D

$$B\,\eta^n + E\,\eta^{-n} = n_1\,(B\,\eta^n - E\,\eta^{-n}), \qquad \text{hence} \qquad E = \frac{n_1 - 1}{n_1 + 1}\,B\,\eta^{2n}.$$

Then (3) and (5) become

$$A\,\eta^{-1} + C\,\eta = B\,\eta^{-n}\left(1 + \frac{n_1 - 1}{n_1 + 1}\,\eta^{4n}\right),$$

$$A\,\eta^{-1} - C\,\eta = n\,B\,\eta^{-n}\left(1 - \frac{n_1 - 1}{n_1 + 1}\,\eta^{4n}\right).$$

By eliminating B, a relation between A and C results which we can write

$$\frac{C}{A} = -\eta^{-2}\,\frac{n-1}{n+1}\,\frac{1 - \nu_1\,\eta^{4n}}{1 - \nu_2\,\eta^{4n}}, \qquad \begin{cases} \nu_1 = \dfrac{n+1}{n-1}\,\dfrac{n_1 - 1}{n_1 + 1}, \\[2ex] \nu_2 = \dfrac{n-1}{n+1}\,\dfrac{n_1 - 1}{n_1 + 1}. \end{cases} \tag{9}$$

Since we are only interested in the reflected *intensity* (or rather its ratio with the incident intensity), we can simplify (9) to

$$\left|\frac{C}{A}\right|^2 = \left(\frac{n-1}{n+1}\right)^2 \left|\frac{1 - \nu_1\,\eta^{4n}}{1 - \nu_2\,\eta^{4n}}\right|^2. \tag{10}$$

In order to discuss this formula we calculate

$$(1 - \nu\,\eta^{4n})\,(1 - \nu\,\eta^{-4n}) = 1 - \nu\,(\eta^{4n} + \eta^{-4n}) + \nu^2 = 1 - 2\,\nu\cos\varphi + \nu^2 \tag{11}$$

($\nu = \nu_1$, ν_2 is real, η has the absolute value 1). The angle φ introduced here is, according to (8), defined by

$$\eta^{4n} = e^{i\varphi}, \qquad \varphi = 4\,n\,k\,h. \tag{12}$$

In section E we shall discuss the physical meaning of this "phase difference" φ in terms of optical path length.

Equation (10) becomes, because of (11)

$$\left|\frac{C}{A}\right|^2 = \left(\frac{n-1}{n+1}\right)^2 \frac{1 + \nu_1^2 - 2\,\nu_1\cos\varphi}{1 + \nu_2^2 - 2\,\nu_2\cos\varphi}. \tag{13}$$

The phase differences between intensity extremals are obtained by differentiating (13) with respect to φ and hence are found from the equation

$$0 = \{2\,\nu_1\,(1 + \nu_2{}^2 - 2\,\nu_2 \cos\varphi) - 2\,\nu_2\,(1 + \nu_1{}^2 - 2\,\nu_1 \cos\varphi)\} \sin\varphi = \\ = 2\,(\nu_1 - \nu_2)\,(1 - \nu_1\,\nu_2) \sin\varphi$$

to be

$$\varphi = z\pi, \qquad z = \text{integer} \tag{14}$$

Substituting this in (13), one finds by elementary computation

$$\begin{aligned} &\varphi = \pi, 3\,\pi, 5\pi \ldots; \left|\frac{C}{A}\right|^2 = \left(\frac{n\,n_1 - 1}{n\,n_1 + 1}\right)^2, \quad \text{maxima,} \\ &\varphi = 2\,\pi, 4\pi, 6\pi \ldots; \left|\frac{C}{A}\right|^2 = \left(\frac{n - n_1}{n + n_1}\right)^2, \quad \text{minima.} \end{aligned} \tag{15}$$

The denotations "maxima" and "minima" refer to the case $n_1 < n$ which prevails in the case of oil on water. In the opposite case the denotations are reversed.

The layer of oil is very thin. Though not monomolecular, it has, however, only a thickness of perhaps a wavelength λ_v at the violet end of the spectrum. If we assume this, i. e. let $2\,h = \lambda_v$, and estimate the index of refraction of oil to be 1.5, then we obtain, according to the definition (12) of φ

$$\varphi = 6 \cdot 2\pi \frac{h}{\lambda_v} = 6\pi.$$

This means, according to (14) and (15), that the reflected *violet* light has a *minimum* for $z = 6$. On the other hand, since $\lambda_r \sim 2\,\lambda_v$, one obtains for the red end of the spectrum

$$\varphi = 6 \cdot 2\pi \frac{h}{\lambda_r} = 3\pi.$$

Hence, according to (14) and (15), one obtains a *maximum* of reflected *red* light for $z = 3$. The middle portion of the spectrum gives rise to a further minimum and maximum corresponding to $z = 4$ and 5. Hence, the light reflected by the oil spot has a mixed color, namely a predominantly blue-green tint under our assumptions. If the thickness of the layer varies locally, the color also varies.

C. Coated (non-reflecting) lenses

The light passing through a system of lenses is weakened by reflection. Even though for central rays (perpendicularly incident light) the attenuation arising from a single reflection is small (4% according to p. 20), it becomes considerable for a system of lenses. Many optical firms strive to eliminate this reflection, which is especially disadvantageous in the design of photographic apparatus. The problem is solved by applying thin layers to all surfaces at which the lenses of the system border on air. Such a layer was originally produced by a *structural modification* of the glass surface (etching or solution of components of the glass flux). Nowadays one prefers to evaporate onto the glass a layer of suitable material whose index of refraction is lower than that of glass, taking care to make the layer as uniform as possible.

If we consider only one boundary layer and because of its thinness neglect the curvature, we are again faced with our problem of three media:

I	air,	index of refraction 1
II	surface layer,	index of refraction n
III	lens,	index of refraction $n_1 = n/n_g$

where n_g is the index of refraction of the lens glass relative to air. Since our n_1 should correspond to the transition $II \to III$ in the formulae (15), the transition $I \to III$ is characterized by $n_g = n/n_1$ in conformity with eq. (3.3 b). When, in view of this, we let $n_1 = n/n_g$ in the first eq. (15), we obtain *zero reflection*, for

$$\frac{n^2}{n_g} = 1, \qquad n = \sqrt{1 \cdot n_g}\,. \tag{16}$$

Hence, the index of refraction n of the evaporated layer II should be the *geometric mean* of the indices of refraction 1 and n_g of media I and III (both relative to air). The optical industry tries to fulfill this requirement together with the first condition (15), namely $\varphi = \pi$, by a choice of a suitable material, e. g. lithium fluoride, and by a suitable thickness of the layer to be precipitated. Since however, n, n_g and therefore also φ depend on wavelength, the condition $\varphi = \pi$, in particular, cannot be satisfied for all wavelengths. One favors the brightest spot of the spectrum ($\lambda = 0.55\ \mu$ in the yellow-green) and suppresses reflection as completely as possible for this wavelength. Then the reflection of the complementary purple is, to be sure, not zero but it is nevertheless small. Indeed a lens prepared in this way has a weak purplish tint.

We have here assumed the layer to be homogeneous, hence n to be a constant for a given wavelength. For literature on reflection in inhomogeneous layers we refer to footnote[1] on p. 19. However, we must still note that reflection is to be avoided not only at the front surface but also at the back surface of the lens as well as at all other surfaces of the lens system which border on air. Because of the interchangeability of media I and III in our condition $n = \sqrt{1 \cdot n_g}$, this is accomplished by the same compensation procedure, that is, by evaporating the same thickness of the same material on the back of the lens as on the front. Our result is true not only for the central ray ($\alpha = \beta = \gamma = 0$) but it is true sufficiently closely also for neighboring ray directions. For, since in our initial formulae (3) and (6) only the cosines of these angles appeared, only a "cosine-error" appears in the reflection of neighboring rays (a second order deviation from zero reflection).

D. Soap Bubbles and Newton's Rings.

The color play of thin *soap bubbles* is explained in the same manner as that of the oil spot. The only difference is that medium *III* (interior of the soap bubble) is now the same as medium *I*, namely air. Hence, we have $n_1 = n$. According to (15) this has the result that the intensity of the minima becomes zero and that, therefore, when the condition for a minimum is fulfilled for a wavelength λ_1, the complementary color λ_2 is seen in particularly pure form. In general however, the reflected colors are mixed colors.

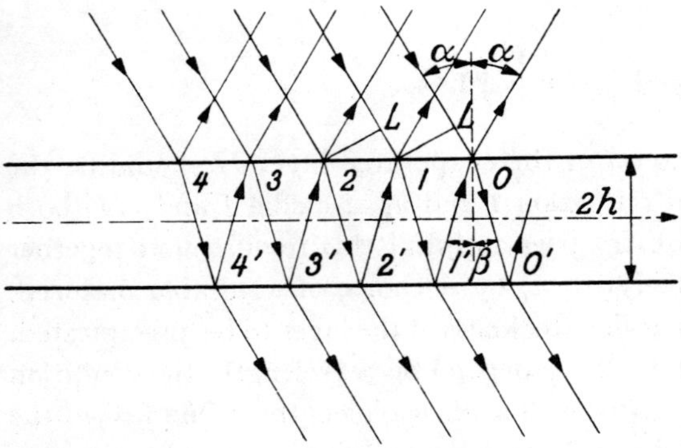

Fig. 9.
Summation process for multiple reflection in a plane-parallel plate.

We estimate the order of magnitude of the thickness of the soap bubble as equal to or smaller than some visible wavelength. This we conclude from the fact that when the bubble is strongly inflated, a dark spot appears at the top which indicates a very small thickness compared to the wavelength. Because the soap solution flows off to the bottom, the thickness of the soap film does indeed become vanishingly small at the top.

Quite similar circumstances prevail in the case of *Newton's rings*: a plano-convex lens is placed with its weakly curved convex surface on a plane glass plate, so that between the two an air gap is created which widens toward the outside. Again media *I* and *III* have the same index of refraction as compared to medium *II* which is now air. If the illumination is monochromatic, one sees a number of dark circles between bright rings. In white light a small number of colored rings appear. These show no pure spectral colors but, rather, mixed colors. The transmitted light is colored complementarily to the reflected light.

E. Comparison of Methods: Summation or Boundary-value Treatment?

Let a plane-parallel glass plate of thickness $2h$ which at the front and back borders on air be obliquely illuminated by parallel, monochromatic light. Ordinarily one argues in the following manner (see fig. 9): at point 0 on the front face there emerges, besides the directly reflected light, also light which entered the plate at 1 and was reflected at 1′. Furthermore, light which entered at 2 and was reflected at 2′, 1 and 1′, etc., will also leave the plate at 0. Generally speaking, the light emerging at an arbitrary point of the front face is a sequence of components which have been refracted *twice* and reflected an *odd* number of times. Correspondingly, the light emerging from an arbitrary point on the back face is composed of a sequence of component waves which have been refracted *twice* and reflected an *even* number of times. We must calculate the differences in *phase* and *amplitude* of these various rays.

The length of the light path 11′0 measured in wavelengths λ_g in the glass (λ = wavelength in air) amounts to

$$\frac{4\,h/\cos\beta}{\lambda_g} = \frac{4\,n\,h/\cos\beta}{\lambda}.$$

When multiplied by 2π, this is the increase in phase which the light attains along the path 11′0

$$\varphi_1 = \frac{2\pi}{\lambda}\frac{4\,n\,h}{\cos\beta} = \frac{4\,k\,n\,h}{\cos\beta}. \tag{17}$$

But, in addition, this light is ahead in phase compared to the light incident at 0 by

$$\varphi_2 = 2\pi\frac{0L}{\lambda} = k\cdot 0L, \tag{18}$$

where L is the point at which the perpendicular form 1 to $0L$ intersects the ray, see fig. 9:

$$0L = \sin\alpha\cdot 01 = \sin\alpha\cdot 2\tan\beta\cdot 2\,h.$$

The *total phase difference* against the light incident at 0 amounts, then, to

$$\varphi = \varphi_1 - \varphi_2 = \frac{4\,n\,k\,h}{\cos\beta}\left(1 - \frac{\sin\alpha\sin\beta}{n}\right) = \frac{4\,n\,k\,h}{\cos\beta}(1-\sin^2\beta) = 4\,n\,k\,h\cos\beta. \tag{18 a}$$

For perpendicular incidence ($\beta = 0$) this φ proves to be identical with the auxiliary angle φ introduced in (12) and explains the physical meaning of the latter. For the light path 22′11′0 the corresponding phase difference obviously amounts to 2φ, for the following path to 3φ, etc.

In order to determine, on the other hand, the *amplitude differences*, we use the energy coefficients r and d from (4.18) which are, as was emphasized there, the same for the front and back side of the plate. Expressed in these terms, the factor which is to multiply the amplitude in the case of one-, three-, five-, . . .fold reflection and, in every case, two-fold transit across the front face amounts to

$$\sqrt{r}\,d, \quad \sqrt{r}\,r\,d, \quad \sqrt{r}\,r^2 d, \ldots \tag{19}$$

From (18 a) and (19) it follows that the summation of the first p rays is given by

$$\sqrt{r}\,e^{i\varphi}\cdot d + \sqrt{r}\,r\,e^{2i\varphi}\,d + \ldots + \sqrt{r}\,r^{p-1}\,e^{pi\varphi}\cdot d. \tag{20}$$

To this is to be added the contribution corresponding to the direct reflection at 0. This is, including the correct phase factor[1], $C/A = -\sqrt{r}$. One obtains finally:

$$\frac{C}{A} = -\sqrt{r}\,\{1 - d\,e^{i\varphi}\,(1 + r\,e^{i\varphi} + \ldots + r^{p-1}\,e^{(p-1)i\varphi})\} = -\sqrt{r}\left\{1 - d\,e^{i\varphi}\,\frac{1 - r^p\,e^{ip\varphi}}{1 - r\,e^{i\varphi}}\right\}. \tag{21}$$

For $p = \infty$, because $r < 1$ and $r + d = 1$, this contracts to:

$$\frac{C}{A} = -\sqrt{r}\left\{1 - \frac{d\,e^{i\varphi}}{1 - r\,e^{i\varphi}}\right\} = -\sqrt{r}\,\frac{1 - e^{i\varphi}}{1 - r\,e^{i\varphi}}. \tag{22}$$

The amplitude factor for the light passing through the plate at 0′, which we shall call D/A, is determined quite similarly. Because of the two-fold passing across the boundary surface (once front, once back) and zero-, two-, four-, . . . fold reflection (see fig. 9), one obtains instead of (19) and (20)

$$d,\, r\,d,\, r^2 d, \ldots \tag{19 a}$$

$$e^{\frac{i\varphi_1}{2}}\,(d + r\,e^{i\varphi}\,d + r^2\,e^{2i\varphi}\,d + \ldots + r^{p-1}\,e^{(p-1)i\varphi}\,d) \tag{20 a}$$

and instead of (21) and (22), respectively

$$\frac{D}{A} = e^{\frac{i\varphi_1}{2}}\,d\,(1 + r\,e^{i\varphi} + \ldots + r^{p-1}\,e^{i(p-1)\varphi}) = e^{\frac{i\varphi_1}{2}}\,d\,\frac{1 - r^p\,e^{ip\varphi}}{1 - r\,e^{i\varphi}} \tag{21 a}$$

[1] For reflection at the denser medium the sign of C/A was chosen opposite from that for reflection at the rarer medium; see, for instance, the comment after formula (5.3).

and for $p = \infty$

$$\frac{D}{A} = \frac{e^{\frac{i\varphi_1}{2}} d}{1 - r e^{i\varphi}}. \tag{22 a}$$

We note here that the expressions (21) and (21 a) become essentially the same if we suppress in (21) the first term on the right which is due to direct reflection. For we then have

$$\frac{C}{A} = \sqrt{r}\, e^{i\left(\varphi - \frac{\varphi_1}{2}\right)} \frac{D}{A} \quad \text{and hence} \quad \left|\frac{C}{A}\right|^2 = r \left|\frac{D}{A}\right|^2. \tag{23}$$

The reflected intensity is thus equal to the intensity transmitted through the plate except for a factor r.

The use of the symbols r and d in (21), (22) already indicate that the results are valid for both polarization cases (parallel and perpendicular to the plane of incidence). r and d only stand for somewhat different expressions in the two cases.

In particular, we consider the special case of perpendicular incidence in which this difference disappears and in which a comparison of (22) with our formula (9) is possible if we specialize the latter by setting $n_1 = n$ (air also at the back face of the plate). Then we have to set in (9)

$$\nu_1 = 1, \qquad \nu_2 = \frac{(n+1)^2}{(n-1)^2} = r. \tag{24}$$

In addition, the following must be taken into account: in (9), just as in our general assumption (I) on p. 41, C and A refer to the center of the plate, namely to $y = 0$ in the choice of coordinates used there, and not to the top surface of the plate $y = h$. Therefore, at the top of the plate A must be multiplied by the factor $\exp(ikh)$ and C by the factor $\exp(-ikh)$ (both for perpendicular incidence: $\alpha = 0$) if we wish to compare (9) with our present amplitude ratio C/A given by (22), since the coordinates of the latter refer to the top surface. This means that we must suppress in (9) the factor $\eta^{-2} = \exp(-2ik_1 h)$. Thereupon (9) becomes, using (24),

$$\frac{C}{A} = -\sqrt{r}\, \frac{1 - e^{i\varphi}}{1 - r e^{i\varphi}} \tag{25}$$

which now indeed agrees with (22).

Thus both of our methods lead to the same result and not only in the special case of perpendicular incidence which was used here for the comparison but quite generally. Both methods have their advantages and their disadvantages. The *boundary value method* saves us the somewhat laborious phase considerations in fig. 9. The *summation method* seems to lend itself more readily to visualization and is **not** limited to the assumption that $p = \infty$. The latter can also be used to treat the case of a plate of finite length and of an incident light bundle of finite width[1]. This problem is inaccessible to the formal hypothesis underlying our boundary value problem which is restricted to the xy-plane. That is why the summation method is preferred in treating the problems on resolving power in Chapter VI. Another reason is that this method fits in better with the usual grating theory. As we shall see, both methods are in fact equivalent for the two types of high resolution interference apparatus which we will now discuss.

F. The Lummer-Gehrke Plate (1902).

In our discussion of total reflection in Sec. 5 the wave was incident in the denser medium and emerged into air. For angles of incidence $\alpha \sim \alpha_{tot}$ we obtained angles of emergence β nearly equal to $\pi/2$ and a reflecting power of almost 1. Lummer's original idea was to let the light impinge on the top

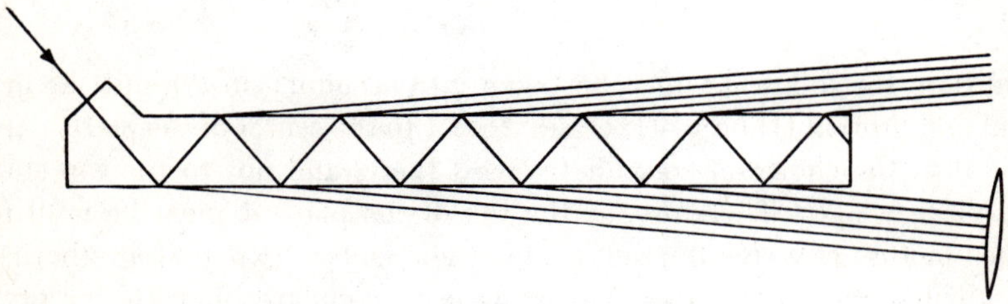

Fig. 10.
The interference of rays in the Lummer Plate.

of the plate at a grazing angle so that the refracted portion would be reflected back and forth at an angle close to α_{tot}. In this way the high reflecting power r of almost 1 would be utilized. Gehrke simplified the procedure by capping the plate with a prism of angle α_{tot} (see fig. 10). Light which is incident perpendicularly upon the face of the prism strikes the lower surface and thence

[1]The fact that the summation method is only an approximation in the sense that it does not account for the diffraction phenomena at the corners of a finite plate and at the boundaries of the light bundle has hardly any practical significance.

alternately the upper and lower surfaces under the desired angle; it emerges at a grazing angle. In this way the first reflection of the incident beam is suppressed as had indeed been assumed in eq. (23).

The number p is not very large for the Lummer Plate because a perfectly homogeneous plane-parallel plate of thickness say 1 cm. cannot be made with a length much greater than 20 cm. Nevertheless we can without hesitation go to the limit $p \to \infty$ and, hence, we can use eq. (22 a) as a starting point. We will write it here in the form

$$\left|\frac{D}{A}\right| = \frac{1-r}{|1-r\,e^{i\varphi}|}. \tag{26}$$

This is justified because the limit $\alpha \to \alpha_{tot}$ cannot be approached arbitrarily closely. If for no other reason, this is so because we never deal with a precisely parallel incident plane wave but always with a wave bundle which has a certain angular distribution. Hence r is never exactly equal to 1, but only to a certain degree of approximation. Therefore, already for $p = 20$, r^p becomes vanishingly small and it is immaterial whether we set p equal to the maximum value occurring in the Lummer Plate or to ∞. Thus it is evident that in spite of the finite value of p there can be no difference between the results of the summation and boundary value methods.

From eq. (26) we find immediately

$$\begin{aligned} &\left|\frac{D}{A}\right| = 1 \text{ for } \varphi = 2\pi z, \;\; z = \text{integer} \\ &\left|\frac{D}{A}\right| \sim 0 \text{ for all } \varphi \text{ appreciably different from } 2\pi z. \end{aligned} \tag{27}$$

The latter expression is correct because the numerator $1-r \sim 0$, the former holds because for $\varphi = 2\pi z$ numerator and denominator of (26) become exactly equal.

In order to see what "appreciably different" means, we rewrite the denominator of (26) in the form

$$\sqrt{(1-r e^{i\varphi})(1-r e^{-i\varphi})} = \sqrt{1+r^2-2r\cos\varphi}. \tag{27 a}$$

We set $\varphi = 2\pi z - \Delta\varphi$, and hence $\cos\varphi = \cos\Delta\varphi = 1 - \dfrac{(\Delta\varphi)^2}{2}$. In particular, we seek those values of $\Delta\varphi$ which correspond to the so-called "half-width" of the intensity maximum 1 occuring at $\varphi = 2\pi z$, that is, those values which satisfy the condition

$$\left|\frac{D}{A}\right|^2 = \frac{1}{2} = \frac{(1-r)^2}{(1-r)^2 + r(\Delta\varphi)^2}. \tag{28}$$

Carrying out the computation one obtains directly

$$(1-r)^2 + r\,(\Delta\varphi)^2 = 2\,(1-r)^2, \qquad \Delta\varphi = \frac{1-r}{\sqrt{r}} \sim \pm\,(1-r).$$

The half-width is twice $|\Delta\varphi|$, hence

$$2\,|\Delta\varphi| \sim 2\,(1-r). \tag{28 a}$$

As expected, this width becomes narrower as r approaches its limiting value 1. This *low value of the half-width* will play a decisive role for the *resolving power of the Lummer Plate.* We shall discuss this problem in greater detail in Chap. VI.

G. The Interferometer of Perot and Fabry (about 1900)

While Lummer attained a high reflecting power r by approaching very closely the limiting angle of total reflection, Perot and Fabry employed the surfaces of a half-silvered glass plate and used an angle of incidence almost perpendicular to the surface. The importance of their method is increased by the fact that the glass plate can be replaced by an "air plate" between two glass surfaces which have been silvered semi-transparently. These surfaces can be spaced by means of invar-steel pieces and in this way a *standard measure* ("etalon") for the exact measurement of wavelengths is created. This standard is entirely independent of temperature, index of refraction or irregularities in the glass.

It is again advantageous to use the boundary value method. But we must alter our former boundary conditions. Assuming p-polarization, that is $\mathbf{E}$ parallel to the z-axis, we consider the z-component of the second Maxwell eq. (2.4). For the displacement current $\dot{\mathbf{D}}$ in this equation we substitute the specific conduction current $\sigma\, E_z$ in the silver layer. We integrate this equation over a rectangle lying in the x, y-plane of length 1 in the x-direction and having the very small thickness of the silver layer as its width in the y-direction. The left-hand side of the integrated equation is then equal to the total current per unit length in the silver layer. The right-hand side yields, according to Stokes' Theorem, the contour integral of $\mathbf{H}$ around the rectangle which equals the *discontinuity of* H_x in passing through the silver layer. Instead of the previous continuity of H_x we have now a discontinuity in H_x which is proportional to E_z. We write

$$\text{Discontinuity of} \quad H_x = -g\sqrt{\frac{\varepsilon_0}{\mu_0}}\,E_z. \tag{29}$$

g is a factor of proportionality which depends on the conductivity and thickness of the silver layer and is dimensionless owing to the factor $(\varepsilon_0/\mu_0)^{1/2}$. Because of the inertia of the electrons g is, in the visible spectrum, actually not a real but a complex number.

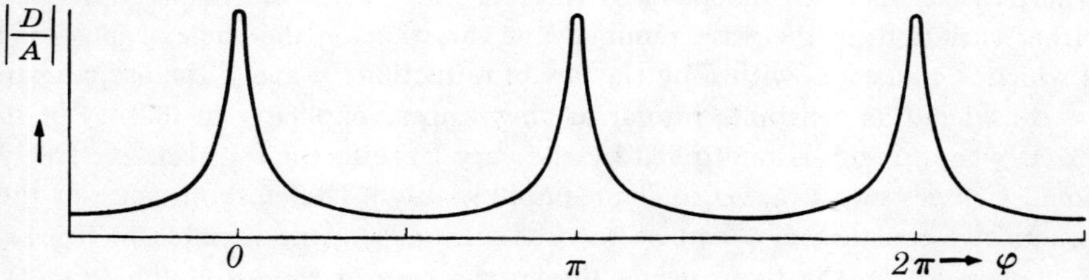

Fig. 11.
The amplitude-ratio $|D/A|$ vs. the phase difference φ for the Lummer Plate ($|D/A|_{max} = 1$) and for the Perot-Fabry Etalon ($|D/A|_{max} \ll 1$).

The continuity of E_z and the resulting boundary conditions (3) and (4) as well as the law of refraction (2) are unchanged by the silvering; they become somewhat simpler, however, because now $n_1 = n$ and $\gamma = \alpha$ (same conditions at the top and bottom plate surfaces). On the other hand, according to (29), the boundary conditions (5) and (6) must be modified in the following manner:

$$(30) \quad \begin{cases} (A\,e^{-ikh\cos\alpha} - C\,e^{+ikh\cos\alpha})\cos\alpha - (B\,e^{-inkh\cos\beta} - E\,e^{+inkh\cos\beta})\,n\cos\beta = \\ \qquad g\,(A\,e^{-ikh\cos\alpha} + C\,e^{+ikh\cos\alpha}) = g\,(B\,e^{-inkh\cos\beta} + E\,e^{inkh\cos\beta}), \end{cases}$$

$$(31) \quad \begin{cases} (B\,e^{+inkh\cos\beta} - E\,e^{-inkh\cos\beta})\,n\cos\beta - D\,e^{+ikh\cos\alpha}\cos\alpha = \\ \qquad = g\,(B\,e^{+inkh\cos\beta} + E\,e^{-inkh\cos\beta}) = g\,D\,e^{+ikh\cos\alpha}. \end{cases}$$

The two forms in which the right-hand sides of these equations are written correspond to the two ways of expressing the value of E_z in (29) in terms of the left- or right-hand sides of eqs. (3) and (4), respectively. Thus the four equations embodied in (30) and (31) represent the complete system of boundary conditions pertaining to the present problem.

For most practical applications only the transmitted light, represented by the quotient D/A, is of interest. By elementary, though somewhat laborious, computation, we obtain from (30) and (31)

$$\frac{D}{A} = \frac{e^{-2ikh\cos\alpha}}{(1+g/\cos\alpha)\cos\frac{\varphi}{2} - \frac{i}{2}\left(\frac{(1+g/\cos\alpha)^2}{n\cos\beta/\cos\alpha} + \frac{n\cos\beta}{\cos\alpha}\right)\sin\frac{\varphi}{2}}; \tag{32}$$

φ represents the phase difference generated by the ray in passing back and forth in the plate and is defined as in (18 a) by

$$\varphi = 4\,n\,k\,h\cos\beta \quad \text{(where } n = 1 \text{ in the case of the air plate).} \tag{32 a}$$

The dependence of the absolute value of (32) on φ is shown in fig. 11, where variations in the independent variable φ may be thought of as expressing either variations in the wave number k or variations in the angle of incidence α which is connected with β by the law of refraction. α and β themselves can be considered as constants insofar as they appear explicitly in (32). But in (32 a) where $\cos\beta$ is multiplied by the very large factor $k\,h$ even extremely small changes in β cause φ to change appreciably. Therefore, in spite of the practically constant β, φ can be used as an independent variable in fig. 11.

Let us check the figure using the special case of the air-étalon ($n = 1$), almost perpendicular incidence ($\alpha = \beta = 0$) and almost real g. Equation (32) then yields simply

$$\left|\frac{D}{A}\right|^{-2} = (1+g)^2\cos^2\frac{\varphi}{2} + \frac{1}{4}([1+g]^2+1)^2\sin^2\frac{\varphi}{2}. \tag{33}$$

The extremal condition which follows from this by differentiation with respect to φ is:

$$\sin\frac{\varphi}{2}\cos\frac{\varphi}{2} = 0.$$

To $\sin\varphi/2 = 0$ corresponds:

$$\left|\frac{D}{A}\right|_{max} = \frac{1}{1+g}, \qquad \varphi = 2\,z\,\pi, \tag{33 a}$$

to $\cos\varphi/2 = 0$:

$$\left|\frac{D}{A}\right|_{min} = \frac{2}{(1+g)^2+1}, \qquad \varphi = (2\,z+1)\,\pi. \tag{33 b}$$

In both cases z is a very large integer.

In (33 a, b) we assumed g to be large. By definition this assumption corresponds to a heavy layer of silver (a strong conduction current). Consequently the incident light is very much weakened even at the maxima. The minima, on the other hand, are weaker than the maxima by a factor of $\frac{1}{1+g}$. The maxima are equidistantly spaced, and so are the minima which lie halfway between the maxima. The maxima are *sharp*; the minima are *very flat.* This is a consequence of (33) because only if the condition for a maximum, i. e. $\sin\varphi/2 = 0$, is *exactly* fulfilled is the maximum value (33 a) of order of magnitude $(1+g)^{-1}$ attained; for all other φ the second term on the right-hand side of (33) dominates because of the fourth power of g which it contains; the resulting magnitude of $|D/A|$ is then about $[(1+g)^2 \sin\varphi/2]^{-1}$ and reaches the minimum as given by (33 b). Thus fig. 11 has been checked for the case of sufficiently large g.

We shall also compute the "half-width" of the intensity maxima. Since, according to (33 a), the latter are equal to $(1+g)^{-2}$, we must substitute on the left-hand side of (33) the value $2\,(1+g)^2$. On the right-hand side we let

$$\frac{\varphi}{2} = z\pi - \Delta\varphi, \qquad \sin^2\frac{\varphi}{2} = (\Delta\varphi)^2, \qquad \cos^2\frac{\varphi}{2} = 1 - (\Delta\varphi)^2$$

and dividing by $(1+g)^2$, we obtain

$$2 = 1 + \frac{1}{4}(1+g)^2(\Delta\varphi)^2, \qquad \Delta\varphi \sim \pm\frac{2}{1+g}.$$

The half-width is therefore

$$2\,|\Delta\varphi| = \frac{4}{1+g}. \tag{34}$$

In exercise I.7 we shall explain these results concerning the positions and half-widths of the interference maxima from the point of view of the electromagnetic characteristic oscillations.

In Chap. VI we shall see that the Perot-Fabry Etalon attains its excellent resolving power only because of a large value of g. Only for large g, i. e. strong silvering, is the half-width of the maxima sufficiently small and thus the prime purpose of this interferometer, namely the resolution of fine structures, is attained. One must, therefore, accept the large loss in intensity which is engendered by heavy silver layers. The Lummer Plate, because of $r \sim 1$, is preferable from the point of view of intensity. But it cannot attain the resolving power of the Perot-Fabry Etalon and is, furthermore, experimentally less convenient than the latter.

It is to be emphasized that the general formula (32) encompasses also the Lummer Plate. The latter is described by the opposite limiting case $g = 0$ (no silvering). For $\varphi = 2\pi z$ (32) gives then immediately

$$\left|\frac{D}{A}\right|_{max} = 1 \tag{35}$$

which agrees with the first equation (27). For all other values of φ, on the other hand,

$$\left|\frac{D}{A}\right| = \left[\cos^2\frac{\varphi}{2} + \frac{1}{4}\left(\frac{\cos\alpha}{n\cos\beta} + \frac{n\cos\beta}{\cos\alpha}\right)^2 \sin^2\frac{\varphi}{2}\right]^{-1/2}.$$

Passing now to grazing incidence, as required by Lummer, that is allowing $\cos\alpha$ to approach 0, the coefficient of $\sin^2\varphi/2$ tends to infinity and one obtains, in agreement with the second eq. (27)

$$\left|\frac{D}{A}\right| \to 0. \tag{35 a}$$

(35) and (35 a) confirm our earlier assertion that the Lummer plate can also be treated by means of the boundary value method.

8. Standing Light Waves

The question of the position of the "light vector" with respect to the plane of polarization was left unanswered by the elastic theory of light. Fresnel was of the opinion that the light vector was perpendicular to the plane of polarization while F. Neumann thought it to be parallel to that plane. But the word light vector could not be clearly defined on the basis of the elastic theory. Electromagnetically we have two light vectors $\mathbf{E}$ and $\mathbf{H}$ (in a crystal there are even four: $\mathbf{E}$, $\mathbf{D}$ and $\mathbf{H}$, $\mathbf{B}$). We saw in Sec. 4 that in the production of polarized light by reflection, the electric vector $\mathbf{E}$ is perpendicular and the magnetic vector is parallel to the plane of incidence. Since in this case the plane of polarization is traditionally identified with the plane of incidence, we have also that $\mathbf{E}$ is perpendicular and $\mathbf{H}$ is parallel to the plane of polarization. Therefore, depending on whether one calls $\mathbf{E}$ or $\mathbf{H}$ the light vector, one decides the question in favor of Fresnel or Neumann. But even in this way only a nominal definition of the word "light vector" is achievable; physical significance, however, can be attributed to it on the force of electromagnetic evidence.

When light acts on a photographic layer, an electron is removed from a silver bromide or chloride molecule and thereby a silver atom is prepared to blacken during the development of the film. Only the electric field strength $\mathbf{E}$ is able to accomplish this. Since, moreover, the processes occurring in the eye's retina are quite similar (both phenomena are without doubt "photoelectric effects"), we have good reason to give the name "light vector" to the field vector $\mathbf{E}$ rather than to the magnetic vector $\mathbf{H}$.

The beautiful experiments by O. Wiener (Ann. d. Physik, 1890) have placed the results of these general considerations on a sound empirical basis. This was accomplished by a thorough study of the photographic process.

A. Monochromatic, linearly polarized light which is incident perpendicularly upon a metal surface

A polished silver mirror is used as a reflector. The normal to this surface shall, as before, be the y-axis. Let the direction of incidence be the negative y-direction and the direction of reflection the positive y-direction. Because of the transversality of light, $E_y = 0$. There is no need to distinguish between E_x and E_z since both directions are equivalent for normal incidence. We can write for either or both of these components:

$$E_i = A\, e^{-iky-i\omega t}, \qquad (1)$$
$$E_r = C\, e^{+iky-i\omega t}.$$

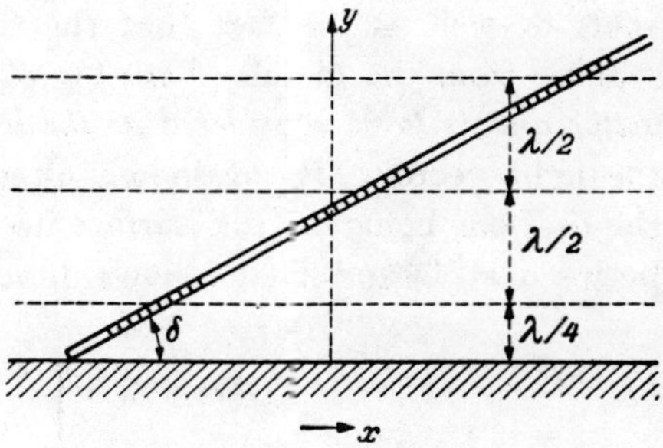

Fig. 12.

Wiener's experiment on standing light waves. The photographic plate placed at an angle δ is blackened at the antinodes of the electric vector (indicated by dotted lines).

As a good conductor ($\sigma \to \infty$) the silver mirror does not permit the existence of an electric field tangential to its surface. Any such field vanishes because of conduction. Hence, we have

$$E_{tan} = E_i + E_r = 0 \quad \text{for} \quad y = 0. \qquad (2)$$

From (1), it follows that

$$C = -A \quad \text{(Phase change during reflection)} \qquad (3)$$

and, writing real parts and letting A be real, we have for $y \geqq 0$

$$E = \operatorname{Re}(E_i + E_r) = 2A \sin k y \cos \omega t. \qquad (4)$$

This is the typical expression for a *standing wave*. The nodes are at

$$k y = n\pi, \qquad y = n\frac{\lambda}{2},$$

the antinodes at

$$k y = \left(n + \frac{1}{2}\right)\pi, \qquad y = \frac{\lambda}{4} + n\frac{\lambda}{2}.$$

We would expect maximum photographic blackening at the antinodes and no blackening at the nodes. The distance of the first blackening from the metal surface should then be equal to half of the spacing between succeeding blackened spots.

To prove the foregoing, Wiener used an age-old method for measuring the water level of rivers. A photographically sensitive film which was spread on the bottom surface of a glass plate was placed against a silver mirror at the extremely small angle δ, as shown in fig. 12. Distances measured perpendicular to the mirror are thus magnified on the film by a factor of $1/\delta$. The distances $\lambda/4$ and $\lambda/2$ are in this way depicted on a macroscopically measurable scale.

The result confirmed completely the expected periodic spacing of blackened spots as well as the fact that the first maximum occurred at 1/2 of that spacing from the metal. *Thus the electric vector* E *is indeed photographically active and is to be considered as the light vector*[1]. The magnetic vector is *not* the light vector. Its antinodes alternate with those of the electric vector, the first one being on the surface itself. Indeed, using Maxwell's relationship between H, D and E, it follows directly from (4) that we obtain for H

$$\mathsf{H} = 2A\sqrt{\frac{\varepsilon_0}{\mu_0}}\cos ky \sin \omega t. \tag{5}$$

B. Obliquely incident light.

The following experiment performed by Wiener is also very revealing. The photographic film was placed in the same position as before but the light was incident upon the silver mirror at an angle of 45° with the normal. When the light was polarized in the plane of incidence, (E perpendicular to that plane), then the film exhibited blackened stripes which were qualitatively in the same positions as in the case of perpendicular incidence. If, however, the plane of polarization was placed perpendicular to the plane of incidence and the angle of incidence was made precisely 45°, then no stripes appeared and rather the blackening was uniformly distributed over the plate.

In exercise I.8 the results of this experiment will be computed for arbitrary angles of incidence α.

---

[1] The fact that, according to H. Jäger, Ann. d. Phys. (Lpz) **34**, 280, 1939, the proof can be based directly on the photoelectric process instead of photography corresponds to the above remark concerning the identity of photographic and photoelectric action.

C. Lippmann's color photography.

Lippmann arranged a very fine grained photographic film in the manner of Wiener's experiment by placing it flat on a mercury surface and shining a *spectrum* perpendicularly upon the surface. At the antinodes of the standing waves thus created, a system of Wiener-type silver layers was formed photochemically. These layers were spaced at distances of $\lambda/2$ where λ was the wavelength of the spectral region which illuminated the particular point in question.

If a film prepared in this way is developed and then illuminated perpendicularly with *white* light, every spot of the film emits the same wavelength λ to which it had been exposed during preparation. Only this wavelength (or a submultiple of it) will fit into the screen formed by the system of Wiener planes. All other λ are destroyed by interference. Thus, looking perpendicularly at the film, one sees the whole spectrum in brilliant interference colors. If the film is breathed upon, it swells and the spectrum is shifted toward the *red* because longer wavelengths fit into the expanded screen. When looked upon obliquely, the spectrum shifts toward the *violet*. This is due to the relationship

$$2\,d\cos\alpha = \lambda_a \tag{6}$$

where λ_a is the wavelength seen at the angle of reflection α and d is the spacing of the planes of the screen. Equation (6) represents the condition that a plane wave incident at the angle α shall be reflected by all planes of the system with the same phase (or phases differing by multiples of 2π). We shall encounter this equation again under the name of *Bragg's Equation* in Chap. V, Sec. 32 where it will be in a somewhat more general form and somewhat different notation. For the present it suffices to make two observations:

1. For $d = \lambda/2$ and $\alpha = 0$ (perpendicular incidence and viewing), $\lambda_a = \lambda$, i. e. the *color is unchanged.*

2. For $d = \lambda/2$ and $\alpha \neq 0$ (oblique incidence and viewing), $\lambda_a = \lambda\cos\alpha < \lambda$, i. e. *shift toward the violet.*

As is well known, the modern practical solution of the problem of color photography is based on entirely different principles. Nevertheless, as the earliest proposal of "photography in natural colors", the Lippmann method has great historical interest.

CHAPTER II

OPTICS OF MOVING MEDIA AND LIGHT SOURCES ASTRONOMICAL TOPICS

The fundamental optical constant is the velocity of light in vacuum. According to the Theory of Relativity this constant governs the scale of time and space. We shall, therefore, discuss the velocity of light before turning, in later chapters, to the optical properties of matter which, though apparently more elementary, are fundamentally really more complicated. We shall discover the most important facts about the velocity of light, c, not from terrestrial experiments but rather by discussing astronomical measurements.

9. Measurement of the velocity of light

The satellites of Jupiter were discovered in 1610 by Galileo. He called them Medicean Stars in honor of his patron Duke Cosimo of Florence. These were the four bright satellites which are close to Jupiter. The periods of their orbital motions amount to several days and are, therefore, very short compared to the period of Jupiter's orbital motion around the sun (twelve years). At the present time twelve satellites of Jupiter are known.

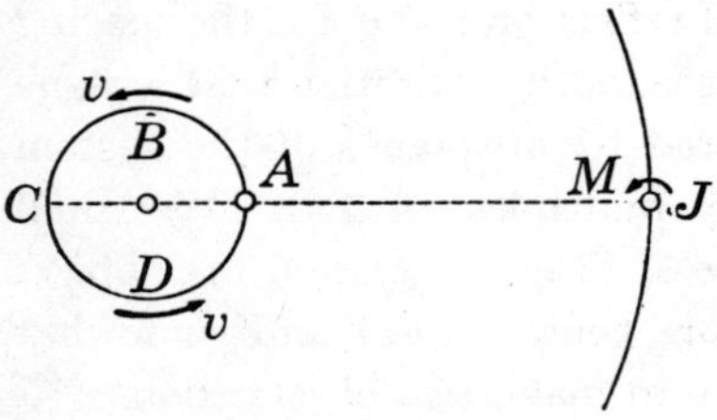

Fig. 13.
Determination of the velocity of light by the method of Olaf Römer. J, Jupiter; M, satellite of Jupiter; ABCD, earth's orbit.

The periods of these satellites can be determined exactly by means of their eclipses (times at which they enter Jupiter's shadow in the sun light). The Dane Olaf Römer (1676) discovered remarkable variations in the records of measurements of these periods: they increased when the earth moved away from Jupiter and decreased when the earth moved towards Jupiter. From this Römer concluded that it must take a finite amount of time for light to traverse the diameter of the earth's orbit. Using the radius of the earth's orbit as it was then known, he computed a fairly accurate value of the velocity of light c.

We can best explain this computation by comparing it with the *Doppler Effect*. This phenomenon will be treated more thoroughly in **11**. For present purposes it will suffice to characterize it by the statement

$$\frac{\Delta\lambda}{\lambda} = \frac{\Delta\tau}{\tau} = \pm\frac{v}{c}. \tag{1}$$

λ = wavelength, τ = period of oscillation, v = relative velocity of observer and source. The signs $\pm$ hold when the distance between source and observer increases or decreases, respectively. The direction of propagation of the observed light is here assumed to be parallel to v, that is in the same direction or opposite to it. Since the Doppler effect is a purely kinematic phenomenon (it is true for sound as well as for light) we can apply it to the satellites of Jupiter with their periodic eclipses. The sunlight reflected by Jupiter itself plays no part in this consideration.

In our case, see fig. 13, τ is the orbital period of a satellite of Jupiter as measured from the moving earth. One of these satellites is denoted by the letter M in fig. 13 and its true orbital period will be called τ_0. $ABCD$ are points on the earth's orbit. At B the earth is moving away from the satellite of Jupiter and the period of the latter appears to be lengthened by $\Delta\tau = v\tau_0/c$. At D the earth is moving toward the satellite and τ appears to be shortened by $v\,\tau_0/c$. At A and C where the velocity of the earth is perpendicular to the direction of the light coming from Jupiter, $\Delta\tau = 0$ and, hence, there the true period τ_0 is observed. The extreme values of τ are found at B and D and, according to (1), they are

$$\tau_{max} = \tau_0 + \frac{v}{c}\tau_0, \qquad \tau_{min} = \tau_0 - \frac{v}{c}\tau_0. \tag{2}$$

From this follows

$$\tau_B - \tau_D = \tau_{max} - \tau_{min} = 2\frac{v}{c}\tau_0 \tag{3}$$

or if the value $2\pi R/T$ is substituted for v:

$$\tau_B - \tau_D = 4\pi\frac{\tau_0}{T}\frac{R}{c}, \tag{4}$$

where T is the time interval of one year and R is the radius of the earth's orbit. Thus, this latter value must be known for the computation of c.

This result can be represented as in fig. 13 a and in this form it has been widely used in popular lectures. As the earth moves from A to C, it moves away from the eclipse-signals coming from Jupiter so that the intervals between receptions of such signals keep increasing. The opposite happens when the earth in its travel from C to A moves toward the eclipse-signals.

The total time difference in each case is equal to the time which light takes in moving through the diameter of the earth's orbit, that is

$$\Sigma \Delta \tau = \frac{2R}{c}. \tag{5}$$

This is fairly close to the area[1] bounded by the sine curve above $A\,C$ (or below $C\,A'$) whose growth (and decay) is plotted separately at the bottom in the step-wise form in which it can be read directly from data tables.

Not until almost 200 years later were the first successful terrestrial determinations of c made. These were done by Fizeau using a rotating toothed wheel (the light passing through a gap between teeth is reflected at a distant point and upon returning is stopped by the following tooth, provided the rotation of the wheel is sufficiently fast) and by Foucault using a rotating mirror (a device used later by Michelson in his much more precise experiments).

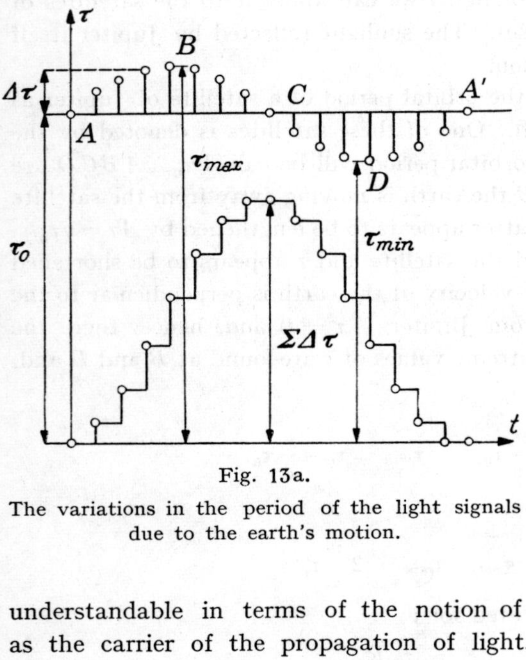

Fig. 13a.

The variations in the period of the light signals due to the earth's motion.

The most important fact about the propagation of light is that it is independent of the state of motion of the emitting source, that the light velocity cannot "remember" the velocity of the source. Only wavelength and period of oscillation have such a "memory" according to eq. (1). This fact seemed to be understandable in terms of the notion of a stationary light ether acting as the carrier of the propagation of light. It is now well known that

[1]This area equals the product of the integral

$$\int_0^{T/2} \sin 2\pi \frac{t}{T}\, dt = \frac{T}{\pi} \tag{a}$$

and the ordinate of the sine curve at C which, according to (2), is

$$\Delta \tau_{max} = \frac{v}{c}\tau_0 = \frac{2\pi R}{c\,T}\tau_0. \tag{b}$$

The product of (a) and (b) divided by τ_0 does, indeed, yield the right hand side of eq. (5).

this theory did not stand up in the face of the experiments to be described in 12 and 14 and in the face of the theory of relativity.

In the following discussion it will be assumed that the reader has a certain amount of knowledge of the special theory of relativity. Such knowledge can be gained, for example, from Sec. 27 of Vol. III of this series. Without such knowledge the reader will have to omit several quantitative proofs in the following paragraphs.

10. Aberration and Parallax

By the *parallax* of a star we will here mean the so-called "yearly parallax", i. e. the solid angle of the cone which is formed by the lines of sight from different points on the earth's orbit to the star. The projection of this cone on the celestial sphere is the parallactic orbit which the star appears to describe in the course of a year. This orbit is generally a small ellipse and, in particular, it is a circle if the fixed star is at the pole of the ecliptic (as is assumed in fig. 14 a) and it is a straight line if the star lies in the ecliptic. Because of its importance as the final confirmation of the Copernican system, the proof of the existence of such an orbit was sought for a long time. While seeking this proof Bradley discovered the *aberration* of light in 1728.

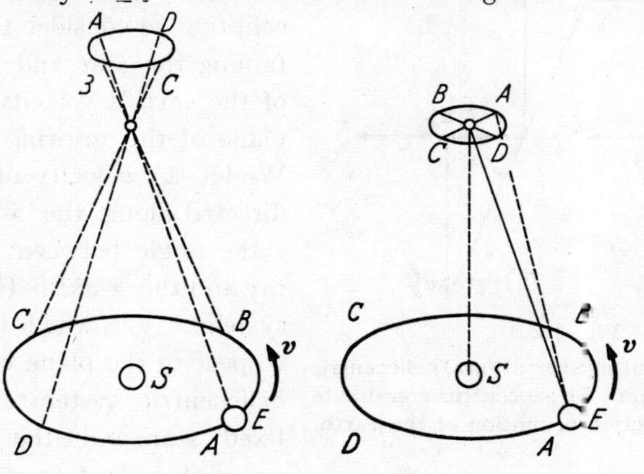

Fig. 14.
Apparent orbits of the star in the case of the North Star. a) The parallactic orbit arising from the star's finite distance from the sun. b) Orbit resulting from aberration of light. The apparent locations of the star ABCD belong to the corresponding points on the earth's orbit.

This phenomenon also causes the locus of a star to describe a small ellipse on the celestial sphere and this ellipse again degenerates into a sphere at the pole and into a straight line in the ecliptic. But the direction and magnitude of the angular deviation are entirely different from those caused by parallax: the two directions are perpendicular to each other, see fig. 14 a, b; the magnitude of the deviation due to aberration is independent of the distance of the star and is much greater than the deviation due to parallax even for the fixed stars which are closest to the sun. The first confirmation of a real parallax of a fixed star was found 100 years later by Bessel.

Lenard's attempt to invoke aberration as a contradiction to the relativity of motion was an incomprehensible misunderstanding; especially so, as Einstein had derived aberration directly from the principle of relativity as early as 1905. Aberration does not reveal the "absolute motion" of the earth in space but rather the earth's relative motion during its yearly orbital period, that is to say, the differences in the direction of motion from one season to the next. Observatories are built (among other reasons) in order to make it possible to determine and measure these differences in motion. If these differences in direction of the earth's motion did not exist, that is, if the earth's motion were linear, then no aberration could be observed.

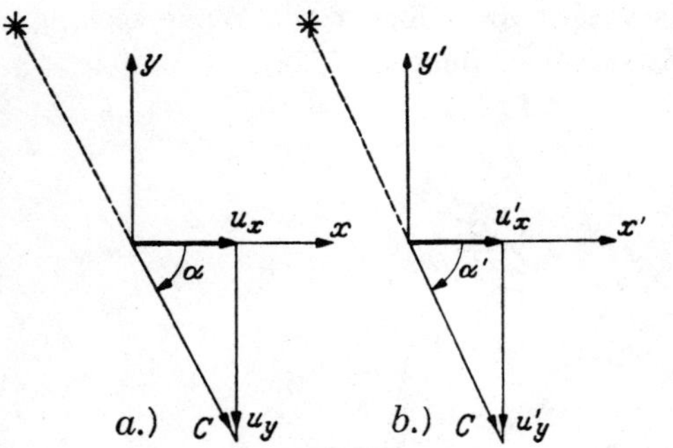

Fig. 15.

Figures for computing aberration a) heliocentric coordinate system, b) geocentric coordinate system x = direction of motion of the earth.

Still referring to fig. 14 b but removing the restriction that the star shall be at the pole of the ecliptic, we consider the plane containing the star and the direction of the earth's velocity. This is the plane of the drawing in fig. 15 a. We let the velocity of the earth be directed along the x-axis and call α the angle between the incoming ray and the x-axis. The coordinate system x, y, z, which is at rest with respect to the plane of fig. 15 a is a heliocentric system since both the fixed star and the sun are at rest with respect to it. If, however, we move with the velocity v of the earth, we have to introduce a geocentric coordinate system x', y', t' (fig. 15 b). The fact that the time as well as the space coordinates are transformed is a characteristic feature of the theory of relativity and is made necessary by the constancy of c. The transformation between the two systems is the Lorentz transformation

$$(1) \qquad x' = \frac{x - v\,t}{\sqrt{1-\beta^2}}, \qquad y' = y, \qquad t' = \frac{t - \beta\, x/c}{\sqrt{1-\beta^2}}, \qquad \beta = \frac{v}{c}.$$

Differentiating, we get

(2) $$dx' = \frac{dx - v\,dt}{\sqrt{1-\beta^2}}, \qquad dy' = dy, \qquad dt' = \frac{dt - \beta\,dx/c}{\sqrt{1-\beta^2}}.$$

Hence, the geocentric velocity components

$$u_x' = \frac{dx'}{dt'}, \qquad u_y' = \frac{dy'}{dt'}$$

are expressed in terms of the heliocentric components

$$u_x = \frac{dx}{dt}, \qquad u_y = \frac{dy}{dt}$$

by

(3) $$u_x' = \frac{u_x - v}{1 - \beta\, u_x/c}, \qquad u_y' = \frac{u_y\sqrt{1-\beta^2}}{1 - \beta\, u_x/c}.$$

Now, instead of considering the velocities of material particles in the two systems we shall consider the velocities u and u' of the light radiated by the star as seen by observers in the respective systems. This is quite admissible. The heliocentric velocity is $u = c$ with the components (see fig. 15 a)

(3 a) $$u_x = c\cos\alpha, \qquad u_y = -c\sin\alpha.$$

We write the corresponding components of u' in the form (see fig. 15 b)

(3 b) $$u_x' = u'\cos\alpha', \qquad u_y' = -u'\sin\alpha'$$

where the magnitude of u' is left undetermined for the present. Taking the ratio of the two equations (3) and using (3 a), we obtain

(3 c) $$\frac{u_y'}{u_x'} = \frac{-c\sin\alpha\sqrt{1-\beta^2}}{c\cos\alpha - v},$$

which, according to (3 b), can be written

(4) $$\tan\alpha' = \frac{\sin\alpha\sqrt{1-\beta^2}}{\cos\alpha - \beta}.$$

Thus the angle of incidence α' as seen in the geocentric system is *different* from the angle α as seen in the heliocentric system. By squaring (3) it is seen that the geocentric velocity of light propagation u' is *equal* to the heliocentric velocity. For, substituting (3 a) we have

$$u'^2 = \frac{c^2\cos^2\alpha - 2\,c\,v\cos\alpha + v^2 + c^2\sin^2\alpha\,(1 - v^2/c^2)}{(1-\beta\cos\alpha)^2}$$

$$= \frac{c^2 - 2\,c\,v\cos\alpha + v^2\cos^2\alpha}{(1-\beta\cos\alpha)^2} = c^2.$$

We could have written down this result without computation merely on the basis of Einstein's theorem on the addition of velocities. According to this theorem the following brief and seemingly paradoxical expression holds (see Vol. III, Sec. 27 F):

$$c + v = c.$$

Returning to the relationship between the angles α and α', we write the first order approximation of (4) by neglecting terms of order β^2

$$\tan\alpha' = \frac{\sin\alpha}{\cos\alpha}\left(1 + \frac{\beta}{\cos\alpha}\right) = \tan\alpha + \frac{\beta\sin\alpha}{\cos^2\alpha}. \tag{5}$$

Letting $\alpha' = \alpha + \Delta\,\alpha$, (5) now becomes

$$\tan(\alpha + \Delta\,\alpha) - \tan\alpha = \frac{\beta\sin\alpha}{\cos^2\alpha}$$

and, expanding the left-hand side in terms of the small quantity $\Delta\,\alpha$, the denominator $\cos^2\alpha$ on both sides cancels to give

$$\Delta\,\alpha = \beta\sin\alpha. \tag{6}$$

β is called the "constant of aberration", and is almost exactly equal to 10^{-4} if $\Delta\,\alpha$ is measured in radians. Hence, in degrees we have

$$\beta = 10^{-4}\,\frac{180^\circ}{\pi} = 20.5''.$$

If the fixed star is at the pole of the ecliptic, then $\alpha = 90^\circ$ all along the earth's orbit. Hence the aberrational orbit is, in this case, a circle about the pole of the ecliptic with radius $\Delta\,\alpha = \beta$. For stars in the plane of the ecliptic, which coincides with the xy-plane in fig. 15 a, α alternates between $\pm\,90^\circ$ and 0. In this case also the aberration is in the plane of the ecliptic and oscillates between $\pm\,\beta$ passing twice through 0. For stars in general the aberrational orbit is an ellipse whose major axis is β and whose minor axis is $\beta\sin\delta$, where δ is the "celestial latitude" of the star (complement of its polar distance).

The above formulae, in particular eq. (2), make it clear that aberration is a direct consequence of the relativistic deviation of the time measure t' from the time measure t. From the point of view of classical kinematics it was difficult to reconcile aberration with the universality of c. Now we recognize it to be a necessary consequence of the fact that the light velocity is independent of the reference system. In the following paragraph aberration will come up again in connection with a more general point of view.

11. The Doppler Effect

The elementary explanation of the Doppler effect is well known. If a light source which is at rest emits waves of period τ, then the number of oscillations which meet a stationary observer during a time interval t is $N = t/\tau$. But if the observer moves toward the wave with a velocity v, thus covering a distance vt in time t, he will meet an additional vt/λ oscillations. Therefore, altogether the moving observer encounters

$$N' = \frac{t}{\tau} + \frac{v\,t}{\lambda} = \frac{t}{\tau}\left(1 + v\frac{\tau}{\lambda}\right) = \frac{t}{\tau}\left(1 + \frac{v}{c}\right) \tag{1}$$

oscillations in the time t.

If, one the other hand, it is the light source which moves with velocity v toward the observer who is now at rest, then the spacing between two successive crests or troughs is no longer λ but, because the light source has progressed a distance $v\tau$ in time τ, this spacing is

$$\lambda' = \lambda - v\,\tau = \lambda\left(1 - \frac{v}{c}\right). \tag{1 a}$$

The corresponding spacing in time is

$$\tau' = \tau\left(1 - \frac{v}{c}\right). \tag{1 b}$$

Therefore, during the time t the observer at rest encounters

$$N'' = \frac{t}{\tau'} = \frac{t}{\tau}\frac{1}{1 - v/c} = \frac{t}{\tau}\left(1 + \frac{v}{c} + \frac{v^2}{c^2} + \dots\right) \tag{2}$$

oscillations. N' and N'' differ by terms of second and higher orders in $\beta = v/c$. They agree only up to terms of first order.

Against this argument can be said: nature knows no absolute motion, be it that of the light source or that of the observer. She gathers both cases (1) and (2) into the same law, which thereby becomes simpler and more beautiful. How this is accomplished we shall discover from the following consideration:

Every physical relation must be invariant with respect to the group of transformations which governs the particular domain being considered. If a relationship is expressed by means of an analytic function, then the argument of that function must be a *dimensionless scalar*. With this in mind, we consider the exponential function in the expression for a plane wave. Its argument is, aside from the factor i, the *phase* of the wave. In particular, this argument may be written in different forms representing various levels of generality as follows:

$$k\,x - \omega\,t, \qquad \mathbf{k}\,\mathbf{r} - \omega\,t, \qquad \vec{\mathsf{K}} \cdot \vec{\mathsf{R}}. \tag{3}$$

In the last of these expressions $\vec{\mathsf{R}}$ is the space-time four-vector

$$\vec{\mathsf{R}} = x_1, x_2, x_3, x_4 \quad \text{where} \quad x_4 = i\,c\,t. \tag{4}$$

$\vec{\mathsf{K}}$ is the wave number four-vector with the dimensions of an inverse length:

(5) $$\vec{\mathsf{K}} = k_1, k_2, k_3, k_4$$

$$\text{where } k_4 = \frac{i\,\omega}{c} = \frac{2\pi\, i}{\tau\, c} = \frac{2\pi\, i}{\lambda},$$

and the space components of $\vec{\mathsf{K}}$ are given by

(5 a) $$k_1, k_2, k_3 = \frac{2\pi}{\lambda}(\cos\alpha_1, \cos\alpha_2, \cos\alpha_3);$$

$\alpha_1\, \alpha_2\, \alpha_3$ are the angles which $\vec{\mathsf{K}}$ makes with the $x_1\, x_2\, x_3$ axes and

$$\cos^2\alpha_1 + \cos^2\alpha_2 + \cos^2\alpha_3 = 1.$$

From this it follows that the absolute value of $\vec{\mathsf{K}}$ is zero:

$$|\mathsf{k}|^2 = k_1{}^2 + k_2{}^2 + k_3{}^2 + k_4{}^2 = \frac{4\pi^2}{\lambda^2}(\cos^2\alpha_1 + \cos^2\alpha_2 + \cos^2\alpha_3 - 1) = 0.$$

We now view the wave from a primed coordinate system which moves with a velocity $v = \beta\, c$ in the x-direction with respect to the unprimed system. The Lorentz transformation (10.1) must then be applied to $\vec{\mathsf{K}}$, where we must replace t by the quantity $x_4/i\, c$ or, in the present case, by k_4/ic. Thus we obtain the following components of the transformed four-vector $\vec{\mathsf{K}}'$:

(6) $$k_1{}' = \frac{k_1 + i\,\beta\, k_4}{\sqrt{1-\beta^2}}, \qquad k_2{}' = k_2, \qquad k_4{}' = \frac{k_4 - i\,\beta\, k_1}{\sqrt{1-\beta^2}}.$$

These expressions are somewhat specialized in that we have taken $\cos\alpha_3 = 0$, that is the wave is assumed to propagate in the $x_1\, x_2$-plane, and we have omitted the equation $k_3{}' = 0$. For $\cos\alpha_3 = 0$ we get $\cos^2\alpha_2 = 1 - \cos^2\alpha_1 = \sin^2\alpha_1$. If $\alpha_1{}'\, \alpha_2{}'$ are the corresponding angles in the $x_1{}'\, x_2{}'$-plane, we also have $\cos^2\alpha_2{}' = 1 - \cos^2\alpha_1{}' = \sin^2\alpha_1{}'$. From now on we shall write α, α' in place of α_1, $\alpha_1{}'$.

Substituting eq. (5) in (6) and applying the definitions (5 a), we get

(7) $$\frac{\cos\alpha'}{\lambda'} = \frac{\cos\alpha - \beta}{\lambda\sqrt{1-\beta^2}}, \qquad \frac{\sin\alpha'}{\lambda'} = \frac{\sin\alpha}{\lambda}, \qquad \frac{1}{\lambda'} = \frac{1-\beta\cos\alpha}{\lambda\sqrt{1-\beta^2}}.$$

Forming the ratio of the first two of these equations, we obtain

(7 a) $$\tan\alpha' = \frac{\sin\alpha\sqrt{1-\beta^2}}{\cos\alpha - \beta},$$

which is identical with the equation for the aberration (10.4), while the third of eqs. (7) *represents the exact relativistic formulation of the Doppler principle.*

We could have derived these equations in an even more elementary manner using the second expression (3) for the phase of the plane wave. We could, namely, require that

$$\mathbf{k}' \cdot \mathbf{r}' - \omega' t' = \mathbf{k} \cdot \mathbf{r} - \omega t,$$

write x_1', x_2', t' in terms of x_1, x_2, t by means of a Lorentz transformation, and then equate the coefficients of the terms x_1, x_2, t on both sides. We have favored the former method (using the covariance of the wave number vector instead of the invariance of the phase) because it expresses more clearly the relativistic four-dimensional origin of the Doppler equation.

For general directions of the velocity of the light source with respect to the observer or vice versa, we let $v \cos \alpha = v_n =$ projection of this velocity on the normal to the wavefront and according to (7), we get

$$\frac{\lambda}{\lambda'} = \frac{1 - v_n/c}{\sqrt{1 - \beta^2}}. \tag{8}$$

Letting $\Delta \lambda = \lambda' - \lambda$, it follows that

$$\frac{\Delta \lambda}{\lambda} = \frac{\sqrt{1 - \beta^2} - 1 + v_n/c}{1 - v_n/c}. \tag{9}$$

In the first order approximation this yields the well-known elementary expression for the Doppler effect

$$\frac{\Delta \lambda}{\lambda} = \frac{v_n}{c}. \tag{10}$$

A more detailed discussion of (9) will show that this equation contains not only the longitudinal Doppler effect in the case $v_n = \pm v$, which is a first order effect, but it also contains the *second* order *transverse Doppler effect*. For if $v_n = 0$, we have

$$\frac{\Delta \lambda}{\lambda} = \sqrt{1 - \beta^2} - 1 = -\frac{\beta^2}{2} + \ldots;$$

recently this transverse effect has been measured accurately by means of the red shift of spectral lines (see Vol. III Sec. 27 D).

12. Fresnel's Coefficient of Drag and Fizeau's Experiment

Regarding the velocity of propagation of light in a moving transparent medium, the most obvious assumption suggested by the classical ether theory would have been that the velocity of light c/n ($n =$ index of refraction of the medium) is added to the velocity v of the medium. However, Fresnel, through ingenious reasoning, found the resulting velocity to be

$$u = \frac{c}{n} + v\left(1 - \frac{1}{n^2}\right). \tag{1}$$

The factor $(1 - 1/n^2)$ is called the *"Fresnel Coefficient of Drag"*. The formula was completely confirmed by performing the *Fizeau experiment* in streaming water.

The light originating from the source L passes in two separate bundles of rays through the two pipes shown in fig. 16. In one of the pipes the light velocity is increased and in the other decreased and the resulting optical path difference can be measured at A by means of an interferometer.

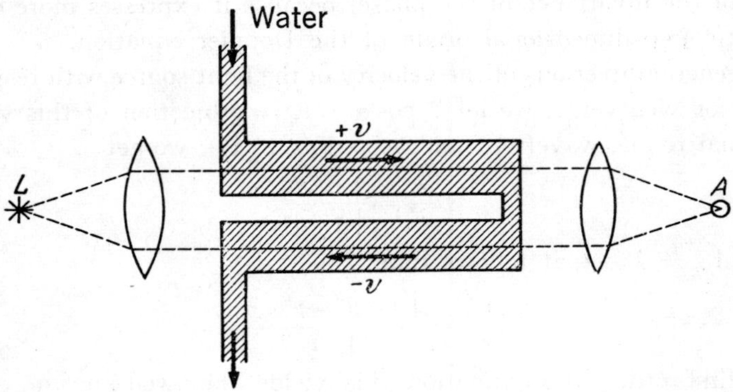

Fig. 16.
Fizeau's experiment for the determination of Fresnel's drag coefficient.

As was first noted by v. Laue[1], Eq. (1) can be explained purely phenomenologically on the basis of the velocity addition theorem as given by formula (27.15) Vol. III

$$u = \frac{v_1 + v_2}{1 + v_1 v_2/c^2} \tag{2}$$

without making any special assumptions regarding the nature of the propagation of light in the moving medium. If in Eq. (2) v_1 is set equal to the phase velocity in water of index of refraction n, i. e. equal to c/n, and v_2 is set equal to the velocity of the water with respect to the reference system which is at rest in the laboratory, i. e. v_2 is $+v$ in the upper pipe and $-v$ in the lower one, then according to the addition theorem, the resulting velocities u are

$$u = \frac{\frac{c}{n} \pm v}{1 \pm \frac{v}{n\,c}}. \tag{3}$$

[1] Ann. d. Phys. **23**, p. 989, 1907.

From this it follows in the first approximation if $v \ll c/n$, that

$$u = \frac{c}{n}\left(1 \pm \frac{v\,n}{c}\right)\left(1 \mp \frac{v}{n\,c}\right) = \frac{c}{n}\left(1 \pm \frac{v\,n}{c} \mp \frac{v}{n\,c}\right),$$

which, indeed, agrees with (1):

(4) $$u = \frac{c}{n} \pm v\left(1 - \frac{1}{n^2}\right).$$

Lorentz[1] showed that this formula can be refined by combining it with the Doppler effect. In this way one obtains

(5) $$u = \frac{c}{n} \pm v\left(1 - \frac{1}{n^2} - \frac{\lambda}{n}\frac{dn}{d\lambda}\right),$$

This equation is derived in the following way: n is not constant but is a function of the wavelength. One now considers a certain spectral line λ which is emitted by the light source L. As seen from a reference frame moving with the water, this spectral line is modified to $\lambda' = \lambda + \Delta\lambda$. One obtains, therefore,

(5 a) $$n(\lambda') = n(\lambda + \Delta\lambda) = n + \frac{dn}{d\lambda}\Delta\lambda.$$

In the upper pipe the water flows away from the light source and in the lower pipe it flows toward it. $\Delta\lambda$ is found from Eq. (11.10) by replacing c by the velocity of propagation c/n in water. Accordingly

$$\frac{\Delta\lambda}{\lambda} = \pm\frac{v}{c/n}; \quad \text{hence from (5 a)} \quad n(\lambda') = n \pm \lambda\frac{dn}{d\lambda}\frac{v}{c}n,$$

and

$$\frac{c}{n(\lambda')} = \frac{c/n}{1 \pm \lambda\frac{dn}{d\lambda}\frac{v}{c}} = \frac{c}{n}\left(1 \mp \lambda\frac{dn}{d\lambda}\frac{v}{c}\right).$$

The numerator in Eq. (3) is thereby changed to

(6) $$\frac{c}{n}\left(1 \mp \lambda\frac{dn}{d\lambda}\frac{v}{c} \pm n\frac{v}{c}\right).$$

The correction in the denominator would amount only to a term of second order in v/c, which may be neglected. As before, the inverse of this denominator reads

(6 a) $$1 \mp \frac{v}{n\,c}.$$

---

[1] Versuch einer Theorie der elektrischen und optischen Erscheinungen von bewegten Körpern, Leiden 1895, p. 101.

Multiplying (6) and (6 a) one obtains

$$u = \frac{c}{n}\left(1 \mp \lambda \frac{dn}{d\lambda}\frac{v}{c} \pm n\frac{v}{c} \mp \frac{v}{n\,c}\right), \tag{7}$$

which agrees with (5). Zeeman's[1] mastery of spectroscopy enabled him to verify this more exact formula experimentally.

The following general conclusion is to be drawn regarding the dragging of light in moving (ponderable and isotropic) bodies: as seen by an observer moving with the body, or at rest with respect to it the light propagates with the velocity c/n uniformly in all directions (first term of eq. (1)], regardless of whether the body is at rest or in a state of uniform motion. For an observer at rest in the laboratory with respect to which the body is moving with velocity v, a first order effect is added in the direction of the motion (this effect is given by the second term in eq. (1) or eq. (5) and is small compared to the first term by the order of magnitude v/c). As in the case of the transverse Doppler effect, an effect of second order exists in the direction perpendicular to the motion. Though not included in the Fresnel formula (1), this second order effect is easily computed by means of the addition theorem of velocities.

When $n = 1$ the first order effect disappears. A medium of index of refraction 1, however fast it may move (for instance, the so-called "ether wind"), has no effect at all on the propagation of light. This fact was once considered to be a proof that the ether was at rest and ponderable matter moved through it. According to that theory, only the charges associated with matter which find their expression in the index of refraction n were to affect the propagation. We know now that no particular assumptions need be made regarding the mechanism of the emission of light. The concepts of electron theory are, to be sure, useful for the visualization of the dragging term but they are in no sense necessary for its derivation.

13. Reflection by a Moving Mirror

The problem to be discussed in this section will serve as a preparation for the experiments to be described in 15 and will, moreover, be of help in connection with the thermodynamics of radiation which will be treated in Vol. V. (Wien's Displacement Law). We distinguish two cases: a) the mirror is moved in a direction tangential to its plane surface and b) it is moved in the direction perpendicular to its surface. In both cases the mirror will be assumed to be perfectly reflecting and its velocity will always be uniform.

---

[1] Amsterd. Akademie Versl. 1914, p. 245 and 1915, p. 18.

a) We use the wave number four-vector defined in (11.5) referred to the "primed" system[1] which moves with the mirror, see fig. 17 a. In this system we call our vector $\mathbf{k}'$ and its components are

(1) k_1', k_2' in the plane of incidence, $k_3' = 0$, $k_4' = i\,\omega'/c$.

We shall call the corresponding quantities describing the reflected ray $\overline{\mathbf{k}}'$ and $\overline{k}_1'$, etc., respectively. Since the mirror is at rest with respect to the primed coordinate system, the ordinary law of reflection holds:

$$\overline{k}_1' = k_1', \quad \overline{k}_2' = -k_2', \quad \overline{k}_3' = 0, \quad \overline{k}_4' = k_4'. \tag{1'}$$

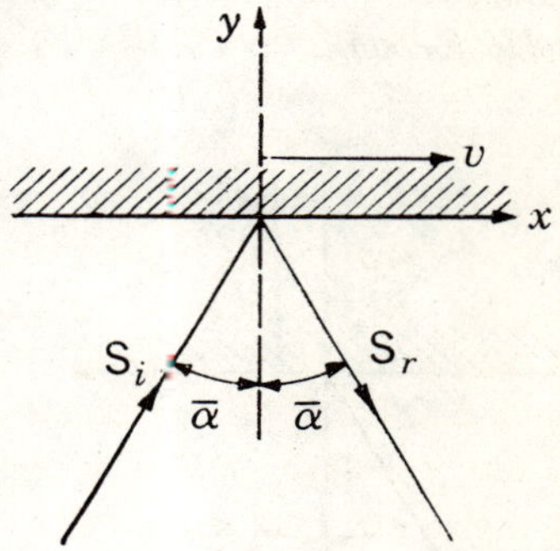

Fig. 17a.
Reflection from a moving mirror. Direction of motion parallel to the plane of the mirror.

When observed from the laboratory with respect to which the mirror moves at a velocity $v = \beta\, c$ in the direction of the x-axis, the four-vectors describing the incident and reflected rays shall be denoted by $\mathbf{k}$ and $\overline{\mathbf{k}}$ and their components by $k_1 \ldots k_4$ and $\overline{k}_1 \ldots \overline{k}_4$, respectively. Inverting the transformation eq. (11.6) by replacing β by $-\beta$ and vice versa, we find for the incident wave

$$k_1 = \frac{k_1' - i\,\beta\, k_4'}{\sqrt{1-\beta^2}}, \qquad k_2 = k_2', \qquad k_4 = \frac{k_4' + i\,\beta\, k_1'}{\sqrt{1-\beta^2}} \tag{2}$$

and for the reflected wave, taking into account (1'),

$$\overline{k}_1 = \frac{k_1' - i\,\beta\, k_4'}{\sqrt{1-\beta^2}}, \qquad \overline{k}_2 = -k_2', \qquad \overline{k}_4 = \frac{k_4' + i\,\beta\, k_1'}{\sqrt{1-\beta^2}}. \tag{2'}$$

If we denote the angles of incidence and reflection as measured in the laboratory system by

α and $\overline{\alpha}$, respectively,

then by definition

$$\tan \alpha = \frac{k_1}{k_2}, \qquad \tan \overline{\alpha} = -\frac{\overline{k}_1}{\overline{k}_2}. \tag{3}$$

[1] The convention by which the primed system is identified with the "moving" body and the un-primed system with the "stationary" laboratory is, though customary, entirely as arbitrary as the words "moving" and "stationary" themselves. Since we shall call the angle of incidence in the primed system α', we must change our former notation for the angle of reflection (Chap. I). We will, therefore, distinguish the latter from the angle of incidence by a superposed bar.

From this it follows, according to eq. (2) and (2′), that

$$\bar{\alpha} = \alpha. \tag{4}$$

From the connection between ω' and k_4' given by (1) and the corresponding relationships between ω and k_4 and between $\bar{\omega}$ and $\bar{k}_4$ it follows that

$$\bar{\omega} = \omega. \tag{5}$$

Thus, *for a mirror moving tangential to its surface the law of reflection which holds for stationary mirrors is preserved,* and, *seen from the laboratory system, the frequency of the light remains unchanged by reflection.* However, α differs from the angle of incidence in the primed system α' by a small term of first order (which we could call the angle of aberration). Also the frequency ω differs somewhat from ω' because of the Doppler effect.

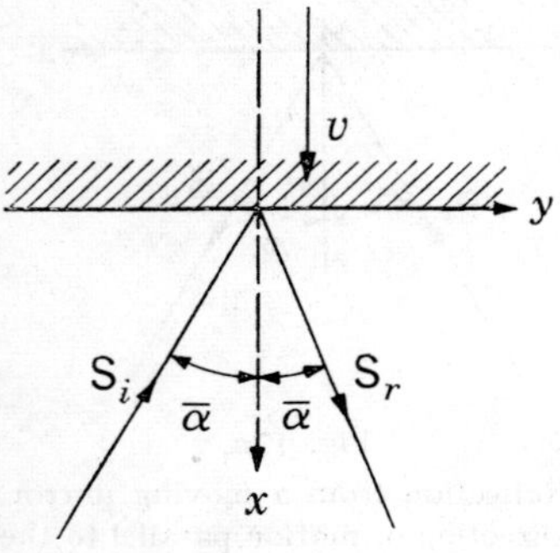

Fig. 17 b.
Reflection from a moving mirror. Direction of motion perpendicular to the plane of the mirror.

b) Now let the mirror move in a direction perpendicular to its surface, for instance, forward against the incident light. The x-axis will again be in the direction of the velocity v, see fig. 17 b, and the y-axis will be in the plane of the mirror. In the primed coordinate system which moves with the mirror one finds, because of the interchange of indices 1 and 2, instead of (1′)

$$\overline{k_1'} = -k_1', \qquad \overline{k_2'} = k_2', \qquad \overline{k_3'} = 0, \qquad \overline{k_4'} = k_4' = i\,\omega'/c, \tag{6}$$

From the Lorentz transformation, taking (6) immediately into account, one obtains instead of (2) and (2′)

$$k_1 = \frac{k_1' - i\,\beta\,k_4'}{\sqrt{1-\beta^2}}, \qquad k_2 = k_2', \qquad k_4 = \frac{k_4' + i\,\beta\,k_1'}{\sqrt{1-\beta^2}}, \tag{7}$$

and

$$\bar{k}_1 = \frac{-k_1' - i\,\beta\,k_4'}{\sqrt{1-\beta^2}}, \qquad \bar{k}_2 = k_2', \qquad \bar{k}_4 = \frac{k_4' - i\,\beta\,k_1'}{\sqrt{1-\beta^2}} \tag{7'}$$

In contrast to (3) the angles of incidence and reflection are now to be defined by

$$\tan\alpha = \frac{k_2}{-k_1} = \frac{k_2'\sqrt{1-\beta^2}}{-k_1' + i\,\beta\,k_4'}, \qquad \tan\bar{\alpha} = \frac{\bar{k}_2}{\bar{k}_1} = \frac{k_2'\sqrt{1-\beta^2}}{-k_1' - i\,\beta\,k_4'}. \tag{8}$$

Thus, *as observed from the laboratory, the angle of reflection differs from the angle of incidence.* The same is true for the *frequencies ω and $\bar{\omega}$.*

In the case illustrated by fig. 17 b, $\overline{\omega} > \omega$; if v were in the opposite direction, then we would find that $\overline{\omega} < \omega$. This is readily understood by temporarily replacing the plane wave by a stationary point source at a finite distance and by considering the image of this source produced by the moving mirror. This image approaches the observer at a velocity $2v$ and hence the wavelength of the reflected light appears shortened due to the Doppler effect, and its frequency appears increased. If the mirror moves in the opposite direction, the whole situation is merely reversed. Correspondingly, $\overline{\alpha} < \alpha$ in the case illustrated by fig. 17 b, while $\overline{\alpha} > \alpha$ when the motion of the mirror is in the opposite direction (or, to speak more precisely when the relative motions of mirror and observer are opposite).

Anticipating the corpuscular considerations of Sec. 16, we may point out a mechanical analogue: a tennis ball which falls obliquely on the racquet is reflected at a smaller angle than that at which it impinges. This is because the perpendicular component of the ball's velocity is increased by the forward motion of the racquet.

14. The Michelson Experiment

The most famous experiment in the field of optics of moving media is that of Michelson. For dates see the historical table in Sec. 1. After its repetition at Jena the negative result of this experiment can be considered as definitely established. The following will serve to indicate the degree of accuracy which was striven for: the apparatus was operated entirely automatically; in order to eliminate every possible temperature effect the apparatus was set up in a cellar of the Zeiss works and was inaccessible to the experimenter. Joos rightfully considered these precautions more important than the measures taken by D. C. Miller, another successor of Michelson. The latter placed his apparatus in a wooden shed on a high mountain in order to provide the "ether wind" with the freest possible passage through the apparatus. The apparatus used by Joos is now in the "Deutsches Museum" in Munich.

Michelson's experimental set-up is sketched in fig. 18. As in the case of other experiments of Michelson as well as of Perot-Fabry, the most important item is the semi-reflecting plate H. This plate allows the light coming from the lamp L to follow two different paths between L and B (observing telescope), namely,

$$L\,H\,S_1\,H\,B \quad \text{and} \quad L\,H\,S_2\,H\,B.$$

Since along both these paths the light passes once through H and is once reflected by H, the attenuation along both paths is the same and is given by the product $r\,d$. It is, therefore, not necessary to obtain exact semi-transparency ($d = r$). Likewise, it is immaterial whether the mirrors S_1 and S_2 are precisely perpendicular to each other or not, a condition which, in any case, would be experimentally unattainable. Hence, we are here not concerned with interference arising from a plane-parallel air plate but rather with the kind of fringe pattern to be expected from an air space which is somewhat wedge shaped. Even the equality of the distances $l_1 = H\,S_1$ and $l_2 = S_2\,H$, though desirable, is not of critical importance (see footnote p. 78) and is never exactly attained. We shall, however, carry out all computations using $l_1 = l_2 = l$. In the experiment of Michelson and Morley the light paths were increased to eleven meters by repeated reflections. The whole apparatus floated on mercury[1]. First it was oriented so that the direction $L\,H\,S$ was parallel to the direction of the earth's motion about the sun. Then the whole apparatus was rotated through 90° and any possible shift in the interference fringes was observed. *According to the theory of relativity no such shift of the fringes can occur.* This is due to the fact that the earth qualifies as a practically unaccelerated reference system and (in contrast to the experiments on aberration) the change of the earth's direction of motion during the duration of the experiment is negligible.

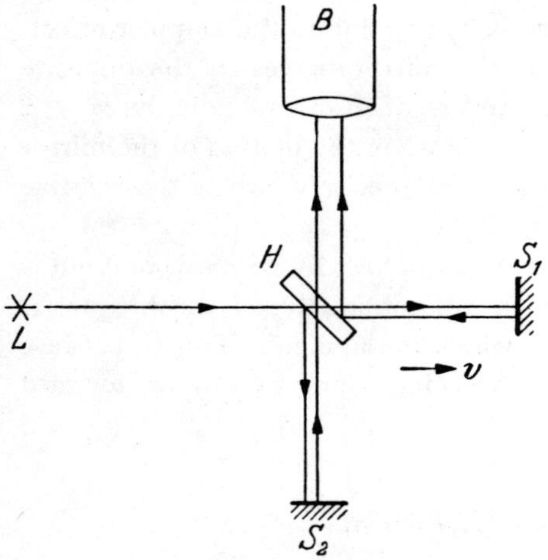

Fig. 18.

Michelson's experiment to prove that the velocity of light is independent of the earth's motion. L is a monochromatic light source. H is a semi-transparent plate. S_1. S_2 are mirrors. B is the telescope used for observation.

This is, however, not true from a non-relativistic viewpoint when the stationary reference system is assumed to be at rest with respect to the sun. For in this case, we would have to assert that the light always moves with a velocity c with respect to *this* reference system. The velocity of propagation of light with respect to the moving apparatus has then to be calculated. For this purpose the positions of H and S_1 for the first part of the experiment

[1] In Joos' set-up the apparatus was suspended by springs. The arms l_1 and l_2 consisted of quartz glass, the light paths amounted to 21 meters and for a source the Hg-line $\lambda = 5461$ A.U. was used. See Ann. d. Phys. (Lpz) 7, p. 385, 1930.

($L\,S_1$ parallel to v) have been drawn in fig. 18 a in the following way. H is the position of the semi-transparent plate at the moment at which it is traversed by a certain phase of the monochromatic wave coming from L, a maximum for instance. S_1 is the simultaneous position of the mirror S_1' is the position of the mirror at the instant at which it reflects the above-mentioned phase. This reflection will occur a time t_1 later. H' is the position of H at that instant. H'' is the position of H at the time t_2 when the same phase returns to the semi-transparent plate. If we use the path lengths $S_1\,S_1' = v\,t_1$, $H'\,H'' = v\,t_2$ shown in the figure and let $H\,S_1 = H'\,S_1' = l$, then ordinary non-relativistic kinematics yields

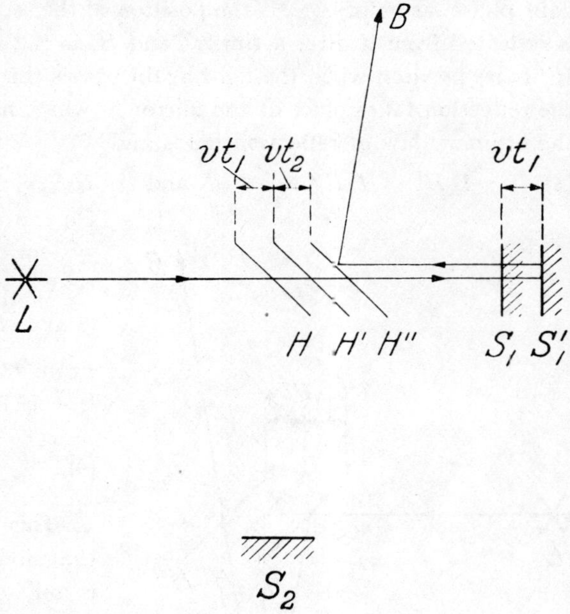

Fig. 18a.
Determination of the light path in Michelson's experiment. Ray parallel to the motion of the earth.

$$\text{(1)} \qquad \begin{cases} l + v\,t_1 = c\,t_1, & t_1 = \dfrac{l}{c-v}, \\[2ex] l - v\,t_2 = c\,t_2, & t_2 = \dfrac{l}{c+v}. \end{cases}$$

Hence, the total time taken by the light is

$$\text{(2)} \qquad t_1 + t_2 = \frac{l}{c-v} + \frac{l}{c+v} = \frac{2\,l\,c}{c^2 - v^2} = \frac{2\,l/c}{1-\beta^2}.$$

The fact that the light actually meets the telescope B, which has itself moved forward, is due to the changed law of reflection at the moving H-mirror when in the position H'' [1].

[1] In this case the reflection is from a mirror which moves neither perpendicular to its surface as in 13a, nor parallel to it as in 13b. This mirror H is inclined at an angle of 45° with respect to its direction of motion. It is evident, however, that also in this case the event (arrival of the reflected light at B) which is observed on earth must be relativistically preserved if the reference system is moved to the sun.

For the computation of the other ray path we use fig. 18 b. Here H is the position of the semi-transparent plate at the time at which it reflects the same light phase, as before. S_2 is the position of the second mirror when this phase is reflected from it after a time t', and H' is the simultaneous position of H. H'' is its position when the light again passes through it. Since, in this case, the reflection takes place at the mirror S which moves parallel to its surface, the ordinary law of reflection holds and

(3) $$H H' = H' H'' = v t' \quad \text{and} \quad H S_2 = S_2 H'' = \sqrt{l^2 + v^2 t'^2}.$$

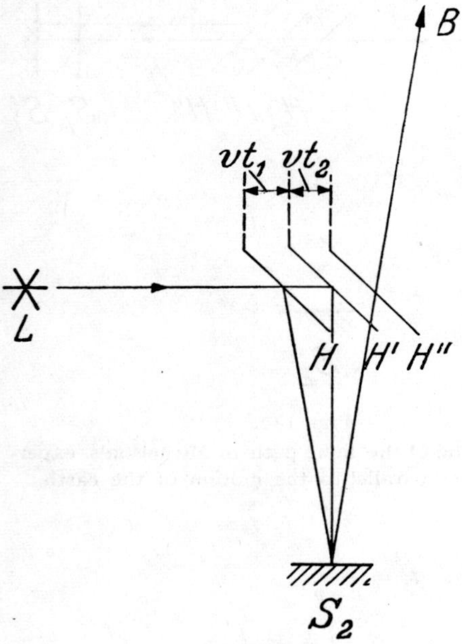

Fig. 18b.

Determination of the light path in Michelson's experiment. Ray perpendicular to the motion of the earth.

On the other hand, in the stationary system of the sun it must be true that

(3 a) $$H S_2 = S_2 H'' = c t'.$$

From (3) and (3 a) follows $c^2 t'^2 = l^2 + v^2 t'^2$ and also

(4) $$2 t' = \frac{2 l/c}{\sqrt{1 - \beta^2}}.$$

This time interval *differs* from that found in (2). The difference is only of second order in β (β = aberration constant = 10^{-4}, see Sec 10), namely,

(5) $$\Delta t = t_1 + t_2 - 2 t' = \frac{l}{c} \beta^2$$

or, expressed as a light path length,

(5 a) $$c \Delta t = l \beta^2.$$

Nevertheless, this difference means that the phase under consideration reaches the observer B noticeably later when it goes by way of S_1 than when it goes by way of S_2.

If we now rotate the apparatus through 90°, then S_2 takes the place of S_1 and vice versa. Thus the time intervals $t_1 + t_2$ and $2 t'$ are interchanged[1]

[1] The arms l_1 and l_2 interchange in the same way even if they differ in length from one another ($l_1 = l$, $l_2 = l + \delta l$). In this case, however the term $2 \delta l (1 + \beta^2/2)$ is added to (5a), and the part independent of β^2 cancels in (5a) and one obtains

(5 c) $$\Delta Z = \frac{2 l}{\lambda} \beta^2 \left(1 - \frac{\delta l}{l}\right)$$

which agrees closely with (5b) as long as $\delta l \ll l$.

so that the expressions (5) and (5 a) change their signs. The difference in the time of arrival at B is thereby doubled. Hence B should observe twice the shift given by (5 a). Expressed in fractions of one whole fringe width, this amounts to

$$\Delta Z = 2\frac{c\Delta t}{\lambda} = 2\frac{l}{\lambda}\beta^2. \tag{5 b}$$

For the above-mentioned values of $l = 21$ meters, $\lambda = 5461 \times 10^{-10}$ m, this gives

$$\Delta Z = 0.4. \tag{6}$$

In contrast to this result Joos summarized the result of the Jena experiments as follows: "We can say with a clear conscience that the upper limit of the effect due to any true ether wind which might still be possible on the basis of these experiments is 1/1000 of a fringe."

In order to reconcile this disagreement between theory and experiment, Lorentz and, independently, Fitzgerald found it necessary to introduce the following bold hypothesis: every moving body contracts in the direction of its motion by a factor $\sqrt{1-\beta^2}$. From Vol. III, Sec. 27 C we know that this "Lorentz contraction" is a general consequence of the principle of relativity and that it holds true, not only for the particular experiment here considered, but for all relative motions and for all space measurements parallel to such motion. Thus the Lorentz contraction is not an "ad hoc hypothesis". We should never have had to mention it had we used the relativistically correct kinematics immediately in (1) and had there replaced l by $l\sqrt{1-\beta^2}$. In eq. (3), where l is the length of the arm which is perpendicular to the direction of the motion, no such change would have been necessary.

15. The Experiments of Harress[1], Sagnac[2] and Michelson-Gale[3]

The negative result of Michelson's experiment has, of course, no bearing on the problem of the propagation of light in *rotating* media. To discuss this problem one must use not the special but rather the general theory of relativity with its additional terms which correspond to the mechanical centrifugal forces. However, in view of the fact that in the following experiments only velocities $v \ll c$ occur and only first order effects in v/c are important, relativity theory can be dispensed with entirely and the computations can be carried out classically.

[1] Diss. Jena, 1912.
[2] Comptes Rendus, 1913.
[3] Astrophys. Journ., 1925.

The *Sagnac experiment* is the easiest to describe. As shown in fig. 19 the semi-reflecting plate H and the three mirrors S are mounted at the corners of a square which is inscribed in a disc. The plate H is mounted in a radial position, the mirrors S are mounted tangentially to the disc. The monochromatic light source L and the photographic plate Ph are, likewise, rigidly attached to the disc. The two rays emitted by L and separated by H are made to interfere at Ph. If the disc is made to rotate, then the ray going in the same direction as the rotation is made to travel a longer path and the ray which is oppositely directed travels a shorter path. Thus interference fringes are formed whose positions differ for oppositely directed rotations. If this shift ΔZ in the fringes is measured, it is found to obey the following theoretical formula:

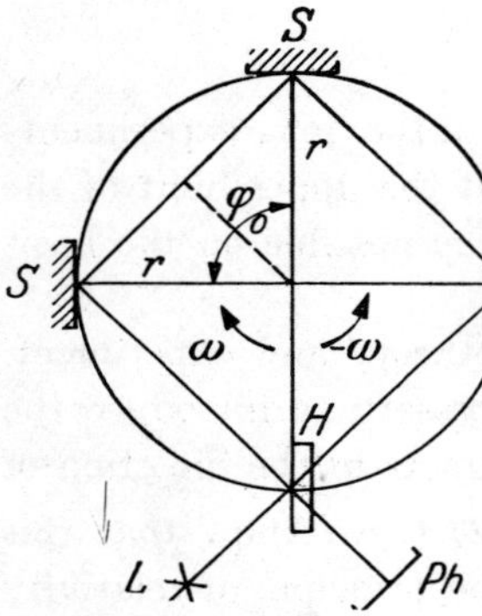

Fig. 19.

Sagnac's experiment. The interference arrangement consists of the light source L, the semi-transparent plate H, three mirrors S, and the photographic plate Ph. The entire set-up rotates with angular velocity $\pm\,\omega$ about a central axis perpendicular to the plane of the paper.

$$\Delta Z = 4\,\beta \frac{F}{r\,\lambda}; \tag{1}$$

F is the surface enclosed by the ray path, i. e. a square in Sagnac's experiment; r is the radius of the disc; v is the velocity of the disc's circumference and $\beta = v/c$.

In order to prove eq. (1) we note that owing to the law of reflection at tangentially moving mirrors, the four sides of the "square" (which for a rotating disc is no longer a closed figure) subtend the same central angle; namely, φ_0 if $\omega = 0$, φ_+ for ω in the same direction, φ_- for ω opposite to the direction of the ray:

$$\varphi_0 = \frac{\pi}{2}, \qquad \varphi_\pm = \frac{\pi}{2} \pm \frac{1}{4}\,\omega\,\tau_\pm,$$

where $\tau_\pm$ is the time which it takes the corresponding ray to travel its path $L \to Ph$. Neglecting the distances $L\,H$ and $H\,Ph$ which are the same for both rays, the length of this path is equal to $c\,\tau_\pm$. It is also equal to four times one side of the square, where this side has been lengthened or shortened by the rotation. Thus:

$$c\,\tau_\pm = 4 \cdot 2\,r \sin\frac{1}{2}\,\varphi_\pm = 8\,r \sin\left(\frac{\pi}{4} \pm \frac{\omega}{8}\,\tau_\pm\right).$$

From this one finds the time difference to be

$$\Delta\tau = \tau_+ - \tau_- = \frac{8r}{c}\left\{\sin\left(\frac{\pi}{4} + \frac{\omega}{8}\tau_+\right) - \sin\left(\frac{\pi}{4} - \frac{\omega}{8}\tau_-\right)\right\}$$

$$(2) \qquad = \frac{16r}{c}\cos\left(\frac{\pi}{4} + \frac{\omega}{16}\Delta\tau\right)\sin\frac{\omega}{16}(\tau_+ + \tau_-).$$

Neglecting small quantities on the right-hand side, one can put

$$\cos\left(\frac{\pi}{4} + \frac{\omega}{16}\Delta\tau\right) \sim \cos\frac{\pi}{4}$$

$$\sin\frac{\omega}{16}(\tau_+ + \tau_-) \sim \sin\frac{\omega}{8}\tau_0 \sim \frac{\omega}{8}\tau_0,$$

where τ_0 is the time taken by the light if the disc is at rest and is given by

$$\tau_0 = \frac{8r}{c}\sin\frac{\pi}{4}.$$

Thus (2) becomes

$$(3) \qquad \Delta\tau = \frac{8\omega r^2}{c^2}.$$

Since $\omega = v/r$ and $F = (r\sqrt{2})^2$, one can write instead

$$(3\text{ a}) \qquad \Delta\tau = 4\beta\frac{F}{rc}.$$

This expression becomes identical with (1) if one computes the shift of fringes ΔZ from the time difference $\Delta\tau$.

We could have shortened the above calculations had we started out by using the Doppler effect. The Doppler effect originates in the semi-reflecting plate H which acts as a moving light source emitting different wavelengths in the forward and backward directions (while no additional Doppler effect is caused by the tangentially moving mirrors). The shift in the fringes ΔZ is due to the difference in wavelength between the ray going with and against the direction of rotation.

In Harress' experiment a number of glass prisms were arranged along the circumference of the disc. The same formula (1) again applies, where, however, F is the area bounded by the polygon that is described by the ray in its path from prism to prism. Formula (1) was entirely confirmed by the results of both experiments.

In the experiment of Michelson and Gale the *earth* plays the role of the rotating disc. The component of the earth's angular velocity along the perpendicular at the place of observation takes the place of ω. Preliminary experiments on Mt. Wilson showed that if the necessarily long light paths were placed in free air, then even under the best atmospheric conditions the interference fringes were too unsteady to make measurements possible. Hence, it was necessary "to resort to a pipe line about one mile long and one foot in diameter which could be exhausted of air". The area F was a rectangle with sides of 340 and 610 meters. The mirrors S and the semi-transparent plate H were attached at the corners of this rectangle. In order to obtain a zero setting for the fringe shift, a comparison path enclosing only a small area was provided. A total of 269 observations yielded a mean shift of $\Delta Z = 0.230 \pm 0.005$ fringes which is again in full agreement with eq. (1).

This experiment is a beautiful analogue to *Foucault's Pendulum experiment.* While the translatory motion of the earth cannot be noticed either mechanically or optically, the earth's rotation is measurable both mechanically according to Foucault and optically according to Michelson-Gale.

16. The Quantum Theory of Light

At the end of the seventeenth century Huygens' wave theory and Newton's corpuscular theory entered upon a period of competition. Although the corpuscular theory prevailed during the eighteenth century, at the beginning of the nineteenth century the interference experiments of Thomas Young brought about the victory of the wave theory. But with the start of the twentieth century a rebirth of the corpuscular theory was brought about by the work of Einstein: *Über einen die Erzeugung und Verwandlung des Lichtes betreffenden heuristischen Gesichtspunkt,* Ann. d. Phys. (Lpz) **17**, 1905. (On a heuristic viewpoint concerning the creation and conversion of light.)

This paper was much more radical than the theory of relativity which had its origin in the same year. While the latter represented the crowning achievement of classical physics, the former revolutionized it.

In 1887 Hertz discovered the photoelectric effect and soon afterwards it was measured electrostatically by Hallwachs. The explanation of this effect on the basis of electron theory as given by Lenard and J. J. Thomson led to the following results: the number of electrons ejected from a metal plate by light depends on the light *intensity* but the kinetic *energy* of these electrons is solely determined by the frequency of the incident light. Einstein, applying Planck's discovery of the quantum of action h and the quantum of energy $h\nu$, interpreted these results in the following way: the upper limit v of the velocity spectrum of the photoelectrons is given by the energy equation

$$h\nu = \frac{m}{2} v^2 + A \tag{1}$$

where A is the minimum energy required to separate the electron from the metal. Only when $h\nu > A$ does a photoelectric effect appear. Ultraviolet light always yields a photoelectric effect while only the alkali metals, which have a small A, are photoelectrically active under red light. In 1916 Millikan confirmed the existence of such an upper bound of the velocity spectrum precisely and used it to determine h.

Thus a new elementary particle, the "photon", was introduced into physics. Its energy is

$$E = h\nu. \tag{1 a}$$

Since this particle always moves with the velocity c, we must ascribe to it the rest mass $\mu_0 = 0$; otherwise its velocity mass $\mu = \mu_0 \sqrt{1 - \beta^2}$ would become infinite. From the general relation between mass and energy

$$E = (\mu - \mu_0)\, c^2$$

the mass is found to be $u = h\nu/c^2$ and the momentum

$$p = \mu c = \frac{h\nu}{c}. \tag{2}$$

In his original paper Einstein also called attention to *Stokes rule for fluorescence*: the frequency of the fluorescent light is always displaced toward the red with respect to that of the exciting light. This rule is also generally true for *phosphorescence bands* (delayed fluorescence) and for the *characteristic radiation frequencies* in the *X-ray spectrum.* To excite, for instance, the K radiation of an atom the exciting radiation must be harder than the hardest line of the K spectrum.

The *continuous X-ray spectrum* is produced by a kind of reversal of the photoelectric effect. While in the ordinary photoelectric effect primary photons produce secondary electrons, we now have the case of primary electrons (cathode rays of energy E) impinging on a target electrode where they excite the secondary photons of the continuous X-ray spectrum. The following Stokes rule is obeyed:

$$h\nu < E, \qquad h\nu_{max} = E. \tag{3}$$

Therefore, the continuous X-ray spectrum has a short-wave limit $\lambda_{min} = c/\nu_{max}$ which can again be used in connection with eq. (3) to determine h. In contrast to this the classical computation of the radiation given off by a decelerating electron, as performed in Vol. III, eq. (19.22) or (30.11), always yields a spectrum which is continuous up to $\nu = \infty$.

The following consideration is to be added: the "accumulation period" which would be necessary if the energy provided by single cathode ray electrons were to accumulate to an X-ray energy of $h\nu$ would become very long — of the order of several hours! But actually the secondary X-rays are emitted simultaneously with the primary cathode rays just as the photoelectric effect starts immediately with the primary illumination. As a last desperate attempt to save the classical theory of radiation, Debye and the author undertook in 1913 to explain the photo effect classically by using a special hypothesis regarding the action integral[1].

Since that time it has, of course, become possible to register directly the discontinuous quantum nature of weak X-rays or ultraviolet light through the amplifying action of the *counter tube.* It is even possible to make the clicks associated with the separate discharges audible. We will not discuss here the *Compton effect* which makes the corpuscular nature of X-rays especially evident. Rather we shall limit our discussion to those effects which have already been explained wave-theoretically in this chapter. In the case of the moving mirror we have already indicated the possibility of a corpuscular explanation by using the example of the tennis ball at the end of Sec. 13. While our previous derivations of the aberration and dragging effects were ultimately based upon the velocity addition theorem, these two phenomena can also be easily explained on the basis of the corpuscular theory. But how about the Doppler effect with its expansion or crowding of wave surfaces which seems to require a definitely wave-theoretical explanation? Schrödinger[2] showed that this effect also can be understood from a photon point of view.

We shall assume that a radiating atom O, instead of emitting a spherical wave, sends out photons of energy $h\nu$ and momentum $h\nu/c$ in random directions. In this way such a photon will occasionally also travel in the direction of the observer P. When this is the case, the atom recoils in the direction PO. We shall assume here that the observer is at rest and the emitting atom is in motion, though we could equally well treat the reverse situation. The recoil momentum $h\nu/c$ combines with the original momentum of the atom Mv_1 to give Mv_2. Let α be the angle between OP and v_1 and $\alpha + d\alpha$ the angle between OP and v_2. We construct the momentum triangle

[1] Ann. d. Phys. **41** (Lpz), 1913, see also: First Solvay Congress, "Theorie du rayonnement et des quanta", p. 344. These efforts had to fail, of course, because of the tremendous length of the required accumulation period.

[2] Physikal. ZS. **23**, 301, 1922.

OAB, as shown in fig. 20, where $OA = Mv_1$, $OB = Mv_2$ $AB = h\nu/c$. We now project OB on OA and find that for the infinitesimal right triangle ABC the following relation holds:

$$M \Delta v = \frac{h\nu}{c} \cos\alpha. \tag{4}$$

This is the law of conservation of momentum. The conservation law of energy is:

$$\frac{M}{2} v_1{}^2 + E_1 = \frac{M}{2} v_2{}^2 + E_2 + h(\nu + \Delta \nu) \tag{5}$$

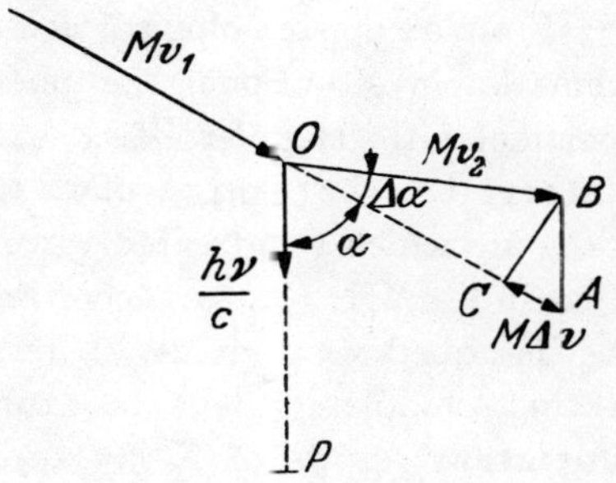

Fig. 20.
The corpuscular explanation of the Doppler effect.

where $h(\nu + \Delta \nu)$ is the energy emitted by the moving atom. Quantum-theoretically the energy of the photon emitted by an atom which is at rest or moving uniformly ($v_1 = v_2$) and which experiences a change in its configuration $E_1 \to E_2$ is given by

$$E_1 - E_2 = h\nu. \tag{5 a}$$

Substituting this in (5) one obtains

$$h\Delta\nu = \frac{M}{2}(v_1{}^2 - v_2{}^2) = M\Delta v \frac{v_1 + v_2}{2}. \tag{6}$$

Neglecting $(\Delta v)^2$ we let $(v_1 + v_2)/2 = v$ and obtain, from (4),

$$h\Delta\nu = h\nu \frac{v}{c} \cos\alpha. \tag{7}$$

Characteristically, h cancels out and we obtain the Doppler formula which is identical with (11.10)

$$\frac{\Delta\nu}{\nu} = \frac{v}{c} \cos\alpha \tag{8}$$

The reader may convince himself that the significance of the sign mentioned in connection with (11.10) is also in agreement with fig. 20.

This derivation appears to be inconsistent in that the recoil momentum was assumed to be $h\nu/c$ instead of $h(\nu + \Delta\nu)/c$. However, if we use the latter value, the result will differ from our present result only in terms of second order, i. e. terms proportional to $(v/c)^2$. If we had considered such terms, however, we should have had to carry out our calculations relativistically from the very beginning. In particular, the kinetic energy of the atom should have been set up differently. As Schrödinger has pointed out, we would then have obtained the relativistically rigorous Doppler formula, that is eq. (11.9).

From an epistemological point of view we are thus faced with an extremely remarkable situation: the phenomena described in this chapter, and in particular the Doppler effect, can be understood in terms of either the wave theory or the corpuscular theory. This is also true of the light pressure which was treated wave-theoretically by means of Poynting's energy flux in Vol. III, Sec. 31. According to the corpuscular theory this pressure can be described very vividly in terms of a "photon-hail". However, the wave theory completely fails to explain the photoelectric effect and the most important results of X-ray spectroscopy. On the other hand, the photon theory, at least in its present state of development, is unable to account precisely for polarization and interference phenomena. *Therefore, we are forced to adopt a dualistic conception of light*: not Huygens *or* Newton, *but* Huygens *and* Newton. Newtons' theory explains the coarse but nevertheless fundamental *energy* problems while Huygens' theory is of use for the much more delicate problems concerning *interference*. Light has a *dual nature*; it presents us with either its corpuscular or its wave aspect depending on the particular question which we are posing. It is wrong to ask which of these aspects is the *true* one. As far as we know today, *they are both on an equal footing* and only both aspects taken together are capable of representing the nature of light completely.

One speaks, therefore, not of a *duality* of light but, more appropriately, of its *complementarity*. This expression, which was coined by Bohr, is all the more appropriate because we know today that also *electrons and all material bodies* possess in addition to their corpuscular character and on an equal basis with it also a wave-mechanical nature. There are *matter waves* as well as *light waves*. Our previous characterization of cathode rays as corpuscles and X-rays as waves was an antiquated way of speaking. The difference between these two lies not in their wave character but rather in their velocities and charges and consequently their very different interactions with the atoms of material bodies.

It is clear that this complementarity overthrows the scholastic ontology. What is truth? We pose Pilate's question not in a skeptical, anti-scientific sense, but rather in the confidence that further work on this new situation will lead to a deeper understanding of the physical and mental world. Indeed, the revised quantum theory and Heisenberg's uncertainty principle on which it is based demonstrate that there can be no logical contradiction between the corpuscular and wave-theoretical viewpoints.

In the following chapters we shall only rarely have occasion to return to these fundamental questions. We shall have to restrict ourselves to a development of the wave theory. However, we should always remember that, though the latter forms the most important practical part of optics, it does not reveal its full content.

CHAPTER III

THEORY OF DISPERSION

So far we have discussed only the *nature of light.* Now we shall investigate more closely the nature of *refractive media.* We have already mentioned in connection with eq. (3.4 a) that the electromagnetic explanation of the index of refraction is inadequate since it even fails to account for the decomposition of white light in a prism. We can understand such phenomena only by learning more about the optical properties of matter.

The electric composition of matter is well known: every atom consists of a positive nucleus and a shell of more or less mobile electrons. However, we need not go into electron theory in the usual sense. We can carry out our calculations using, instead of individual electrons, an *electron fluid*[1] which is spread uniformly throughout the whole body. The situation here is analogous to the theory of hydrodynamics in which a continuous density replaces the individual molecules. In the same way the electrons will be thought of as "smeared-out"[2] into a continuum.

We shall treat the charges of the positive ions in the same way. They will serve the purpose of neutralizing the enormous electrostatic repulsive forces which would otherwise act within the electron fluid, and conversely the latter will neutralize the electrostatic repulsion of the ions. This point of view, which is usually adopted in dispersion, is entirely justified in the optical spectrum where a cube one wavelength in dimension contains a tremendously large number of atoms; in the X-ray region, however, this mode of attack fails.

In hydrodynamics we use a volume element to define displacement and velocity; the definition of displacement in the electron fluid will be explained in eq. (17.2); the corresponding definition for ions will be found in the beginning of Sec. 18.

---

[1] Using a currently popular word one could speak of an electron "plasma".

[2] I am afraid that this ugly word which I used in my lecture in 1912 has come into general usage. At that time I posed to P. Ewald, as a theme for his dissertation, the problem of explaining double refraction and dispersion in crystals in terms of their lattice structures, that is not to smear out the electrons but to assume them to be bound to the individual building stones of the crystal. There was a close causal connection between this dissertation and the ingenious idea of M. von Laue to investigate the lattice structure of crystals with X-rays. See also Chap. V, Sec. 32 C.

17. Ultraviolet Resonance Oscillations of the Electrons

Let us now investigate a transparent isotropic non-conducting material. The optical field will be described by the two light vectors $\mathbf{E}$ and $\mathbf{H}$. Excluding, for the present, magnetizable materials we set $\mathbf{B} = \mu_0 \mathbf{H}$. However, we do not set $\mathbf{D} = \varepsilon \mathbf{E}$ as in the case of slowly varying fields, but rather more generally we let

$$\mathbf{D} = \varepsilon_0 \mathbf{E} + \mathbf{P}. \tag{1}$$

$\mathbf{P}$ is the *polarization* vector which was introduced in Vol. III, Sec. 11. Polarization means that the field $\mathbf{E}$ displaces the electrons from their rest positions. We shall call the displacement vector $\mathbf{s}$ and set

$$\mathbf{P} = -N e\, \mathbf{s}. \tag{2}$$

$-e$ is the charge of the electron and N is the number of dispersion electrons per unit volume. These definitions of $\mathbf{P}$ and $\mathbf{s}$ obviously presuppose individual electrons rather than a continuum. Equation (2), therefore, refers to the state of affairs before the "smearing-out".

Let us note that this assumption is dimensionally correct. The dimension of $\mathbf{P}$ is, like that of $\mathbf{D}$ and $\varepsilon_0 \mathbf{E}$, $Q\,M^{-2}$ (see table at the end of Sec. 2). And indeed, since N has the dimensions M^{-3}, $N\mathbf{s}$ has the dimensions M^{-2}. Let us note further that the sign in (2) is chosen correctly. Figure 21 represents the action of the field $\mathbf{E}$ upon the electronic charge $-e$. This charge is separated from the ionic charge $+e$, with which it originally coincided, by being displaced in a direction opposite to the displacement of the positive charge. We can assume the ionic charge $+e$ which carries the ionic mass M to be stationary. Thus an electric moment $(+e, -e)$ with moment arm $|\mathbf{s}|$ and directed parallel to $\mathbf{E}$ is created as represented in eq. (2).

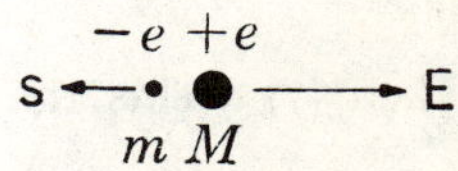

Fig. 21.
Separation of the electron charge (mass m) and ion charge (mass M) in the electric field $\mathbf{E}$.

In fig. 21 we assume that the electrons are bound "quasi-elastically" to their rest positions, so that they will seek to return to those positions when displaced from them by a field $\mathbf{E}$. Therefore $\mathbf{s}$ satisfies the following differential equation:

$$m\ddot{\mathbf{s}} + f\mathbf{s} = -e\,\mathbf{E}. \tag{3}$$

Here the restoring force $-f\mathbf{s}$ has been put on the left-hand side with the opposite sign. On the right-hand side is the field force acting on $-e$. Instead of (3) we shall write

$$\ddot{\mathbf{s}} + \omega_0^2 \mathbf{s} = -\frac{e}{m}\mathbf{E} \qquad \text{where} \qquad \omega_0^2 = \frac{f}{m}. \tag{3 a}$$

ω_0 is the characteristic frequency of the electron (and therefore also of the electronfluid) which the title of this paragraph asserts to be in the far *ultraviolet.* This is particularly the case for the gases H_2, N_2, O_2, etc. Written in terms of the polarization eq. (3 a) becomes:

$$\left(\frac{\partial^2}{\partial t^2} + \omega_0{}^2\right) \mathbf{P} = \frac{N e^2}{m} \mathbf{E}. \tag{4}$$

Thus we have not two, but *three* light vectors, $\mathbf{E}$, $\mathbf{H}$ and $\mathbf{P}$ which are connected by three vector differential equations. For we have, besides (4), the two Maxwell equations

$$\mu_0 \dot{\mathbf{H}} = -\operatorname{curl} \mathbf{E} \qquad \text{and} \qquad \varepsilon_0 \dot{\mathbf{E}} + \dot{\mathbf{P}} = \operatorname{curl} \mathbf{H}. \tag{5}$$

In the second of these equations $\dot{\mathbf{D}}$ is already expressed in terms of $\mathbf{E}$ and $\mathbf{P}$ by means of (1). Eliminating $\mathbf{H}$ from the two eqs. (5), we obtain, taking account of $\operatorname{div} \mathbf{E} = 0$ (uncharged dielectric)

$$\varepsilon_0 \mu_0 \ddot{\mathbf{E}} + \mu_0 \ddot{\mathbf{P}} = -\operatorname{curl}\operatorname{curl} \mathbf{E} = \Delta \mathbf{E}, \tag{6}$$

see Vol. III eq. (6.2). We need now only eliminate $\mathbf{P}$ from eqs. (4) and (6) to obtain a pure differential equation for $\mathbf{E}$. For this purpose we operate on (6) and (4) with the operators $\partial^2/\partial t^2 + \omega_0{}^2$ and $\mu_0\, \partial^2/\partial t^2$, respectively. Thus we get the following fourth order differential equation:

$$\left(\frac{\partial^2}{\partial t^2} + \omega_0{}^2\right)\left(\frac{1}{c^2} \ddot{\mathbf{E}} - \Delta \mathbf{E}\right) + \frac{\mu_0 N e^2}{m} \ddot{\mathbf{E}} = 0. \tag{7}$$

We specialize this immediately to the case of a linearly polarized plane wave of frequency ω and wave number k

$$E_y = A\, e^{i(kx - \omega t)}, \qquad E_x = E_z = 0. \tag{8}$$

Then (7) yields the following algebraic relation between ω and k:

$$(-\omega^2 + \omega_0{}^2)\left(-\frac{\omega^2}{c^2} + k^2\right) = \frac{\mu_0 N e^2}{m} \omega^2,$$

which, when solved for k^2, gives

$$k^2 = \frac{\omega^2}{c^2}\left(1 + \frac{\mu_0 c^2 N e^2/m}{\omega_0{}^2 - \omega^2}\right). \tag{9}$$

Now, $u = \omega/k$ is the phase velocity of our plane wave (8) in the dispersive medium. The index of refraction of this medium relative to the vacuum is, according to the definition (3.4),

$$n = \frac{c}{u} = \frac{c\,k}{\omega}. \tag{9 a}$$

Setting $\mu_0 c^2$ equal to $1/\varepsilon_0$ we write instead of (9)

$$n^2 = 1 + \frac{N e^2/m\,\varepsilon_0}{\omega_0{}^2 - \omega^2}. \tag{10}$$

Thus, the *index of refraction has become frequency dependent.* This is what is meant by *dispersion.* Because

$$\omega_{\text{red}} < \omega < \omega_{\text{violet}} < \omega_0$$

the denominator $\omega_0{}^2 - \omega^2$ is positive in the entire visible spectrum and is *larger* at the red end than at the violet end. *Blue light is refracted more than red light.* This is normal dispersion.

Assuming ω_0 very large, we now develop (10) in powers of ω/ω_0 and retain only the first two terms:

$$n^2 = 1 + \frac{N e^2}{m\,\varepsilon_0\,\omega_0{}^2}\left(1 + \frac{\omega^2}{\omega_0{}^2}\right). \tag{11}$$

Setting $\omega = \dfrac{2\pi c}{\lambda}$, where λ is the wavelength in vacuum, we are led to a formula which corresponds to an old molecular elastic theory of Cauchy (ca. 1830). Abbreviating, we write

$$n^2 = 1 + A\left(1 + \frac{B}{\lambda^2}\right), \qquad A = \frac{N e^2}{m\,\varepsilon_0\,\omega_0{}^2}, \qquad B = \frac{4\pi^2 c^2}{\omega_0{}^2}. \tag{12}$$

A and B are called the *coefficients of refraction* and *dispersion,* respectively. We note that the ratio B/A does not contain the characteristic frequency ω_0, so that the value of this ratio is the *universal* number

$$\frac{B}{A} = \frac{4\pi^2 c^2\,\varepsilon_0}{N e^2/m}. \tag{12 a}$$

Its dimension is M^2, as can be ascertained directly from (12).

We shall now compare (12) and (12 a) with very exact measurements[1] of the dispersion in hydrogen which yield

$$n^2 = 1 + 2.721 \times 10^{-4} + \frac{2.11}{\lambda^2}\,10^{-18}. \tag{13}$$

This gives the values

$$B = \frac{2.11}{2.721}\,10^{-14}\,\text{M}^2, \qquad \frac{B}{A} = \frac{2.11}{(2.721)^2}\,10^{-10} = 0.2\text{[illegible]} \times 10^{-10}\ \text{[illegible]}$$

[1] J. Koch, Nova Acta Upsal. 2, 1909. The measurements refer t[illegible] In conformity with our system of units we have written (13) so [illegible] meters.

We substitute this in the left-hand side of (12 a). In accordance with the table at the end of Sec. 2, we put $4\pi c^2 \varepsilon_0 = 10^7 \text{ M S}^{-1} \Omega^{-1}$ on the right-hand side and obtain

$$\text{(14)} \qquad \frac{N e^2}{m} = \frac{\pi}{0.29} 10^{17} = 1.1 \times 10^{18} \text{ M}^{-1} \text{ S}^{-1} \Omega^{-1}.$$

N is found from the density of H_2 which at 0° C and 760 mm Hg is almost exactly equal to $9.00 \times 10^{-2} \text{ KM}^{-3}$. Hence, the mass per unit volume is 9.00×10^{-2} K. But the mass per unit volume is also equal to $2 N_0 m_H$ where m_H is the mass of a hydrogen atom and N_0 the number of molecules per unit volume. Hence

$$N_0 = \frac{9.00 \times 10^{-2}}{2 m_H}.$$

Furthermore, since each H_2 molecule has two electrons,

$$N = 2 N_0, \qquad N e = 9.00 \times 10^{-2} \frac{e}{m_H}.$$

Now e/m_H, which is Faraday's *electrochemical equivalent* (the charge of one gram-atom in electrolysis), is also a very precisely known number, namely 9649 in abs. e. m. u. and hence $9649 \times 10^4 = 9.65 \times 10^7 \text{ QK}^{-1}$ in our Q, K units. Equation (14) becomes, therefore,

$$N e \cdot \frac{e}{m} = 9.00 \times 10^{-2} \times 9.65 \times 10^7 \times \frac{e}{m} = 1.1 \times 10^{18}$$

which yields

$$\text{(15)} \qquad \frac{e}{m} \sim 1.4 \times 10^{11} \text{ QK}^{-1}.$$

This is of the same order of magnitude as the "specific charge of the electron" $e/m = 1.76 \times 10^{11} \text{ QK}^{-1}$. The resonance frequeny ω_0 which is easily computed by comparing (12) and (13) confirms our basic assumption by being in the far *ultraviolet*. Clearly, our theory is still very crude in comparison to the detailed results on molecular structure and emission of light which atomic physics has provided.

We shall not justify a once much debated rule (Drude, Natanson) according to which the number of "dispersion electrons" of an ideal gas is the same as the "valence number" of the particular molecule (namely 2×2 for O_2, 2×3 for N_2). From the point of view of our present atomic models this is less understandable than the number 2×1 in the case of H_2, because cases of O and N the valence is not the number of electrons *present* e case of H, but rather it is the number of electrons *missing* from the shell of eight electrons. Empirically, however, this rule is satisfied to a good degree of approximation, and this is also understandable from the point of view of atomic physics because in many respects (for instance, in the Pauli principle) present and missing electrons play the same role.

We shall now mention, though again only briefly, a refinement of the dispersion formula which yields the dependence of the index of refraction on pressure in the case of gases and which is known as the Lorenz-Lorentz formula. This formula results from a more exact calculation of the polarization **P**, which depends not only on the external field **E**, as was assumed in (2) and (3), but is also influenced by the electric moments of the neighboring molecules. We investigated this dependence more closely in Vol. III, Sec. 11 and derived there the Clausius-Mosotti formula (11.8) for the dielectric constant, in which we now have to replace ε_{rel} by n^2 and the molecular constant as defined there

$$N\alpha \text{ by } \frac{N e^2/m\,\varepsilon_0}{\omega_0^{\,2}-\omega^2}.$$

Thus we obtain

$$\frac{n^2-1}{n^2+2}=\frac{1}{3}\,\frac{N e^2/m\,\varepsilon_0}{\omega_0^{\,2}-\omega^2}. \tag{16}$$

When n^2 differs only slightly from 1, which is true in particular for ideal gases, then it can be seen that (16) goes over into the above eq. (10).

18. Infrared Resonance Oscillations of the Ions in Addition to Ultraviolet Electron Resonance Oscillations

If there is a considerable difference between the indices of refraction for visible light and for Hertzian waves, one must expect that other resonance oscillations besides the ultraviolet one contribute to the index of refraction. If, furthermore, the material in question is transparent so that there are no resonances in the visible spectrum, these additional resonances must be *infrared resonance oscillations* (perhaps rotational resonances). It is reasonable to ascribe these not to the mobile electrons, but rather to the much more inert ions. These, too, we shall consider to be "smeared out"; thus we shall not use individual ions, but rather a continuous ion-fluid.

The polarization **P** is then the sum of the two contributions $\mathbf{P}_1$ (electrons) and $\mathbf{P}_2$ (ions):

$$\mathbf{P}=\mathbf{P}_1+\mathbf{P}_2. \tag{1}$$

Again, as in the case of $\mathbf{P}_1$ in eq. (17.2), the definition of $\mathbf{P}_2$ refers to the st[illegible] before the "smearing-out", namely to the dipole moments forme[illegible] separations of the ions from ions of opposite charge.

The *resonance frequency of the electrons* which we called ω_0 in Sec. 17 will now be called ω_1. The number of electrons per unit volume will again be N. The differential equation for the oscillations of the electrons brought about by the alternating field E is, according to the model of (17.4)

$$\left(\frac{\partial^2}{\partial t^2} + \omega_1{}^2\right) \mathsf{P}_1 = \frac{N e^2}{m} \mathsf{E}. \tag{2}$$

where the displacements have already been replaced by the electric moment P.

The *resonance oscillation of the ions*, whose frequency will be called ω_2, consists of a relative oscillation of oppositely charged components. If only one pair of such components is present in each molecule, and we shall here limit ourselves to this case, then the relative displacement of these components takes the place of an absolute displacement and the so-called "reduced mass" M takes the place of the individual masses M_1 and M_2. As will be shown in exercise III. 1, the reduced mass is given by

$$\frac{1}{M} = \frac{1}{M_1} + \frac{1}{M_2}. \tag{3}$$

This mass M enters into the differential equation for the *forced* ion oscillations in place of the mass m in eq. (2). Let p denote the valence of the ion as determined by electrolysis, e. g. $p = 1$ for $Na^+ Cl^-$, $p = 2$ for $Ca^{++} F_2{}^{--}$ etc. Because the optical material is electrically neutral, the number of ions per unit volume is equal to the number of electrons N *divided* by p, while the charge of each individual ion is the electron charge e *multiplied* by p. In this way we have in place of $N e^2/m$

$$\frac{N}{p} \frac{(p\, e)^2}{M} = \frac{N p\, e^2}{M}.$$

Hence, we obtain the following differential equation for P_2:

$$\left(\frac{\partial^2}{\partial t^2} + \omega_2{}^2\right) \mathsf{P}_2 = \frac{N p\, e^2}{M} \mathsf{E}. \tag{4}$$

The Maxwell relationship between P and E remains the same as in (17.6), where however, P must now be replaced by $\mathsf{P}_1 + \mathsf{P}_2$ in accordance with (1):

$$\varepsilon_0 \mu_0 \ddot{\mathsf{E}} + \mu_0 (\ddot{\mathsf{P}}_1 + \ddot{\mathsf{P}}_2) = \Delta \mathsf{E}. \tag{5}$$

By eliminating P_1 and P_2 from (2), (4) and (5), we obtain in place of (17.7) a somewhat complicated sixth order differential equation for E. We do not need to write out this equation, but can immediately treat the pure harmonic state of a linearly polarized wave of frequency ω of the type (17.8); that is, we assume that the electrons as well as the ions have attained their steady state in this field of frequency ω. Then, according to (2) and (4)

$$\ddot{\mathsf{P}}_1 \text{ is proportional to } \frac{-\omega^2}{\omega_1{}^2-\omega^2}, \qquad \ddot{\mathsf{P}}_2 \text{ is proportional to } \frac{-\omega^2}{\omega_2{}^2-\omega^2},$$

and by (5) k^2 becomes an algebraic function of ω^2:

$$k^2 - \frac{\omega^2}{c^2} = \mu_0\,\omega^2 \left(\frac{N\,e^2/m}{\omega_1{}^2-\omega^2} + \frac{p\,N\,e^2/M}{\omega_2{}^2-\omega^2}\right).$$

According to the definition of the index of refraction in (17.9 a), the left-hand side equals $(n^2-1)\,\omega^2/c^2$. Hence, we obtain immediately the following generalization of (17.10):

(6) $$n^2 = 1 + \frac{N\,e^2/m\,\varepsilon_0}{\omega_1{}^2-\omega^2} + \frac{p\,N\,e^2/M\,\varepsilon_0}{\omega_2{}^2-\omega^2}.$$

Evidently when a larger number of resonance oscillations are present, whether they be in the ultraviolet or in the infrared or perhaps in the visible spectrum, a formula of the same structure is obtained. The summation on the right-hand side must then include a term for every one of the resonance oscillations.

In order to make eq. (6) more convenient for purposes of comparison with observations, we express ω in terms of the wave length in vacuum λ, and similarly we express ω_1 and ω_2 in terms of λ_1 and λ_2:

$$\omega = \frac{2\pi\,c}{\lambda}, \qquad \omega_1 = \frac{2\pi\,c}{\lambda_1}, \qquad \omega_2 = \frac{2\pi\,c}{\lambda_2}.$$

Using the abbreviations

(7) $$C_1 = \frac{N\,e^2}{4\pi^2\,c^2\,\varepsilon_0\,m}, \qquad C_2 = \frac{p\,N\,e^2}{4\pi^2\,c^2\,\varepsilon_0 M}$$

we obtain

(7 a) $$n^2 - 1 = C_1 \frac{\lambda^2\,\lambda_1{}^2}{\lambda^2-\lambda_1{}^2} + C_2 \frac{\lambda^2\,\lambda_2{}^2}{\lambda^2-\lambda_2{}^2}$$

or, eliminating λ^2 from the numerators,

(8) $$n^2 = 1 + \lambda_1{}^2\,C_1 + \lambda_2{}^2\,C_2 + \frac{\lambda_1{}^4\,C_1}{\lambda^2-\lambda_1{}^2} + \frac{\lambda_2{}^4\,C_2}{\lambda^2-\lambda_2{}^2}.$$

Next we consider the limiting case as $\lambda \to \infty$. Then the last two terms on the right-hand side vanish and we obtain

(9) $$n_\infty{}^2 = 1 + \lambda_1{}^2\,C_1 + \lambda_2{}^2\,C_2.$$

Only in this limiting case when the actual resonance terms vanish is the Maxwell relation (3.4 a) fulfilled exactly. Thus, *the failure of the Maxwell relation to hold in the visible spectrum can be understood in terms of the existence of infrared resonance oscillations* (because $\lambda_1^2 \ll \lambda_2^2$ the term in (9) which contains λ_2^2 is clearly the decisive one). Hence the Maxwell relation should be corrected to read

$$n_\infty = \sqrt{\varepsilon} \qquad \text{instead of} \qquad n = \sqrt{\varepsilon} \tag{9 a}$$

(by ε is meant in both cases the dielectric constant relative to vacuum).

We turn now to the visible part of the spectrum. In this range very exact measurements for some cubic halide crystals are available. (As we shall see in Chap. IV, cubic crystals, strangely enough, behave *optically isotropic,* while elastically, thermally, etc. they reveal their anisotropy). From among these we select fluorspar CaF_2 (or fluorite, from which substance the phenomenon of fluorescence received its name). According to measurements by Paschen[1] the following holds for fluorite:

$$n^2 = 6.09 + \frac{6.12 \times 10^{-15}}{\lambda^2 - 8.88 \times 10^{-15}} + \frac{5.10 \times 10^{-9}}{\lambda^2 - 1.26 \times 10^{-9}}. \tag{10}$$

Comparing this with (8) we obtain

$$\frac{C_2}{C_1} = \frac{\lambda_1^4}{\lambda_2^4} \cdot \frac{5.10 \times 10^{-9}}{6.12 \times 10^{-15}} = \left(\frac{8.88 \times 10^{-15}}{1.26 \times 10^{-9}}\right)^2 \cdot \frac{5.10 \times 10^{-9}}{6.12 \times 10^{-15}} = 4.15 \times 10^{-5}.$$

On the other hand (7) yields

$$\frac{C_2}{C_1} = p\,\frac{m}{M}.$$

Because $p = 2$ (the Ca^{++} ion has given two electrons to the two F ions) we conclude that

$$\frac{M}{m} = \frac{2 \times 10^5}{4.15}. \tag{11}$$

To calculate the reduced mass M we set in (3) $M_1 = 40\, m_H$ (40 = atomic weight of Ca, m_H = mass of the H atom), $M_2 = 2 \times 19\, m_H$ (19 = atomic weight of F, therefore 2×19 = molecular weight of the negative component F_2). Thus we obtain

$$\frac{1}{M} = \left(\frac{1}{40} + \frac{1}{38}\right)\frac{1}{m_H}, \qquad \text{hence} \qquad M = 19.5\, m_H.$$

[1] Ann. d. Phys. (Lpz.) **54**, p. 672, 1895.

From this and (11) we find

$$\frac{m_H}{m} = 2450. \tag{12}$$

This value is of the same order of magnitude as that obtainable from the values for e/m and e/m_H given in Sec. 17. Our original assumption: ultraviolet resonance oscillations = electron oscillations, infrared resonance oscillations = ion oscillations, is thus confirmed.

The relationship between $n_\infty{}^2$ and the dispersion constants C_1, C_2, λ_1, λ_2 required by eq. (9) is also tolerably well fulfilled. For, according to (9), one obtains

$$n_\infty{}^2 = 6.09, \qquad \lambda_1{}^2 C_1 = \frac{6.12}{8.88} = 0.7, \;\; \lambda_2{}^2 C_2 = \frac{5.10}{1.26} = 4.06,$$

hence

$$1 + \lambda_1{}^2 C_1 + \lambda_2{}^2 C_2 = 5.76.$$

The electric determination of the dielectric constant gives

$$\varepsilon = 6.7 \quad \text{to} \quad 6.9.$$

When Drude arrived at these and many similar results around 1900, he made the incidental remark to the author: "We live in a grandiose era; we are beginning to get a glimpse of the electric composition of matter." Had he lived to witness the developments of the following decades, he would have seen his boldest hopes surpassed.

From a practical point of view the shape of the dispersion curve for glass is clearly of paramount importance to the problem of designing achromatic lenses and other optical apparatus. In exercise III. 3 we shall treat the achromatic prism and in the same connection also the direct vision prism. Exercise III. 2 serves as preparation for these problems.

19. Anomalous Dispersion

We shall now investigate the dispersion in the immediate vicinity of a resonance frequency $\omega = \omega_0$. We assume that the latter lies in the visible spectrum, because only in the visible spectrum can sufficiently precise measurements be made with which to check the theory. Hence, our body is no longer transparent as had previously been assumed, but, as we shall see, it is *colored* and the coloration depends on the value of ω_0

Since for $\omega = \omega_0$ the equation for the forced oscillations (17.3 a) would result in an oscillation of infinite amplitude, we must add a *damping term*. This same procedure is used in all other resonance problems of mech and electrodynamics. We shall write this term in the form $g\,\omega_0\,\dot{\mathbf{s}}$, because it is convenient to make the term proportional to the velocity $\dot{\mathbf{s}}$ and because inclusion of the factor ω_0 makes the *damping constant* g a dimensionless number. In order for the resonance to be sharp, g must be $\ll 1$. Equation (17.4) is thus changed into

$$\left(\frac{\partial^2}{\partial t^2} + g\,\omega_0\,\frac{\partial}{\partial t} + \omega_0{}^2\right)\mathbf{P} = \frac{N\,e^2}{m}\,\mathbf{E}. \tag{1}$$

Since ω_0 is in the visible spectrum, and the oscillating particles are therefore *ions,* m is a reduced mass and also includes the valence number p of the ion as in (18.4).

For a field $\mathbf{E}$ of frequency ω and for steady harmonic oscillations of the ions, (1) gives

$$\mathbf{P} = \frac{N\,e^2}{m}\,\frac{\mathbf{E}}{(\omega_0{}^2 - \omega^2 - i\,g\,\omega_0\,\omega)}$$

and, corresponding to (17.10) and (18.6), the resulting index of refraction is

$$n^2 = n_m{}^2 + \frac{N\,e^2/m\,\varepsilon_0}{\omega_0{}^2 - \omega^2 - i\,g\,\omega_0\,\omega.}\,. \tag{2}$$

n_m is the average contribution in the vicinity of $\omega = \omega_0$ of all other resonances which add to the dispersion in the visible range (also included in n_m is the contribution 1 of the pure displacement current which has heretofore been written down separately). As in the case of metallic reflection the index of refraction n has now become *complex.* As in eq. (6.2) we again replace n by $n\,(1 + i\,\varkappa)$; by separating real and imaginary parts we obtain from (2)

$$n^2\,(1 - \varkappa^2) = n_m{}^2 + \frac{N\,e^2}{m\,\varepsilon_0}\,\frac{\omega_0{}^2 - \omega^2}{(\omega_0{}^2 - \omega^2)^2 + g^2\,\omega_0{}^2\,\omega^2}, \tag{3}$$

$$2\,n^2\,\varkappa = \frac{N\,e^2}{m\,\varepsilon_0}\,\frac{g\,\omega_0\,\omega}{(\omega_0{}^2 - \omega^2)^2 + g^2\,\omega_0{}^2\,\omega^2}. \tag{4}$$

We introduce the abbreviation

$$a^2 = \frac{N\,e^2}{m\,\varepsilon_0\,\omega_0{}^2} \quad \text{(dimensionless number)} \tag{5}$$

and use as variables

$$x = \frac{\omega^2 - \omega_0{}^2}{\omega_0{}^2}, \qquad y = \frac{n^2\,(1 - \varkappa^2)}{a^2}, \qquad z = \frac{2\,n^2\,\varkappa}{a^2}. \tag{6}$$

Then eqs. (3) and (4) become

$$y = \frac{n_m{}^2}{a^2} - \frac{x}{x^2 + g^2\,(1 + x)}, \qquad z = \frac{g\sqrt{1 + x}}{x^2 + g^2\,(1 + x)}. \tag{7}$$

The extrema of y are found in the following way:

$$\frac{dy}{dx} = -\frac{1}{x^2 + g^2(1+x)}\left(1 - \frac{2x^2 + g^2 x}{x^2 + g^2(1+x)}\right) = 0, \qquad \text{hence} \qquad x^2 = g^2,$$

$$x = +g, \qquad y = y_{min} = \frac{n_m^2}{a^2} - \frac{1}{2g + g^2},$$

$$x = -g, \qquad y = y_{max} = \frac{n_m^2}{a^2} + \frac{1}{2g - g^2}.$$

We now visualize the content of our formulae by means of a table and a graph. For the sake of clarity we shall treat g as a small number. Then we can neglect g^2 as compared to g and, where they occur together, g as compared to 1. In the first line of the table x is used as a spectral measure and in the last line ω is given.

x	-1	$-g$	0	$+g$	∞
$y - n_m^2/a^2$	$+1$	$1/2g$	0	$-1/2g$	0
z	0	$1/2g$	$1/g$	$+1/2g$	0
ω	0	$\omega_0\sqrt{1-g}$	ω_0	$\omega_0\sqrt{1+g}$	∞

In agreement with this table fig. 22 shows that the extrema of the y-curve occur at $x = \pm g$ and that the curve cuts the line $y = n_m{}^2/a^2$ at $x = 0$. Furthermore, the narrow bell-shaped z curve attains its maximum at $x \sim 0$ (more precisely at $x = -g^2/4 + \ldots$) and its half-value width is $2g$. The scales to which the y- and z-curves are drawn in fig. 22 are not comparable. The scale of the z-curve is indicated on the right of the figure.

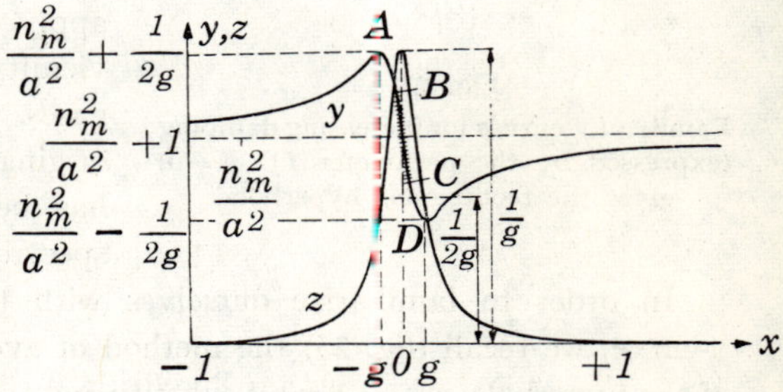

Fig. 22

Representation of anomalous dispersion.

$$x = \frac{\omega^2 - \omega_0{}^2}{\omega_0{}^2}, \qquad y = \frac{n^2(1-\varkappa^2)}{a^2}, \qquad z = \frac{2n^2\varkappa}{a^2}$$

The y-curve (scale on the left) indicates substantially the course of the index of refraction. The z-curve (scale on the right) indicates the coefficient of absorption.

Besides representing y and z, fig. 22 also serves as a qualitative illustration of the behavior of n^2 and $\varkappa$. Only in the vicinity of $\varkappa = 0$ are the curves somewhat distorted from the exact representations of n and $\varkappa$ as given by eq. (3) and (4).

In fig. 22 the portion of the curve for the index of refraction lying between A and D is of greatest interest to us. While before A and after D this curve *rises* with *increasing* ω (normal dispersion), between A and D it descends with increasing x. This is called *anomalous dispersion*, i. e. *the short waves are refracted less than the longer ones*. The Dane Christiansen first observed this phenomenon in fuchsin around 1870. Almost at the same time and independently Kundt detected it in various dyes.

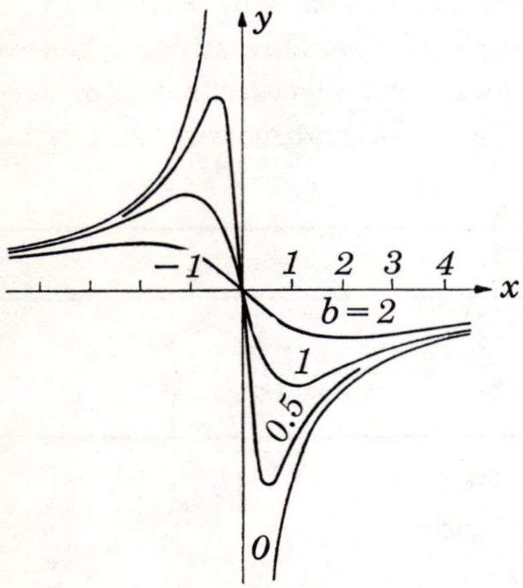

Fig. 23.

Family of y curves for increasing damping (expressed by the parameter b). $b = 0$ gives the rectangular hyperbola.

The portion BC of the descending curve AD is marked with hatch lines to indicate that in this range the spectrum is extinguished by absorption. Hence the anomalous dispersion is observable only along the short segments AB and CD. In the case of fuchsin the absorption band is in the yellow-green. The transmitted unabsorbed light therefore, has the complementary intensely red color.

The above remarks regarding dyes apply even more strongly to the vicinity of all the spectral lines of gases. By virtue of this fact Kundt's original "method of crossed prisms" has been developed into a powerful spectroscopic method.

In order to familiarize ourselves with the characteristic shape of the y-curve, we recall (fig. 23) the method of avoiding the singularity, at $x = 0$, of the hyperbola $y = -1/x$ by substituting the continuous function

$$y = -\frac{x}{x^2 + b^2},$$

which, in the limit as $b \to 0$, approaches the rectangular hyperbola $y = -1/x$. The latter function corresponds to the index of refraction for the case of an undamped resonance denominator. The continuous function corresponds to the index of refraction for the case of the damped resonance denominator in eq. (7).

Figure 22 furthermore demonstrates that in passing through the resonance from short wavelengths to longer wavelengths the y-curve (n^2/a^2) is raised by unity. We can also verify this by comparing the second and last columns in the above table. Every resonance frequency contributes such an increase in the value of n^2/a^2. Every time n^2 passes through a resonance it is increased by a^2. This enables us to understand why for solids with complex molecules, e. g. glass, the indices of refraction in the optical range differ from those in the range of Hertz waves.

The particularly striking case of water ($n = 4/3$ in the visible spectrum, $n_\infty = \sqrt{\varepsilon} = \sqrt{80}$ in the limiting case of electrostatics $\lambda = \infty$) does not belong in the same category, but is rather due to the polar nature of the H_2O-molecule; see Vol. III, p. 74, footnote 1. Owing to their triangular structure the water molecules possess a permanent electric moment which can follow the external field oscillations only at long wavelengths. A sort of molecular stiffness prevents them from being set into oscillation by short waves. The transition between the two regions of behavior is in the vicinity of $\lambda = 1.7$ cm; and it is here also that the main portion of the jump between the optical value of n^2 and the electrostatic dielectric constant lies.

20. Magnetic Rotation of the Plane of Polarization

One of the important contributions to the development of the electromagnetic theory of light was Faraday's discovery in 1845 (see historical table) of a connection between optics and magnetism, two up to that time entirely distinct fields. Although this connection does not apply to processes in free space, but is limited to ponderable bodies and is related to the motions of the dispersion electrons in such bodies, Faraday's discovery was, nevertheless, an impressively strong hint of the electromagnetic nature of light.

We begin with eq. (17.3) where the force of the field on the electron was described solely by $-e\mathbf{E}$. On the other hand we know from Vol. III that the force of the field on a *moving* charge (which we shall again denote by $-e$) is given generally by the Lorentz force expression

$$\mathbf{F} = -e(\mathbf{E} + \mathbf{v} \times \mathbf{B}). \tag{1}$$

Setting $\mathbf{B} = \mu\mathbf{H}$ and noting that according to (2.5) the relation

$$|\mathbf{H}| = \sqrt{\frac{\varepsilon}{\mu}}\,|\mathbf{E}|$$

holds for a light field, we remark that the $\mathbf{B}$ term in (1) differs from the $\mathbf{E}$ term by the factor

$$\sqrt{\varepsilon\mu}\,v = \frac{v}{u} = n\frac{v}{c} = n\beta.$$

The magnetic force is, therefore, only a correction of first order in β and can be neglected in comparison to the electric force $-e\,\mathbf{E}$.

But now we shall suppose that the material is placed in an external field $\mathbf{B}$ which can be made much stronger than the optical $\mathbf{B}$-field. In that case the equation of motion (17.3) must be provided with a possibly large correction term, and it then reads:

$$m\ddot{\mathbf{s}} + f\,\mathbf{s} = -e\,(\mathbf{E} + \dot{\mathbf{s}} \times \mathbf{B}). \tag{2}$$

We choose the direction of $\mathbf{B}$ as the positive z-direction and assume that the light wave propagates in this same direction. With these assumptions $\mathbf{E}$ and $\mathbf{s}$ are vectors in the xy-plane[1]. Separating the components of (2), dividing by m, and letting $\omega_0^2 = f/m$, we obtain the two simultaneous equations.

$$\begin{aligned} \ddot{s}_x + \frac{e}{m} B\,\dot{s}_y + \omega_0^2\,s_x &= -\frac{e}{m} E_x, && \Big|\ 1 \\ \ddot{s}_y - \frac{e}{m} B\,\dot{s}_x + \omega_0^2\,s_y &= -\frac{e}{m} E_y. && \Big|\ \pm i \end{aligned} \tag{3}$$

Multiplying by the factors indicated on the right and adding we get

$$\ddot{S} \mp \frac{e}{m} i\,B\,\dot{S} + \omega_0^2 S = -\frac{e}{m} E \tag{4}$$

where the following abbreviations have been used:

$$S = s_x \pm i\,s_y, \qquad E = E_x \pm i\,E_y. \tag{4 a}$$

However, it must be emphasized that this method of deriving (4) by way of (3) is unnecessarily indirect. For (4) is nothing else than the initial eq. (2), provided the latter is interpreted in the following way: two-dimensional vectors are complex numbers of the form $a + i\,b$ and therefore the vectors $\mathbf{s}$ and $\mathbf{E}$ in (2) are identical with the complex quantities S and E as defined in (4 a). Moreover, since multiplication by i means a right-handed rotation from x to y around z, and since the vector product is also defined by the right-handed screw rule, it follows that $\dot{\mathbf{s}} \times \mathbf{B}$ in (2) is simply the complex number $\mp i\,B\,S$ which appears multiplied by $-e/m$ on the left-hand side in eq. (4).

This point of view is quite generally valid. It suggests that we treat the complex quantities E themselves as the basic field variables, rather than the components E_x, E_y. Physically this implies a *transition from linear to circular*

---

[1] Like all quantities which we use to represent monochromatic states, $\mathbf{E}$ and $\mathbf{s}$ are, of course, everywhere multiplied by the same time factor $\exp(-i\omega t)$. While heretofore we could almost always omit writing this factor it improves the clarity of the calculations to retain it occasionally and we shall do so starting with eq. (5) The transition to real parts is again postponed until the final formula (14).

polarization. If originally (i. e. without magnetic field or where the light enters this field) we have a linearly polarized wave **E**: $E_x = A \cos \omega t$, $E_y = 0$, then we decompose this wave into the two right and left circularly polarized waves, say E_+ and E_-, and in agreement with (4 a) set

$$E_x = \frac{1}{2}(E_+ + E_-), \qquad E_y = \frac{1}{2i}(E_+ - E_-). \tag{4 b}$$

Then we carry out all calculations using $E_\pm$ as the simplest elements of the field and find expressions for their velocities of propagation and indices of refraction (which will differ somewhat for the two components). After the light has passed through the magnetic field, we again combine $E_\pm$ into a linear oscillation **E** which will have been turned through a certain angle χ in the xy-plane compared to the original oscillation E_x. The resulting law for χ is rather complicated and becomes clear only because of our simpler assumptions about E.

We assume the wave to be monochromatic, i. e. simply periodic in t,

$$E_\pm = A\, e^{i(k_\pm z - \omega t)} \tag{5}$$

where A is a real number. We thereby fulfill our initial condition at $z = 0$ (entrance of linearly polarized light into the magnetic field):

$$E_x = A \cos \omega t, \qquad E_y = 0, \tag{5 a}$$

and at the same time we take into account the circular nature of $E_\pm$ in the magnetic field:

$$|E_\pm| = A\, |e^{i(k_\pm z - \omega t)}| = A, \tag{5 b}$$

since, as we shall show below, this expression describes two circular oscillations, provided only that $k_\pm$ are real.

Using (5) and the corresponding expression for S we obtain from (4) for the steady state, that is, for the purely periodic state of the electron fluid

$$S_\pm = \frac{-e/m}{-\omega^2 \mp \dfrac{e}{m} B\omega + \omega_0^2} E_\pm . \tag{6}$$

It should be remarked that the denominator here is *real* and not complex as in the case of anomalous dispersion, eq. (19.2). This is due to the fact that the magnetic field *does no work* on the electrons. We can disregard the *absorption* which takes place when $\omega = \omega_0$ since that effect is not caused by B.

The vector **P** which is proportional to **s** behaves just like the latter. If we set

$$P_\pm = P_x \pm i\, P_y$$

then, according to (17.2), we get

(6 a) $$P_\pm = \frac{N e^2/m}{-\omega^2 \mp \frac{e}{m} B\,\omega + \omega_0{}^2} E_\pm .$$

Therefore, for the periodic state the differential eq. (17.6) gives in place of (17.9)

(7) $$k_\pm{}^2 = \frac{\omega^2}{c^2}\left(1 + \frac{\mu_0 c^2 N e^2/m}{\omega_0{}^2 \mp \frac{e}{m} B\,\omega - \omega^2}\right).$$

Hence there are two different indices of refraction corresponding to the two wave numbers $k_\pm$, namely

(7 a) $$n_\pm = \frac{c\,k_\pm}{\omega}$$

and their values are given by (compare with (17.10))

(8) $$n_\pm{}^2 = 1 + \frac{N e^2/m\,\varepsilon_0}{\omega_0{}^2 \mp \frac{e}{m} B\,\omega - \omega^2}.$$

This then, is the result of our dispersion calculation in its simplest form, i. e. using $E_\pm$. We see that n_+ and n_- *differ from one another*, and n_+ is somewhat larger than n_-.

This difference is small, however, because, as was remarked in connection with (2), the middle term in the denominator is only a correction term. By neglecting the square of this correction term we get from (8)

(9) $$n_+{}^2 - n_-{}^2 = \frac{N e^2}{m\,\varepsilon_0} \cdot 2\,\frac{e}{m}\,\frac{B\,\omega}{(\omega_0{}^2 - \omega^2)^2}$$

or, if we introduce the mean index of refraction $n = 1/2\,(n_+ + n_-)$,

(9 a) $$n_+ - n_- = \frac{N e^3}{n\,m^2\,\varepsilon_0}\,\frac{B\,\omega}{(\omega_0{}^2 - \omega^2)^2}.$$

We now turn to the method of measuring this effect. First we shall justify the title of this section: "*Rotation of the Plane of Polarization.*" Let the path of the light in the magnetic field extend from $z = 0$ to $z = l$. Let us determine the values of $E_\pm$ at $z = l$. For this it is advantageous to decompose $k_\pm$ into two terms which are symmetric and antisymmetric, respectively, with respect to an interchange of k_+ and k_-. Thus we set

(10) $$k_\pm = \frac{1}{2}(k_+ + k_-) \pm \frac{1}{2}(k_+ - k_-).$$

We also introduce the abbreviations

(11) $$\varphi = \frac{l}{2}(k_+ + k_-) - \omega\, t, \qquad \chi = \frac{l}{2}(k_+ - k_-).$$

As we shall see, φ is a *phase difference* and χ is an *angle of rotation.* We obtain then

(12) $$E_\pm = A \exp i \left\{\frac{l}{2}(k_+ + k_-) \pm \frac{l}{2}(k_+ - k_-) - \omega\, t\right\},$$

and applying (11)

(13) $$E_+ = A\, e^{i\varphi} e^{i\chi}, \qquad E_- = A\, e^{i\varphi} e^{-i\chi},$$

from which follows, according to (4 b),

(14) $$E_x = A\, e^{i\varphi} \cos\chi, \qquad E_y = e^{i\varphi} \sin\chi.$$

Since E_x and E_y oscillate with the same phase, they combine into a *linear oscillation which is rotated in a positive sense (right-handed screw direction around the magnetic field B) by an angle χ with respect to the incident oscillation* (5 a). At the same time the phase φ is changed from its original value at $z = 0$ in (5 b) by the amount

$$\frac{k_+ + k_-}{2}\, l.$$

The angle χ can be measured very precisely. One sets

(15) $$\chi = V\, l\, H$$

where V is called *Verdet's Constant.* By (11) and the relation (7 a) between $k_\pm$ and $n_\pm$ this constant is

$$V = \frac{\omega}{2\,c}\,\frac{n_+ - n_-}{H}.$$

From this follows by (9 a)

(16) $$V = \frac{N\, e^3}{2\, n\, m^2\, c}\,\frac{\mu}{\varepsilon_0}\,\frac{\omega^2}{(\omega_0{}^2 - \omega^2)^2}.$$

At first sight it might seem that the strong rotation of the plane of polarization in ferromagnetic materials is due to the factor μ in (16). However, this is not the case. For μ plays only a formal role in (16). It appears there only because of the conventional definition (15) of χ (which postulates proportionality to H, rather than to B, which would actually be better). As a matter of fact, our theory cannot encompass ferromagnetism at all because it fails to take the electron spin[1] into account.

[1] The extent to which the electron spin is able to explain the rotation of polarization in ferromagnetic media has been investigated by H. R. Hulme, Proc. Roy. Soc. London **135**, 237, 1935.

Clearly, (16) is frequency dependent. Therefore, as with refraction, the magnetic rotation of the plane of polarization is connected with *dispersion*. One could develop V in powers of ω^2/ω_0^2, which is the procedure used for n in Sec. 17, and by eliminating ω_0 from the first two coefficients of the resulting expansion one could derive a universal relationship between e, m and N. However, the accuracy with which the dispersion factor in (16) can be measured is hardly sufficient to justify this procedure. A better method is to use the first terms of the expansions of both n and V; the resulting relation is satisfied sufficiently well for the gases H_2, O_2, N_2 [1].

More important and more interesting than the magnetic rotation is the *natural rotation* of the plane of polarization in crystals which have helical structure (quartz, sodium chlorate, etc.) and in liquids which have an asymmetrically bound carbon atom (sugar solutions). This rotation of the plane of polarization is an indispensable reagent in the entire sugar industry. We shall take up these phenomena in Chap. IV.

At this stage a basic difference between natural and magnetic rotation should be emphasized: if at the end of the path l, one reflects the light ray backwards, then the *natural* rotation is cancelled while the magnetic rotation is *doubled*. The magnetic effect is due to the fact that for the return path not only k_+ and k_- are interchanged in formula (11), but also i and $-i$ must be interchanged in formulae (12) and (13). For after the reflection the vector direction of a positive rotation in the Gaussian plane is opposite to the direction of the magnetic field. Because of this Faraday was able to multiply his very minute rotation effect by repeated back and forth reflections.

21. The Normal Zeeman Effect and Some Remarks on the Anomalous Zeeman Effect

The above considerations provide us with a very simple approach to the *Zeeman effect,* even if only to the *normal* Zeeman effect in which the spin of the electron plays no essential role. Even for the hydrogen atom with its single electron the Zeeman effect is actually *anomalous*.

Strictly, the normal Zeeman effect occurs only for *singlet lines*, i. e. when the spins of the contributing electrons add up to zero. The simplest example is parahelium (two electrons with opposing spins). On the other hand, the hydrogen lines, as well as the alkali lines, are *doublet lines*. However, the Zeeman effect of the hydrogen atom approaches the normal Zeeman effect very closely even for weak magnetic fields. In the case of the alkalis, which

[1] According to observations by Siertsema. See A. Sommerfeld, Ann. d. Phys. (Lpz.) **57**, 513, 1917.

have the same anomalous Zeeman effect as hydrogen, this approach to the normal Zeeman effect, the so-called *Paschen-Back effect*, takes place only in very much stronger fields. Latter, we shall give the necessary field strengths for each case. H. A. Lorentz worked out the theory of the normal Zeeman effect on a classical basis. The spin of the electron, and hence the anomalous Zeeman effect, can be understood only quantum-mechanically.

The treatment of the Zeeman effect for *absorption*, that is of the interaction of an incoming light field with a magnetic field, remains entirely within the framework of the basic concepts of the theory of dispersion. Woldemar Voigt used this method very successfully to treat the D-lines of sodium. He called it the *method of the inverse Zeeman effect*. Experimentally one is usually concerned with the Zeeman effect in emission, the "direct Zeeman effect", which we prefer to discuss here because of its mathematically simpler theory.

We start out from the equation of motion of the electron (20.2) in which we must, however, set $\mathbf{E} = 0$, since we are concerned only with the *magnetic* action upon the emitted radiation. This equation then reads

$$\ddot{\mathbf{s}} + \omega_0^2\,\mathbf{s} = -\frac{e}{m}\,\dot{\mathbf{s}} \times \mathbf{B}. \tag{1}$$

ω_0 is the frequency of the light emitted by the atom without magnetic field. The term $\omega_0^2\,\mathbf{s}$ is due to the retarding "quasielastic force"; see fig. 21 and eq. (17.3).

As in Sec. 20 let the z-axis lie along the direction of the magnetic field $\mathbf{B}$. In this direction $\dot{\mathbf{s}} \times \mathbf{B} = 0$, hence

$$\ddot{s}_z + \omega_0^2\,s_z = 0. \tag{2}$$

Therefore, the z-oscillation of the electron has the original frequency ω_0. It is *not influenced* by the magnetic field.

As before we use complex notation in the xy-plane; that is, as in (20.4 a) we set

$$S = s_x \pm i\,s_y. \tag{3}$$

Hence we get

$$\ddot{S} \mp i\frac{e}{m}B\,\dot{S} + \omega_0^2\,S = 0 \tag{4}$$

which corresponds to (20.4). This equation is integrated by assuming

$$S = a\,e^{i\omega t}, \tag{5}$$

where the factor a is due to the original excitation producing the oscillation and remains, therefore, undetermined. Equation (5) implies a *circular oscillation*. Substitution of (5) in (4) yields:

$$-\omega^2 \pm \frac{e}{m}B\,\omega + \omega_0^2 = 0. \tag{6}$$

The middle term is small compared to the other two. Hence, assuming $\Delta\,\omega$ small, we set

$$\omega = \omega_0 + \Delta\,\omega, \qquad \omega^2 = \omega_0{}^2 + 2\,\omega_0\,\Delta\,\omega, \qquad B\,\omega = B\,\omega_0,$$

and find from (6)

$$-2\,\omega_0\,\Delta\,\omega \pm \frac{e}{m}\,B\,\omega_0 = 0.$$

Hence

$$(7) \qquad \Delta\,\omega = \pm\,\frac{1}{2}\,\frac{e}{m}\,B.$$

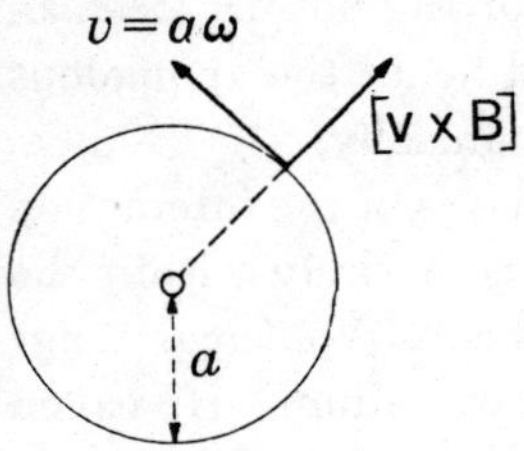

a.)

Fig. 24 a.

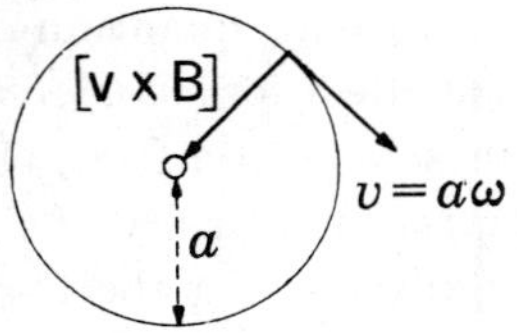

b.)

Fig. 24 b.

Direction of the Lorentz force $\mathbf{v} \times \mathbf{B}$ as centrifugal force for left-handed rotation (a), and as centripetal force for right-handed rotation (b). The vector $\mathbf{B}$ is directed out of the paper.

We verify (7) by the following elementary considerations: the circular oscillation must be such as to maintain equilibrium between the centrifugal inertia force on the one hand, and the sum of the centripetal quasielastic force and the magnetic force on the other. Figure 24a represents the oscillation $S_+ = s_x + i s_y$. For the radius $r = a$ and the velocity $v = a\,\omega$ (ω = angular frequency and also the angular velocity), the centrifugal force is

$$(8) \qquad m\,\frac{v^2}{a} = m\,a\,\omega^2 = m\,a\left(\omega_0 + \frac{1}{2}\,\frac{e}{m}\,B\right)^2 = m\,a\,\omega_0{}^2 + a\,\omega_0\,e\,B.$$

The first term in the last of these expressions is balanced by the quasielastic force. The second term is balanced by the magnetic force $-e\,\mathbf{v} \times \mathbf{B}$. As shown in the figure, $\mathbf{v} \times \mathbf{B}$ is directed centrifugally, hence $-e\,\mathbf{v} \times \mathbf{B}$ is centripetally directed like the quasielastic force.

Figure 24b shows the same for the oscillation $S_- = s_x - i\,s_y$ which we can describe, according to (5), by reversing the sign of i:

$$s_x - i\,s_y = a\,e^{-i\omega t}.$$

Hence, we are now dealing with a circular path with the same radius a but the opposite direction of motion and with its $\Delta\,\omega$ given by the lower sign in (7). The centrifugal force is now

$$(8\text{ a}) \qquad m\,\frac{v^2}{a} = m\,a\,\omega^2 = m\,a\left(\omega_0 - \frac{1}{2}\,\frac{e}{m}\,B\right)^2 = m\,a\,\omega_0{}^2 - a\,\omega_0\,e\,B.$$

$\mathbf{v} \times \mathbf{B}$ is here centripetally directed and, therefore, $-e\,\mathbf{v} \times \mathbf{B}$ is directed centrifugally. The magnetic force acts oppositely to the quasielectric centripetal force which it holds in equilibrium together with the now reduced inertial force.

How do the theoretically expected spectra look now? 1. *Longitudinal observation,* i. e. observation in the z-direction. The linear oscillation (2) has the magnetically uninfluenced frequency ω_0 and *does not radiate* in the z-direction; just as a radio antenna does not radiate in its own direction of oscillation. On the other hand, the two circular oscillations with their magnetically influenced frequencies, eq. (7), radiate two *circularly polarized electromagnetic waves,* one of which is *left* polarized and the other *right* polarized. We define these directions of polarization as those seen by an observer who is looking in the direction of **B** (that is, looking out of the picture in fig. 24). Thus we obtain in fig. 25a the picture which shows the spectrogram seen by an observer looking in the direction of the **B**-field: no light is seen at the position of the original spectral line. To the right and left of that position there are equally intense magnetically displaced lines.

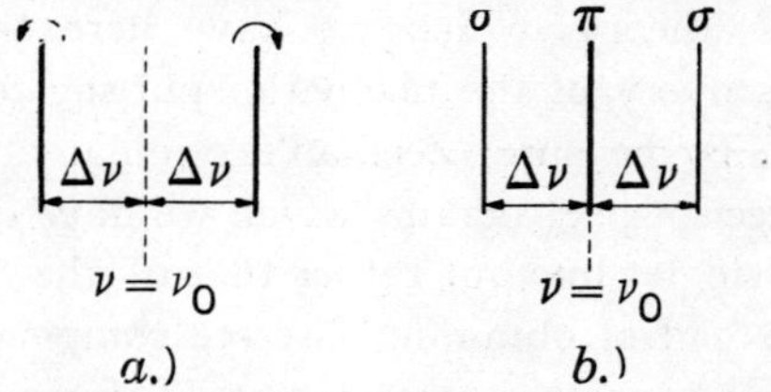

Fig. 25 a. Fig. 25 b.
Normal Zeeman Effect for (a) longitudinal and (b) transverse observation.

The quantitative connection between the primary electron oscillation and the emitted radiation which has here been implicity assumed rests upon the treatment in Vol. III, Sec. 19.

We also note that in fig. 25a, b the angular frequency ω has been replaced by the frequency $\nu = \omega/2\pi$, as is usual in spectroscopy Thus we get instead of $\Delta\,\omega$ as given by (7)

$$\Delta\,\nu = \frac{\mu_0}{4\,\pi}\frac{e}{m}H, \tag{9}$$

where, following common usage, B has been replaced by $H = B/\mu_0$.

2. *Transverse observation,* i. e. observation in a direction perpendicular to the magnetic field, as, for example, in the y-direction. The components s_y of the circular oscillations do not radiate in this direction and can be omitted. The components s_z emit their strongest radiation in this direction, just as the radiation of an antenna or a Hertzian dipole has a maximum in the transverse direction. The frequency ν_0 belonging to s_z is now present in the spectrogram. The frequencies $\nu_0 \pm \Delta\,\nu$ of the two circular components are also present, but their intensities are only half as great[1] because only s_x contributes to

[1]For statistical excitation the intensity of the linear oscillation s_z is of the same magnitude as that of each of the circular oscillations $s_x \pm i\, s_y$ which were denoted by a^2 in (5). Since on the average $s_x{}^2 = s_y{}^2$, it follows that $s_x{}^2 = a^2/2$ as stated in the text.

them. Since s_x oscillates perpendicularly to the magnetic field, the transverse **E**-field emitted by s_x is also perpendicular to **H**. On the other hand, the **E**-field emitted by s_z is directed parallel to **H**. In fig. 25b the corresponding lines are denoted by the usual symbols π (parallel) and σ (perpendicular). The intensity ratio 2 : 1 is indicated by the widths of the lines. The resulting picture is called a "normal Lorentz triplet". Indeed, H. A. Lorentz formulated the theory, which we have here sketched, immediately upon Zeeman's discovery of the magnetic splitting of lines in 1896.

To be sure, Zeeman's original observations were far from yielding the precise spectrograms which we have drawn here. He did not use the light of a singlet line but rather that of the (unresolved) sodium D-line doublet and, instead of obtaining discrete components, he saw only a general broadening of the spectroscopic picture. This was, nevertheless, sufficient to demonstrate the existence of a new fundamental effect; an effect, by the way, for which even Faraday had searched in vain. Furthermore, Zeeman's result sufficed to indicate a qualitative similarity to Lorentz's theory of the effect. For, the outer edges of the spot of light were *linearly polarized* with the direction of oscillation of the **E**-vector, *perpendicular* to **H**, if the observation was made transverse to the magnetic field. For longitudinal observation, on the other hand, the edges were *circularly polarized* with the rotational sense the same as that indicated in fig. 25a. This latter fact was of special significance to the electron theory which was then being formulated because it indicated the *negative* charge of the oscillating particles. Indeed, if these particles had positive charges,, the sign of $\Delta\nu$ and thereby the sense of the circular oscillations would be reversed in all the above formulae and figures.

The following circumstance, which could not be known at that time, was essential to this comparison between experiment and theory: *Also in the anomalous Zeeman effects, the σ-components lie near the edges of the pattern while the π-components are nearer its center, as shown by fig. 25b. In these effects the short wave components are circularly polarized in a right-handed screw sense around the magnetic field lines while the long wave components are left circularly polarized, as in fig. 25a.*

We shall confirm this by considering the complete transverse decomposition of the two D-lines as it was later measured by Zeeman and others. The two lines are the D_2 line, $\lambda = 5896$ Å, fig. 26a, and the D_1 line, $\lambda = 5890$ Å, whose intensity is half that of the D_2 line; fig. 26b. In both drawings the distance of each component from the position $\nu = \nu_0$ is a multiple of *one third of the normal $\Delta\nu$.* In both cases the center position is unoccupied and is indicated by a broken line. The positions $\pm\Delta\nu$ are occupied by strong σ-components in fig. 26a and are unoccupied in fig. 26b. The π-components lie nearest to the center and are displaced from it by $\frac{\Delta \nu}{3}$ in fig. 26a and by $2 \Delta \nu/3$ in fig. 26 b. Instead of being normal Lorentz triplets as shown in fig. 25b, fig. 26a is a sextet and fig. 26b a quartet.

Runge's rule says: for all anomalous Zeeman effects the displacements of the components from the position of the original line measured in wave numbers, are *rational multiples* of the Lorentz $\Delta \nu$. The denominator of these rational multiples is called Runge's denominator. It's value is 3 for the principal series of sodium and all other alkalis. A general formula which is due to Landé enables one to compute the complete splitting diagram, including the denominator, for every series character. *Preston's rule* says that spectral lines with the same series character have the same Zeeman splitting.

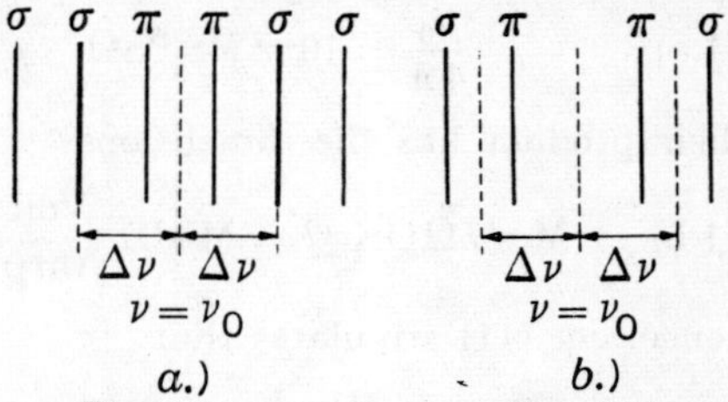

Fig. 26 a. Fig. 26 b.
Splitting of the Na D-lines under transverse observation.
D_1: $\lambda = 5890$ Å; D_2: $\lambda = 5896$ Å.

However, to these rules must be added the reservation "provided the magnetic field is not too strong". What is meant by "not too strong"? Paschen and Back found the answer to this question to be:

$$\Delta \nu \ll \Delta \nu_0. \tag{10}$$

where $\Delta \nu$ is the magnetic splitting arising from the normal Zeeman effect as given by (9), and $\Delta \nu_0$ is, in the case of a doublet line such as the D-lines, the spacing of the original two lines. In the case of a "multiplet" $\Delta \nu_0$ is the smallest spacing which occurs between two individual lines. If, with increasing H, $\Delta \nu$ approaches $\Delta \nu_0$ in magnitude, then $\Delta \nu$ no longer increases proportionally to H. As $\Delta \nu$ becomes much larger than $\Delta \nu_0$, the multiplet shrinks, so to speak, into a singlet compared to the strong magnetic field; the Zeeman effect becomes more and more *normal*. This degeneration phenomenon is called the *Paschen-Back effect*. One consequence of this effect is that the hydrogen lines with their extremely small doublet spacing $\Delta \nu_0$ exhibit a normal Zeeman effect even for very weak fields. Therefore these hydrogen lines as well as the helium lines (and not only the singlet lines of parahelium, but also the close triplet lines of orthohelium) were for a long time considered to be typical examples of the normal Zeeman effect.

We shall compute the critical value of H for which $\Delta \nu = \Delta \nu_0$ in the case of hydrogen. The magnitude $\Delta \nu_0$ for the hydrogen doublet is given by the formula $R \alpha^2/2^4$ where R is the "Rydberg frequency" in reciprocal seconds and $\alpha \sim 1/137$ is the "fine structure constant". In agreement with spectroscopic results this formula yields $\Delta\,\nu_0 = 1.08 \times 10^{10}\ \mathrm{S}^{-1}$. Hence, according to (9), we set

(11) $$\frac{\mu_0}{4\pi}\frac{e}{m}H = 1.08 \times 10^{10}\ \mathrm{S}^{-1}.$$

The factors on the left are (see table at the end of Sec. 2):

(11 a) $$\frac{\mu_0}{4\pi} = 10^{-7}\,\mathrm{M}^{-1}\,\mathrm{S}\Omega, \qquad \frac{e}{m} = 1.76 \times 10^{11}\ \mathrm{QK}^{-1}.$$

Their product has the dimensions

(11 b) $$\mathrm{M^{-1}S\Omega QK^{-1}} = \mathrm{M^{-1}S}\,\frac{\mathrm{Volt}}{\mathrm{Amp}}\,\mathrm{QK^{-1}} = \mathrm{M^{-1}SK^{-1}}\frac{\mathrm{Joule}}{\mathrm{Amp}} = \frac{\mathrm{MS^{-1}}}{\mathrm{Amp}}$$

Hence, eq. (11) stipulates that

$$1.76 \times 10^4\, H\,\frac{\mathrm{MS^{-1}}}{\mathrm{Amp}} = 1.08 \times 10^{10}\ \mathrm{S}^{-1}, \qquad H = 5.8 \times 10^5\,\frac{\mathrm{Amp}}{\mathrm{M}}.$$

Since (see Vol. III, Sec. 8, eq. (5a))

(11 c) $$1\,\frac{\mathrm{Amp}}{\mathrm{M}} = 4\pi \cdot 10^{-3}\ \mathrm{Oerstedt}$$

the desired value turns out to be

(12) $$H = 4\pi \times 5.8 \times 10^2 = 7200\ \mathrm{Oerstedt}.$$

This agrees well with very precise experiments made by Försterling and Hansen[1] with a Lummer plate. They observed the beginning of the Paschen-Back effect at 4000 Oerstedt and found that the π-components of the hydrogen doublet merged at 10,000 Oerstedt. In the case of the D-lines whose $\Delta\,\nu_0$ is fifty times that of the hydrogen doublet, the critical field is, instead of (12),

(12 a) $$H = 50 \times 7200 = 360{,}000\ \mathrm{Oerstedt},$$

a field strength which even today is not easy to attain.

Before ending our brief description of the extremely interesting subject of the anomalous Zeeman effect, we wish to reproduce a photometer curve taken by Zeeman which he very kindly contributed to the fifth edition of "Atombau und Spektrallinien" Vol. I, p. 523. This result will serve to demonstrate the progress in technique which has been made in this field. The line in question is the chromium line $\lambda = 4254$ Å from the septet system of that element. In agreement with Landé's theory the splitting consists of seven π-components ($\Delta\,\nu > \Delta\,\nu_{\mathrm{norm}}$) and twice seven σ-components ($\Delta\,\nu \gtreqless \Delta\,\nu_{\mathrm{norm}}$).

[1] Z. f. Phys. **18**, p. 26, 1923. A precise comparison of these observations with the theory of the Paschen-Back effect can be found in Sommerfeld and Unsöld, ibid. **36**, p. 268, 1926.

Runge's denominator is 4. All 21 components are beautifully discernible on the photometric curve which represents an automatic thirty-six fold enlargement of the original photographic pattern.

Returning once more to the *normal* Zeeman effect and its splitting $\Delta\nu_{\text{norm}}$, we evaluate eq. (9) numerically using the numerical values and dimensions given in eqs. (11 a, b, c). Thus we obtain

$$\Delta\nu_{\text{norm}} = 1.76 \times 10^4 H_{\text{Amp/M}} = 1.76 \times 10^4 \frac{10^3}{4\pi} H_{\text{Oerstedt}} \tag{13}$$

where $\Delta\nu$ has the dimensions $\sec^{-1}$, in agreement with the definition of ν as frequency. However, we want to express this result in the dimensions cm^{-1} as is customary in spectroscopy (reciprocal wave length instead of reciprocal period of oscillation). To do this, we must divide (13) by $c = 3 \times 10^{10}$ cm/sec. Then we obtain

$$\Delta\nu_{\text{norm}} = 4.67 \times 10^{-5} H. \tag{13 a}$$

Since the units Gauss and Oerstedt are defined as the basic units of the absolute *cgs*-system, H in (13 a) means H_{Oerst} as well as H_{abs}.

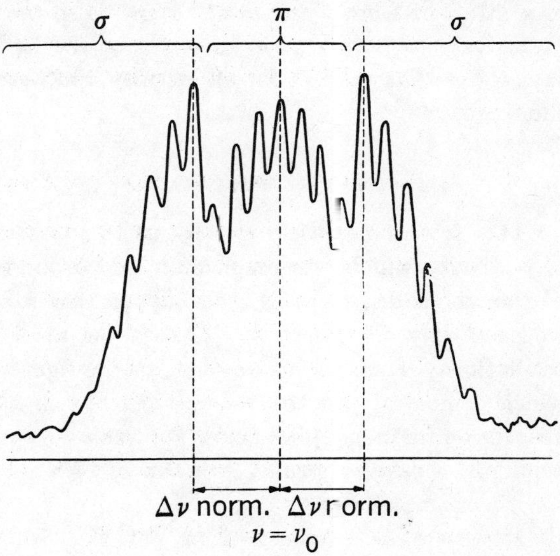

Fig. 27.
Photometer curve of the anomalous Zeeman effect for the chromium line λ 4254 Å.

The entire discussion of this section has remained within the realm of *classical* mechanics and electrodynamics. That these results still remain valid in the *quantum theory* is due to the fact that Planck's constant h, which is characteristic of the quantum theory, accidentally, so to say, drops out of the quantum conditions for magnetically influenced spectral lines. A somewhat similar statement can be made with regard to the anomalous Zeeman effect. The introduction of electron spin and the "vector model", which is constructed from the spins and the orbital angular momenta of the electrons, made it also possible to formulate a theory of the anomalous Zeeman effect even before the introduction of a definitive quantum theory. In that way Runge's rule, the Paschen-Back effect, etc. could be understood. The complete theory of spin and the anomalous Zeeman effect, however, had to await the relativistic Dirac theory. We want to point out briefly the interesting interpretation of $\Delta\,\omega$ in (7) as the *Larmor frequency.* According to this point of view the additional *frequency* defined in (7) can be considered as the *angular velocity* of an additional rotation which the emitting atom experiences in a magnetic field which is switched on infinitely slowly (adiabatically). See exercise III. 4.

22. Phase Velocity, Signal Velocity, Group Velocity

In our discussion of dispersion we considered only the steady, purely sinusoidal state of the electrons and ions. It was made clear that without these induced oscillations no dispersion and refraction, and therefore no value of n different from 1 can exist. Hence also the phase velocity $u = c/n$ refers exclusively to purely periodic states of the light and of matter, which is to say, states that will last for all eternity and were established an infinitely long time ago.

A. Fourier representation of a bounded wave train

This fact immediately enables us to overcome an objection to the theory of relativity which was much discussed around the year 1910[1]. In a region of anomalous dispersion it may happen that $n < 1$, hence $u > c$. To see this we need only assume in fig. 22 that the medium has no infrared resonance oscillations. Then the value of n_m in the figure is equal to 1 and the y-curve which coincides with the n-curve (except in the immediate vicinity of the absorption frequency) lies below the line $n = n_m = 1$ to the right of D. Thus u would be a *velocity greater than that of light* which cannot exist according to the theory of relativity.

However, we emphasized in Vol. III, Sec. 27 F, that this prohibition is limited to processes which can serve as a *signal* and are able to initiate material events. A monochromatic light wave without beginning or end can do no such thing. The Morse signals used in wireless telegraphy are interrupted wave trains. So far, our considerations in no way imply that the front of such a Morse signal propagates with the phase velocity u. In order to be able to apply our previous results to such a signal we must decompose the interrupted signal into a sum of purely periodic waves without beginning or end. We do this with the aid of the *Fourier Integral.*

[1] Gesellschaft der Naturforscher 1907, Physikalische Zeitschrift 8, p. 841, and Weber Festschrift 1912 (publ. by Teubner). Further discussed in Ann. d. Phys. (Lpz.) **44**, 1914: A. Sommerfeld, p. 177, L. Brillouin, p. 203.

The resulting spectrum of partial waves is calculated in Vol. VI, exercise I.4, and is represented there in fig. 33 c. It can be described as a "fluted spectrum" which has a pronounced maximum at $\omega = 2\pi/\tau$ and a half-width which decreases as the length of the wave train contained in the signal increases. This result pertains to a signal which consists of a finite sequence of identical sine oscillations of period τ. It is noted in Vol. VI that such a wave train which is bounded on two sides can be treated as the difference between two wave trains each of which is bounded at only one end.

However, for a signal bounded on only one side, such as

$$f(t) = \begin{cases} 0 & t < 0 \\ \sin 2\pi t/\tau & t > 0 \end{cases} \tag{1}$$

the usual form of the Fourier integral fails because the latter obviously diverges since $f(t)$ does not vanish as $t \to \infty$. In Vol. VI the Fourier integral is, therefore, replaced by a converging contour integral in the *complex* plane.

We repeat[1] the same procedure here. The original path of integration shall be the upper curve in fig. 28 a:

$$f(t) = -\frac{1}{\tau}\int e^{-i\omega t}\frac{d\omega}{\omega^2 - (2\pi/\tau)^2}. \tag{2}$$

One sees immediately that for negative values of t, $-i\omega t$ has a negative real part in the upper half of the complex ω-plane, and that this real part goes to $-\infty$ as the distance from the real axis increases; that is, $\exp(-i\omega t)$ becomes vanishingly small. Since there is nothing to prevent us from shifting the path of integration to infinity in the upper half-plane (indicated by the ↑ arrows), $f(t)$ vanishes as required by the first line in (1). For $t > 0$, however, $\exp(-i\omega t)$ vanishes at infinity when approached through the lower half-plane. If the path of integration is pushed out to infinity in the lower half plane (indicated by the ↓ arrows), it is left hanging on the poles $\omega = \pm 2\pi/\tau$. Integration around these poles yields $-2\pi i$ times the sum of the residues at the two poles (direction of integration in a negative sense in the complex

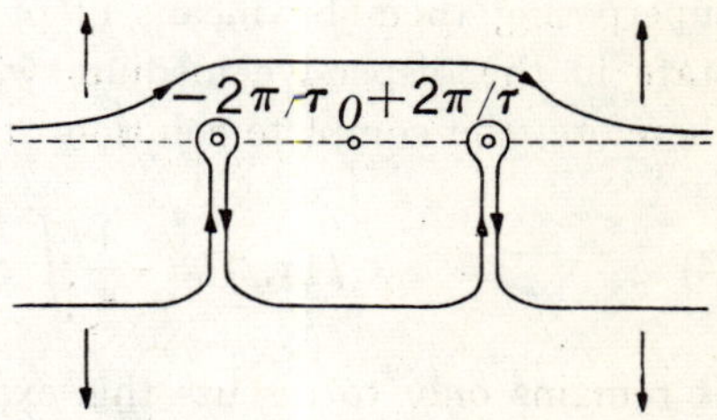

Fig. 28 a.

Representation of a wave train bounded at one end. Path of integration in the complex ω-plane.

[1] In a somewhat changed form since we write $\exp(-i\omega t)$ instead of $\exp(+i\omega t)$. Correspondingly, the upper and lower half planes of the complex ω-plane are interchanged in the present figure in comparison to fig. 24 b in Vol. VI.

plane). Since the contributions to the integral of the paths from the poles to infinity and back cancel each other, it follows from (2) that

$$f(t) = \frac{2\pi i}{\tau} \frac{e^{-2\pi i t/\tau} - e^{+2\pi i t/\tau}}{4\pi/\tau} = \sin 2\pi\, t/\tau \tag{2 a}$$

as required by the second line of (1).

B. Propagation of the Wave Front in a Dispersive Medium

We now consider one of the individual component oscillations of (2) having the time dependence $\exp(-i\,\omega\,t)$ and complement it to form the wave $\exp i\,(k\,x - \omega\,t)$ which propagates in the positive x-direction. At time $t = 0$ let the front of the resulting wave train fall upon the boundary plane $x = 0$ of a dispersive medium which extends from $x = 0$ to $x = \infty$. Each component wave train knows nothing, so to speak, of its origin in the bounded wave train but behaves exactly like the plane wave in a dispersive medium which we treated in Sec. 17. Therefore, we can use for k the value obtained in Sec. 17:

$$k = \frac{\omega\, n}{c}, \qquad n^2 = 1 + \frac{a^2\,\omega_0^2}{\omega_0^2 - \omega^2}, \qquad a^2 = \frac{N\,e^2}{m\,\varepsilon_0\,\omega_0^2}. \tag{3}$$

The abbreviation a^2 is the same as that used in eq. (19.5).

Treating all component waves of the aggregate (2) in this same way and superposing them by means of our complex integral, we obtain a possible state in the dispersive medium which has the form (2) at $x = 0$ and is, therefore, the complete solution of our problem, namely

$$f(x, t) = -\frac{1}{\tau} \int e^{i\,(k\,x - \omega\,t)} \frac{d\omega}{\omega^2 - (2\pi/\tau)^2}. \tag{4}$$

It remains only to discuss this expression for values of $x > 0$.

To do this we must know its singularities in the ω-plane. These are, besides the poles $\omega = \pm\, 2\,\pi/\tau$, the singularities of k. (3) yields

$$k = \frac{\omega}{c} \sqrt{\frac{\omega_0^2\,(1 + a^2) - \omega^2}{\omega_0^2 - \omega^2}} = \frac{\omega}{c} \sqrt{\frac{\omega - \omega_1}{\omega - \omega_0}} \sqrt{\frac{\omega + \omega_1}{\omega + \omega_0}}, \quad \omega_1 = \omega_0 \sqrt{1 + a^2}. \tag{5}$$

Thus k has two pairs of branch points. For small values of a ($\omega_1 \sim \omega_0$, $n \sim 1$) it is best to treat ω_0 and ω_1 together as one pair, and $-\omega_0$ and $-\omega_1$ as another. Each of these pairs of branch points is joined by a branch cut which the path of integration must not cross. Since damping was neglected in (3) so that

ω_0 and ω_1 in (5) are real, we consider the cuts to be along the real axis in fig. 28 b. In any case the upper half-plane is free of singularities. According to (3), k approaches ω/c asymptotically at infinity in the upper half-plane. Hence, there we can replace

$$\text{(6)} \qquad \exp\{i(k x - \omega t)\} \quad \text{by} \quad \exp\left\{i\omega\left(\frac{x}{c} - t\right)\right\}.$$

From this it follows that for $t < x/c$ the argument of the exponential function has a negative real part in the upper half-plane of the complex ω-plane. Hence, we can shift the integral (4) into the upper half-plane and thus obtain

$$\text{(7)} \qquad f(x, t) = 0 \quad \text{for} \quad t < \frac{x}{c}.$$

The wave front penetrates to the depth x in the medium only after a time $t \geqq x/c$. *It certainly does not propagate with a velocity greater than that of light. If any light at all is noticeable at the time* $t = x/c$, (see C.), *then it must have propagated with the velocity* c *of light in vacuum.*

This is also made clear by the following consideration: the dispersion electrons are originally at rest (their thermal agitation which is in no way related to the rhythm of the light wave can obviously be disregarded). But according to our theory, refraction and dispersion are due entirely to the induced periodic oscillations of the electrons or ions. Thus, to begin with, the medium is *optically void* like a vacuum. The propagation velocity is equal to c and the index of refraction, if one still cares to speak of one, is equal to 1.

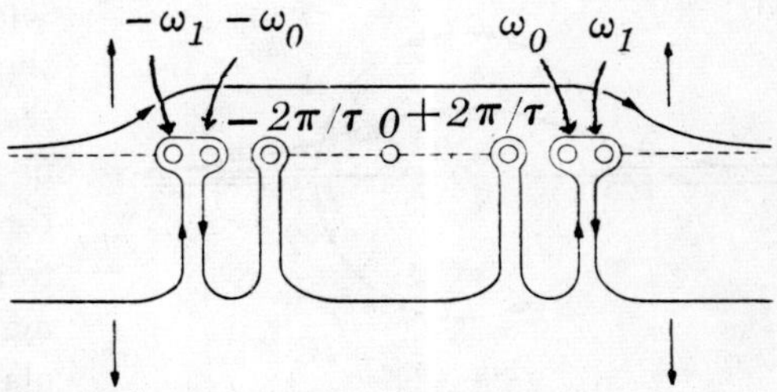

Fig. 28 b.

Propagation of the wave train in a dispersive medium. In deforming the integration path downward the poles and branch points must be taken into consideration.

So far we have assumed that the wave train falls *perpendicularly* upon the surface $x = 0$. If we let it enter obliquely, then at first it is neither *refracted* nor *reflected*. *The law of refraction takes effect only as the electrons are brought into forced oscillations*. Accordingly, on a photographic plate placed behind a dispersive medium, the gap between the light spot corresponding to the regular refraction and the rectilinear projection of the incident beam should be bridged by an extremely weak line of light.

So far we have assumed our medium to be *isotropic*. If the medium is a crystal, calcite for instance, then *no double refraction* should appear at the moment of incidence of the wave train. Such crystals also must be traversed by the initially incident wave train in an undeviated straight line.

It is obvious, however, that the above paradoxes depend on a practically unattainable degree of monochromatism, straightness of direction, and regularity of the wave train.

C. The precursors

We shall use this name, adopted from seismology, to denote the events observed at a depth x immediately following the arrival of the initial wave front. We introduce the time interval

$$\mathrm{t} = t - \frac{x}{c}$$

which according to the above discussion is positive and which we shall assume to be very small. We deform the original integration path of fig. 28 b into a semicircle of very large radius R in the upper half-plane plus the segments of the real axis, as shown in fig. 29. Because of the denominator $\omega^2 - (2\pi/\tau)^2$, the integrand goes to zero as $1/\omega^2$ on the real axis. We may add the path in the lower half plane which is shown as a dotted line in the figure, for if the radius of the semicircular portion of this lower path is increased to infinity, the integrand vanishes exponentially because $\mathrm{t} > 0$. Therefore, we may replace our original path of integration by the entire circle. Expressing t in terms of t we obtain, instead of (4),

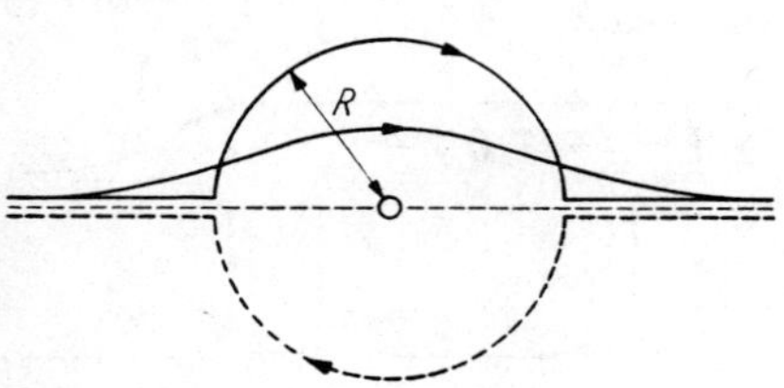

Fig. 29.

Transformation of the path of integration of fig. 28b into a very large semicircle in the upper half-plane (for the calculation of the precursors).

$$f(x,t) = -\frac{1}{\tau}\oint \exp i\left\{\left(k - \frac{\omega}{c}\right)x - \omega \mathrm{t}\right\}\frac{d\omega}{\omega^2 - (2\pi/\tau)^2}. \tag{8}$$

Now, according to (3), for large $|\omega|$

$$k - \frac{\omega}{c} = (n-1)\frac{\omega}{c} = \left(\sqrt{1 - \frac{a^2\,\omega_0^{\,2}}{\omega^2}} - 1\right)\frac{\omega}{c} = -\frac{a^2\,\omega_0^{\,2}}{2\,c\,\omega}. \tag{8 a}$$

Using the abbreviation

$$\xi = \frac{a^2 \omega_0^2}{2c} x, \tag{9}$$

and henceforth neglecting $2\pi/\tau$ as compared to ω, we obtain from (8)

$$\begin{aligned} f(x,t) = f_1(\xi,\mathrm{t}) &= \frac{-1}{\tau} \oint \exp i \left\{ -\frac{\xi}{\omega} - \omega\, \mathrm{t} \right\} \frac{d\omega}{\omega^2} = \\ &= \frac{-1}{\tau} \oint \exp - i \left\{ \sqrt{\xi\, \mathrm{t}} \left(\frac{1}{\omega} \sqrt{\frac{\xi}{\mathrm{t}}} + \omega \sqrt{\frac{\mathrm{t}}{\xi}} \right) \right\} \frac{d\omega}{\omega^2}. \end{aligned} \tag{10}$$

This integral can be transformed into a known form by making the substitution

$$\omega \sqrt{\frac{\mathrm{t}}{\xi}} = e^{iw}, \qquad \frac{d\omega}{\omega} = i\, dw, \qquad \frac{d\omega}{\omega^2} = i \sqrt{\frac{\mathrm{t}}{\xi}}\, e^{-iw}\, dw. \tag{11}$$

Then (10) becomes

$$f_1(\xi,\mathrm{t}) = \frac{-i}{\tau} \sqrt{\frac{\mathrm{t}}{\xi}} \oint \exp\{-2i\sqrt{\xi\,\mathrm{t}} \cos w\}\, e^{-iw}\, dw. \tag{12}$$

Taking our radius R equal to $\sqrt{\xi/\mathrm{t}}$ (since $\mathrm{t} \ll 1$, this is indeed a very large radius), w becomes, by (11), the central angle of our circle, and its value therefore goes from 0 to 2π along the path of integration. Now we compare (12) with the familiar integral representation of the Bessel function of order 1 (see, for instance, Vol. VI, eq. (19.8):

$$J_1(\varrho) = \frac{1}{2\pi} \oint_0^{2\pi} \exp(i\,\varrho \cos w)\, e^{i\left(w - \frac{\pi}{2}\right)}\, dw. \tag{12 a}$$

Because J_1 is real for real ϱ, we can change the sign of i. Then we see that (12) can simply be written as

$$f_1(\xi,\mathrm{t}) = \frac{2\pi}{\tau} \sqrt{\frac{\mathrm{t}}{\xi}}\, J_1(2\sqrt{\xi\,\mathrm{t}}). \tag{13}$$

From the behavior of $J_1(\varrho)$ for small ϱ (where J_1 becomes equal to $\varrho/2$) we find the state of the signal immediately upon its arrival at the depth x to be the following: the initial amplitude is very small compared to 1, that is to the amplitude of the incident oscillation. The initial period of oscillation is extremely small compared to the incident period τ. The amplitude and period both increase with increasing t, the former because of the factor $\sqrt{\mathrm{t}}$, the latter because of the positions of the roots $J_1(\varrho) = 0$ which are spaced at distances of approximately π apart. This gives the following time interval for the mth half-period of the precursor:

$$\Delta\, \mathrm{t}_m \sim \frac{m\pi^2}{2\xi}.$$

According to (9) this value is independent of the incident period of oscillation τ. It depends only on the depth x and on the dispersive power of the medium. For not too small values of x the first precursors are perhaps in the x-ray region. Figure 30 illustrates this phenomenon qualitatively by means of a very coarsely scaled graph.

D. The signal in its final steady state

In contrast to C, we now assume t to be so large that the electrons have already attained their final state of oscillation of period τ. The *process by which the electrons attain this state* is, clearly, represented by the integration around the two pairs of branch points in the lower part of fig. 28 b. The positions of these branch points depend on the nature of the electron binding forces and on the resonance frequencies of the electrons. But now the previously neglected damping term must be taken into consideration, so that the resonance oscillations can die out. If we do this, the branch points which fig. 28 b show to lie on the real axis are shifted somewhat downward into the lower half plane. This means that for very large values of t the factor $\exp(-i\,\omega\,t)$ which occurs in (8) becomes very small. Therefore, the contributions to the integral from the paths around the branch points vanish. All that is left, then, are the loop integrals around the two poles $\pm 2\pi/\tau$ on the real axis in fig. 28 b and these can be evaluated directly by the method of residues.

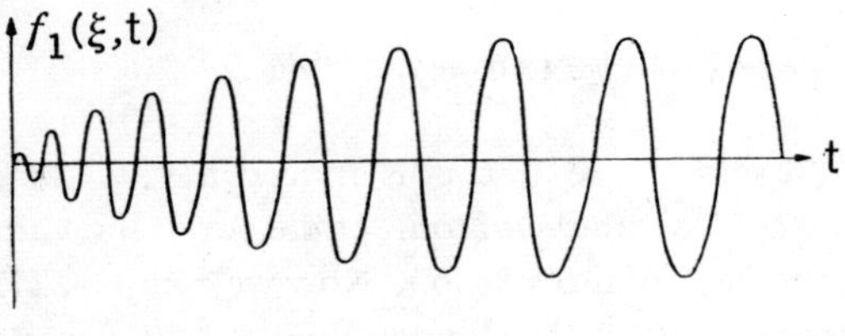

Fig. 30.

Schematic sketch of the excitation immediately upon arrival of the precursors.

At the two poles we have, according to (3),

$$k = \pm \frac{2\pi}{\tau}\frac{n}{c} = \pm \frac{2\pi}{\tau}\frac{1}{u}$$

where n and u are the index of refraction and the phase velocity, respectively, which belong to the period of oscillation τ. Then we get from (4)

$$f(x,t) = \frac{2\pi i}{\tau}\left\{\exp\left[\frac{2\pi i}{\tau}\left(\frac{x}{u}-t\right)\right] - \exp\left[\frac{-2\pi i}{\tau}\left(\frac{x}{u}-t\right)\right]\right\}\frac{1}{4\pi/\tau} =$$

$$= \sin\left\{\frac{2\pi}{\tau}\left(t-\frac{x}{u}\right)\right\}. \tag{14}$$

But this is precisely the wave pattern that results from the incident wave (2 a) when it is displaced towards increasing x with *phase velocity u*.

In fig. 31 we have plotted t horizontally to the right and the depth x in the medium divided by c vertically downward. The straight line $t = x/c$ makes an angle of 45° with the horizontal and it marks for each depth x the time of arrival of the precursors. The line $t = x/u$ makes a smaller angle with

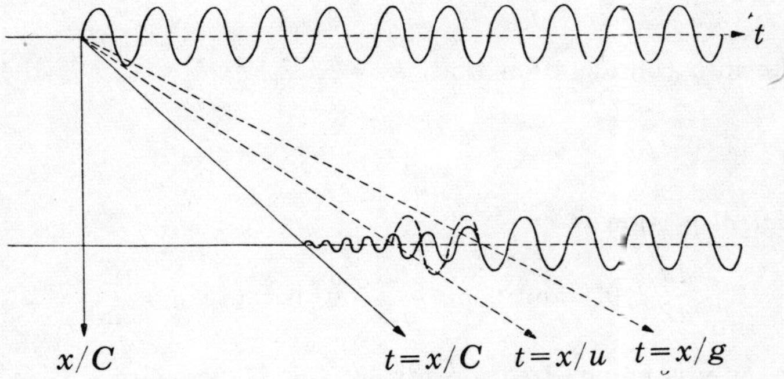

Fig. 31.

Scheme of propagation of a wave to a depth x in a dispersive medium. Transition of the precursors to the steady state.

the horizontal because $u < c$ and it indicates how the amplitude and phase are transmitted to the depth x: The wave train starting at time $t = 0$ on the surface reproduces itself identically at the depth x, its phase being merely shifted by x/u. Any other result would be disastrous to the theory of interference phenomena which rests upon the precise transmission of the phase through dispersive media.

To be sure, the validity of eq. (14) is assumed only if a sufficiently long time $t = t - x/c$ has elapsed. This condition is not necessarily fulfilled at $t = x/u$. Therefore, we have shown the beginning of the wave train which starts at $t = x/u$ as a dotted curve and have drawn it as a solid curve only after a later time $t = x/g$. The locus of points $t = x/g$ is drawn in the figure as a dotted straight line (just like the locus of points $t = x/u$). If $g < u$, this line makes a smaller angle with the horizontal than the line $t = x/u$.

E. Group velocity and energy transport

The concept of group velocity is familiar from hydrodynamics, Vol. II, Sec. 26. This velocity refers not to the propagation of the phase but rather to the propagation of energy (or amplitude). We denote[1] the group velocity by g and compare its formal definition with that of the phase velocity;

$$g = \frac{d\omega}{dk}, \qquad u = \frac{\omega}{k}. \tag{15}$$

Because $\omega = u\,k$, $d\omega = u\,dk + k\,du$, it follows that

$$g = u + k\frac{du}{dk}.$$

If we take into consideration that $k = 2\pi/\lambda$, $dk/k = -d\lambda/\lambda$, then we can write

$$g = u - \lambda\frac{du}{d\lambda}. \tag{16}$$

For *normal* dispersion

$$\frac{dn}{d\lambda} < 0, \quad \text{and hence } \frac{du}{d\lambda} > 0 \text{ because } u = \frac{c}{n}.$$

Therefore, as was emphasized already in Vol. II, loc. cit.,

$$g < u. \tag{16 a}$$

This fact accounts for the smaller slope of the line $t = x/g$ in fig. 31. In the case of anomalous dispersion this line would be *steeper* than the line $t = u/x$.

As in hydrodynamics we expect that the full amplitude 1 of the incident wave will be attained not at $t = x/u$, but at $t = x/g$. This has been confirmed by L. Brillouin in the paper referred to on p. 114. The previously neglected contour integrals around the branch points in fig. 28 b are here essential and are discussed precisely by Brillouin by means of the saddle-point method (Vol. VI, Sec. 19 E and Sec. 21 D). The evaluation of these integrals is not too simple and will be omitted here. The result shows that the precursors are followed by a *transition state* which corresponds to the gradual building up of the electron oscillations to the point where these oscillations correspond to the incoming frequency and amplitude. Thus the ultimate steady state of amplitude 1 is reached not at the time $t = x/u$ but rather (for normal dispersion) at the later time $t = x/g$. In this final state the free oscillations of the electrons have been damped out and only the forced oscillation of period τ remains. The wave train which is drawn dotted in fig. 31 is to be replaced by

[1] In Vol. II we wrote V, U instead of the present notation u, g.

this transition state[1]. Beyond $t = x/g$ the solid wave train correctly represents the final amplitude and phase. The formula $t = x/g$ also gives the time required for the transport of light energy through a distance x in the dispersive medium.

These results can also be derived directly by the *method of stationary phase*[2]. The phase of the exponential function under the integral in (4)

$$\varphi(\omega) = k\,x - \omega\,t$$

is "stationary" with respect to a displacement along the real ω-axis if

$$\frac{d\varphi}{d\omega} = \frac{dk}{d\omega}x - t = 0. \tag{17}$$

While neighboring oscillations generally cancel each other when integrated with respect to ω because of the changes in sign of the exponential function, this is not the case for the "stationary" ω given by (17). In the vicinity of this ω the contributions to the integral have the same sign and add. Therefore, the energy propagation is determined essentially by the stationary ω, that is, by eq. (17). According to the definition (15) of g, (17) does indeed yield

$$t = \frac{x}{g}\,.$$

We conclude with a remark, due to Lord Rayleigh, concerning the measurement of the velocity of light c. Fizeau's toothed wheel, as well as Foucault's mirror, uses cut-off wave trains of the type we have considered here. Therefore, it is the group velocity and not the phase velocity which determines the time interval necessary for the light to pass through the required distances in air. Thus, these experiments really measure g, and not u or c. It is only because of the small dispersion and refraction in air that $g \sim u$ and that c can be computed from u by applying a small correction.

23. The Wave-Mechanical Theory of Dispersion

So far we have not made use of an atomic model. We shall now show how according to Schrödinger[3] we can obtain a deeper understanding of the theory of dispersion if we replace our previous rough assumption of a "quasielastic binding" by well-defined wave-mechanical binding energies.

---

[1] According to fig. 20 in Brillouin, loc. cit., the transition from $t = x/u$ to $t = x/g$ is by no means as simple as we have sketched it in fig. 31.

[2] See Vol. II Sec. 27 under 3, where we also used this method as a substitute for the mathematically precise saddle-point method.

[3] In his fourth communication, Ann. d. Phys. (Lpz.) 81, 1927.

Of course, we cannot here develop the formalism of wave mechanics with any degree of completeness. Rather, we must limit ourselves to describing the progress made by the wave-mechanical treatment as compared to our previous discussion. This will be done in subsection A. In subsection B we shall only indicate how the dispersion formula used in A may be derived.

We shall use the particularly simple case of the Na spectrum as an example (qualitatively the same results will hold for the other alkalis). We consider, therefore, a gas of Na atoms mixed with one of the noble gases (the latter will not further enter into our considerations). If we illuminate this gas with the prismatically decomposed continuous spectrum of a hot flame, then in the light which has passed through the gas, the *principal series* of the Na atom appears as an absorption spectrum. The first of these lines is the yellow *D*-line[1]. The spectrum has a series limit in the near ultraviolet where the lines of higher frequency converge. By means of extremely refined resolving apparatus spectroscopists have been able to find and measure more than 50 lines in the principal series.

We denote the angular frequencies of these series lines

$$\omega_1 \ (D\text{-line}),\ \omega_2,\ \omega_3,\ \ldots,\ \omega_\infty .$$

Wave mechanics associates with these frequencies the energy levels of the atom

$$W_0,\ W_1,\ W_2,\ \ldots,\ W_\infty .$$

Let W_0 be the energy of the atom in its ground state and let $W_1, W_2, \ldots$ be the energies of excited states when the valence electron of the Na-atom is lifted out of its original orbit[2] into a higher orbit which is more distant from the atom. $W_\infty = W_J$ is the ionization energy needed to separate the electron from the atom and leave a Na^+ ion behind. Above the series limit there is a continuous spectrum of ω-values, or energy levels W, which we will not need to go into further here. The connection between ω_j and W_j is

$$\omega_j = \frac{W_j - W_0}{\hbar} \tag{1}$$

where $\hbar$ is Planck's quantum of action divided by 2π. See fig. 32.

[1] We can here neglect the doublet nature of the *D*-line.

[2] Abbreviation for "eigenfunction".

A. Comparison of the Older Dispersion Formula with the Wave-Mechanical Formula

If we take into account not just a single resonance frequency ω_0 but a series of such frequencies $\omega_1, \ldots, \omega_j, \ldots,$ our dispersion formula (17.10) becomes

$$n^2 - 1 = \frac{e^2}{m\,\varepsilon_0} \sum_j \frac{N_j}{\omega_j^2 - \omega^2}. \tag{2}$$

N_j is the number of electrons per unit volume which have a resonance frequency ω_j. The numbers N_j themselves are unknown. But they must collectively satisfy

$$\sum_j N_j = N \tag{2 a}$$

where N is the number of Na-atoms and therefore also (for single valency Na) the total number of valence electrons per unit volume.

In place of this, wave mechanics yields

$$n^2 - 1 = \frac{e^2}{m\,\varepsilon_0} \sum_j \frac{N f_j}{\omega_j^2 - \omega^2}. \tag{3}$$

f_j is called the "transition probability", or the "oscillator strength", and is a definite number which can be calculated from the atomic model by wave-mechanical methods. It is subject to the "sum rule" which is analogous to the requirement (2 a), namely,

$$\sum f_j = 1. \tag{3 a}$$

The difference between eq. (2) and (3) lies not so much in the greater definiteness which distinguishes the latter because the f_j occurring in it can be computed. The difference lies principally in the meaning of the ω_j. In eq. (2) which we have taken over from Sec. 17 the ω_j are *resonance frequencies* of different electrons with different binding energies. In (3) the ω_j are *frequencies of transition* of one and the same valence electron from the excited state W_j to the ground state W_0. In (2) the oscillations ω_j occur side by side and independently of one another. In (3) the transitions take place one after the other, depending on the excitation just preceding each, so that they mutually exclude one another. Thus, in spite of their formal similarity, the meanings of eq. (2) and (3) are quite different. This new interpretation of the ω_j as energy differences is equivalent to Ritz's *combination principle* which, since Bohr, has been the foundation of the theory of spectral lines. The ideas on which eq. (3) is based are represented schematically in fig. 32. The energy levels W_j are plotted vertically upward from the ground state W_0 to the ionization energy W_J. The transition frequencies ω_j with their limiting value ω_∞ are plotted horizontally to the right.

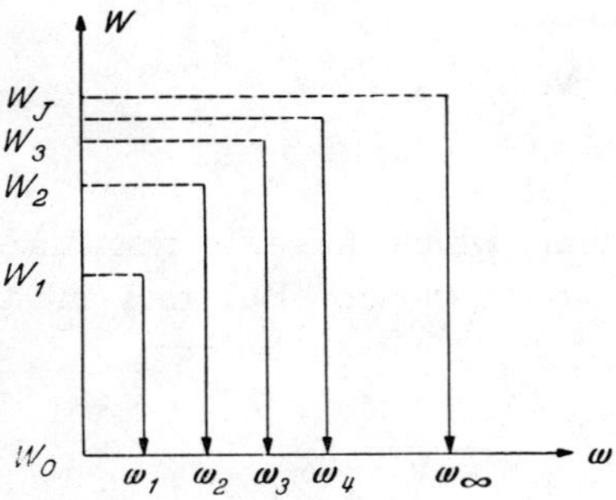

Fig. 32.

Correspondence between emission frequencies ω_1, ω_2, ... and energy level W_1, W_2, ... and the ground level W_0.

The wave-mechanical scheme can be extended considerably. Instead of starting with the ground state, we could investigate the dispersion formula for any desired excited state of energy W_k. Then we would have to draw the arrows originating above W_k only down to W_k. Arrows pointing up from the levels below W_k would also enter the picture and these would contribute negative dispersion terms. See fig. 32 a in which we have chosen $k = 2$. In such a case the transition frequencies must be provided with double indices ω_{jk}. To be consistent the ω_j in fig. 32 should, then, be denoted as ω_{j0}.

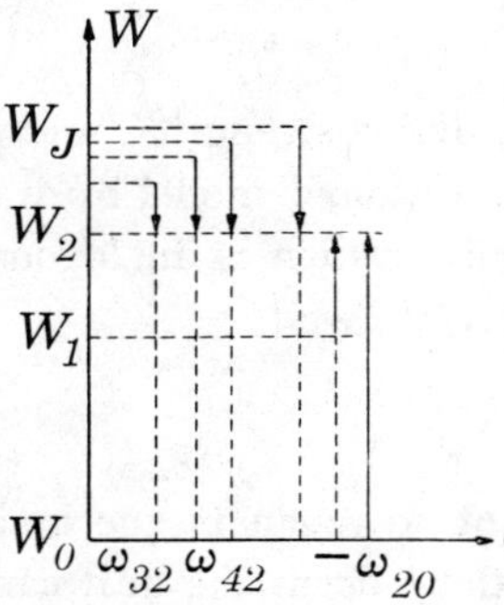

Fig. 32 a.

Dispersion of light due to an excited atom. Besides the positive dispersion terms also negative terms appear which correspond to transitions from deeper lying levels.

B. Outline[1] of the Derivation of Eq. (3)

The Schrödinger equation for the wave function ψ of our valence electron reads

$$\Delta \psi + \frac{2m}{\hbar^2}(W - V)\psi = 0. \tag{4}$$

V is the potential of the force field and takes into account not only the attraction of the nucleus but also the mean repulsions of the remaining atomic electrons. Equation (4) has continuous, normalizable solutions only for the discrete set of values $W = W_0, W_1, \ldots, W_k, \ldots$ These solutions are the "eigenfunctions" of the atom. The kth of these functions, completed so as to include its time dependence, is:

[1] For further details see any text book on wave mechanics, e. g. "Atombau und Spektrallinien", Vol. II, p. 360.

(5) $$u_k = \psi_k \exp\left(-\frac{i\,W_k\,t}{\hbar}\right).$$

Let the atom be perturbed by an incident light wave of angular frequency ω which propagates in the x-direction and is polarized in the y-direction. We represent the space-time dependence of this wave by

$$e^{i\omega\left(t-\frac{x}{c}\right)} + e^{-i\omega\left(t-\frac{x}{c}\right)}.$$

The perturbed state u of the valence electron satisfies the time-dependent Schrödinger equation

(6) $$\varDelta u + \frac{2\,i\,m}{\hbar}\frac{\partial u}{\partial t} - \frac{2\,m}{\hbar^2} V\,u = a\left\{e^{i\omega\left(t-\frac{x}{c}\right)} + e^{-i\omega\left(t-\frac{x}{c}\right)}\right\}\frac{\partial u}{\partial y}.$$

a is a constant factor which is proportional to the amplitude of the incident wave. The reader may convince himself that for the case of no perturbation ($a = 0,\ u = u_k$) eq. (6) reduces to eq. (4). The factor $\partial u/\partial y$ on the right corresponds to the term ($\mathbf{A}\cdot\operatorname{grad} u$) in the general time dependent Schrödinger equation and takes that form because the "vector potential" $\mathbf{A}$ has the direction of the light vector $\mathbf{E}$ which, according to our assumption, is the y-direction.

Owing to the perturbation the state (5) becomes

(7) $$u = u_k + a\left\{w_+ \exp\left(-\frac{i}{\hbar}W_k\,t - i\,\omega\,t\right) + w_- \exp\left(-\frac{i}{\hbar}W_k\,t + i\,\omega\,t\right)\right\}$$

where the perturbation factors $w_\pm$ must satisfy a time-independent differential equation which is derived from (6), namely

(7 a) $$\varDelta w_\pm + \frac{2\,m}{\hbar^2}(W_k \pm \hbar\,\omega - V)\,w_\pm = \frac{\partial\psi_k}{\partial y}\,e^{\pm\frac{i\,\omega\,x}{c}}$$

This equation can be integrated by the general methods of perturbation theory. The right-hand side of (7 a) which is to be considered as a known function of the position x, y, z must be expanded into a series in terms of the complete system of eigenfunctions ψ_j; that is, it is written in the form[1]

(8) $$\sum_j A_j\,\psi_j.$$

---

[1] Actually we should have denoted the coefficients A_j of the series (8) by $A_j\pm$, corresponding to the $\pm$ sign on the right-hand side of eq. (7 a). But because the wave length $\lambda = 2\pi\,c/\omega$ of the incident light is very large compared to the size of the atom, the exponent $\pm\,i\,\omega\,x/c$ is very small for all values of x which come into consideration. Therefore, the $\pm$ sign can be suppressed for A_j.

In the same way we write on the left-hand side of (7 a)

(8 a)
$$w_{\pm} = \sum_j B_j^{\pm} \psi_j$$

and obtain

(9)
$$\sum_j B_j^{\pm} \left\{ \Delta \psi_j + \frac{2\,m}{\hbar^2} (W_k \pm \hbar\,\omega - V)\, \psi_j \right\} = \sum_j A_j \psi_j.$$

If we substitute here the value of $\Delta\,\psi_j$ as given by (4), the position dependent quantity V is eliminated form the left-hand side and (9) simplifies to

(9 a)
$$\frac{2\,m}{\hbar^2} \sum_j B_j^{\pm} (W_k - W_j \pm \hbar\,\omega)\, \psi_j = \sum_j A_j \psi_j.$$

By comparing coefficients one finds

$$B_j^{\pm} = \frac{\hbar^2}{2\,m} \frac{A_j}{W_k - W_j \pm \hbar\,\omega}.$$

Applying eq. (1) and recalling the remarks made in connection with fig. 32 a, we may write this as

(10)
$$B_j^{\pm} = -\frac{\hbar}{2\,m} \frac{A_j}{\omega_{jk} \mp \omega}.$$

Thus the values of the *transition frequencies* ω_{jk} (which in (1) are defined for the ground state) result automatically from the perturbation calculation and take the place of the *oscillation frequencies* ω in the dispersion formula (2). With this result we have attained the essential purpose of our wave-mechanical considerations. The following remarks will serve only to show how this result leads to eq. (3) which is analogous to the classical dispersion formula (2).

It follows from (7) that together with $w_{\pm}$ the function u, which describes the perturbed state, may also be represented as a series in terms of the ψ_j. We need not go into the calculation of the coefficients A_j of this series which is performed in the Fourier manner and requires that the system of eigenfunctions ψ_j be known. From u one obtains the density distribution $\varrho = u\,u^*$ and from it is found P_y, the component of the electric moment of this distribution in the direction of polarization y of the incident wave. From $\overline{P}$ which is the average of P_y taken over all possible orientations of the atom the value of $n^2 - 1$ is found. In this way one obtains precisely eq. (3) with a definite expression for f which takes the form of a space integral over the eigenfunctions.

CHAPTER IV

CRYSTAL OPTICS

So far, we have assumed all optical media to be isotropic. But the complete range of optical refinement is revealed only by anisotropic media. The interference patterns of crystal plates in polarized light are among the most beautiful and splendidly colored phenomena of nature. They indicate the regular structure of crystals even more clearly than the outward shape does. Moreover, calcite, mica, and quartz are essential components of some of the most important optical apparatus.

However, we shall in general not deal with the atomistic structure of anisotropic media but shall treat these only from a phenomenological point of view, just as we have done with isotropic media. The simplest assumptions regarding directional dependence and symmetry suffice for a fairly complete description of the phenomena. The condition required by such methods, i. e. that "the wave length of the light shall be large compared to the interatomic distances", is certainly fulfilled in the visible spectrum.

24. Fresnel's Ellipsoid, Index Ellipsoid, Principal Dielectric Axes

A medium is electrically anisotropic if the relationship between the excitation $\mathbf{D}$ and the field strength $\mathbf{E}$ is determined by a "linear vector function"

$$
\begin{aligned}
D_1 &= \varepsilon_{11} E_1 + \varepsilon_{12} E_2 + \varepsilon_{13} E_3, \\
D_2 &= \varepsilon_{21} E_1 + \varepsilon_{22} E_2 + \varepsilon_{23} E_3, \\
D_3 &= \varepsilon_{31} E_1 + \varepsilon_{32} E_2 + \varepsilon_{33} E_3
\end{aligned} \tag{1}
$$

rather than by the simple proportionality $\mathbf{D} = \varepsilon\, \mathbf{E}$ as in the isotropic case. 1, 2, 3 in (1) are three mutually perpendicular coordinate directions which are fixed in the crystal in some particular way. The dielectric constant is now not a scalar but a symmetric tensor of rank two. The symmetry condition

$$\varepsilon_{ik} = \varepsilon_{ki} \tag{1 a}$$

follows from the requirement that the work done per unit volume $(\mathbf{E} \cdot d\mathbf{D})$ in building up a field must be a total differential. See Vol. III, eq. (5.6 d) and footnote 2. Only if this symmetry condition is fulfilled does there exist an "electric energy per unit volume" as a variable of state which is independent of previous events, namely

$$W_e = \frac{1}{2}(\mathbf{E}\cdot\mathbf{D}) = \frac{1}{2}\sum_i \sum_k \varepsilon_{ik} E_i E_k. \tag{2}$$

Because of (1), **E** and **D** are not parallel but have, in general, different directions.

We have met linear vector functions before on various occasions. There was, for instance, the relationship between the angular velocity and the angular momentum of a rigid body in Vol. I, eq. (24.9). This relationship led to Poinsot's construction: to find the angular momentum **M** corresponding to a rotation ω, place a plane tangential to the inertia ellipsoid ($f =$ const.) through the tip of the ω vector, which is in that ellipsoid. Then draw a perpendicular to this tangential plane from the center of the ellipsoid. This perpendicular is in direction and magnitude the desired angular momentum. The same thing written as formulae[1] reads:

$$M_1 = \frac{\partial f}{\partial x_1}, \qquad M_2 = \frac{\partial f}{\partial x_2}, \qquad M_3 = \frac{\partial f}{\partial x_3}, \qquad f = \frac{1}{2}\sum_i \sum_k I_{ik}' \, x_i \, x_k \tag{3}$$

where x_1, x_2, x_3 are the rectangular coordinates of the tip of $\boldsymbol{\omega}$ as measured in our 1, 2, 3 system. We may also say: "**M** is the normal to that polar plane of the inertia ellipsoid which belongs to the $\boldsymbol{\omega}$-direction."

We have already emphasized in Vol. I that this same construction is valid for any linear vector function which is derived from a symmetric tensor. Indeed, one obtains our eq. (1) for **D** from the eq. (3) for **M** by replacing in the latter the tensor (I_{ik}') by (ε_{ik}) and the vector $\boldsymbol{\omega}$ by **E**. Then the characteristic "tensor surface" to be used in the construction

$$\sum \sum \varepsilon_{ik} \, x_i \, x_k = \text{const.}, \qquad \text{const.} = 2\,W_e \tag{4}$$

is called *Fresnel's ellipsoid*. That this surface is in fact an ellipsoid and not a general second order surface follows, as in the case of the inertia ellipsoid, from the fact that the left-hand side of (4) represents an energy. This left-hand side must therefore be a positive definite quadratic form.

The polar planes belonging to the three principal axes of the ellipsoid are perpendicular to these axes. Hence along these axes and only along them are **D** and **E** parallel. We call these axes "principal dielectric axes" (in contrast

[1] The I_{ik}' are related to the usual products of inertia I_{ik} of Vol. I in the following way: $I_{ii}' = I_{ii}$, $I_{ik}' = -I_{ik}$. This changed notation is clearly convenient for the comparison with the ε_{ik} in eq. (1) and (2).

to the "optic axes" to be introduced later). If these axes are chosen as the coordinate axes, one obtains instead of (1)

$$D_1 = \varepsilon_1 E_1, \qquad D_2 = \varepsilon_2 E_2, \qquad D_3 = \varepsilon_3 E_3. \tag{5}$$

These ε_i are called "*principal dielectric constants*". Just as in mechanics where the products of inertia I_{ik} vanish in the coordinate system of the principal moments of inertia, so here the mixed ε_{ik} vanish and Fresnel's ellipsoid assumes the form

$$\varepsilon_1 x_1^2 + \varepsilon_2 x_2^2 + \varepsilon_3 x_3^2 = \text{const.}, \quad \text{const.} = 2 W_e. \tag{6}$$

Making use of Maxwell's relationship for non-magnetic media $n = \sqrt{\varepsilon/\varepsilon_0}$, we can write instead of (6)

$$n_1^2 x_1^2 + n_2^2 x_2^2 + n_3^2 x_3^2 = \text{const.}, \quad \text{const.} = \frac{2 W_e}{\varepsilon_0} \tag{6 a}$$

The n_i which are here introduced are called the "*principal indices of refraction*". Equation (6 a) shows that *the lengths of the principal axes of Fresnel's ellipsoid are the reciprocals of the three principal indices of refraction.* For later use we also define the three "principal light velocities" as

$$u_i = \frac{c}{n_i} = \frac{1}{\sqrt{\varepsilon_i \mu_0}}. \tag{6 b}$$

We now take the opposite viewpoint and assume **D** to be given and express **E** as a linear vector function of **D**. This is done by solving eq. (1). We write this solution as

$$\begin{aligned} E_1 &= \eta_{11} D_1 + \eta_{12} D_2 + \eta_{13} D_3, \\ E_2 &= \eta_{21} D_1 + \eta_{22} D_2 + \eta_{23} D_3, \\ E_3 &= \eta_{31} D_1 + \eta_{32} D_2 + \eta_{33} D_3. \end{aligned} \tag{7}$$

The η are the minors of the previous ε, divided by the determinant of ε:

$$\eta_{mn} = \frac{|\varepsilon_{ik}|_{mn}}{|\varepsilon_{ik}|}. \tag{7 a}$$

The symmetry of the η-tensor follows from that of the ε-tensor as expressed in eq. (1 a) (our present notation obviously has nothing to do with the electric susceptibility in Vol. III, Sec. 11 C).

Using (7), eq. (2) can be rewritten

$$W_e = \frac{1}{2} (\mathbf{D} \cdot \mathbf{E}) = \frac{1}{2} \sum_m \sum_n \eta_{mn} D_m D_n. \tag{8}$$

The corresponding tensor surface becomes

$$\sum \sum \eta_{mn} x_m x_n = \text{const.} \tag{9}$$

This equation is different from (4), but because of its connection with the electric energy it also represents an ellipsoid. Transformed to its principal axes system this equation assumes the form

$$\eta_1 x_1^2 + \eta_2 x_2^2 + \eta_3 x_3^2 = \text{const.} \tag{9 a}$$

The principal axes of the η-tensor (9 a) are in the same direction as those of the ε-tensor (6) because both are determined by the condition that **D** and **E** shall be parallel.

Furthermore, it can easily be shown that

$$\eta_i = \frac{1}{\varepsilon_i}. \tag{10}$$

For by substituting into (7 a) the principal axis values for the ε_i, one obtains

$$\eta_1 = \begin{vmatrix} \varepsilon_2 & 0 \\ 0 & \varepsilon_3 \end{vmatrix} \div \begin{vmatrix} \varepsilon_1 & 0 & 0 \\ 0 & \varepsilon_2 & 0 \\ 0 & 0 & \varepsilon_3 \end{vmatrix} = \frac{1}{\varepsilon_1} \quad \text{etc.}$$

If we now express the ε_i in terms of the principal indices of refraction n_i, as in (6 a), then we obtain from (9 a) and (10)

$$\frac{x_1^2}{n_1^2} + \frac{x_2^2}{n_2^2} + \frac{x_3^2}{n_3^2} = \text{const.}, \qquad \text{const.} = 2\, W_e\, \varepsilon_0. \tag{11}$$

Therefore, the lengths of the principal axes of our present ellipsoid are *equal* to the principal indices of refraction and not to their reciprocals as in the Fresnel ellipsoid. (11) is, therefore, called the *index ellipsoid* (also Fletcher's ellipsoid or "reciprocal ellipsoid").

The positions of the principal dielectric axes in the crystal change a little with temperature and also differ somewhat for different frequencies. Therefore, one speaks of a "dispersion of the principal axes". Only the *symmetry* of the crystal lattice, if it exists, which controls all physical phenomena, completely fixes the principal axes. We shall go further into this in Sec. 28.

All these considerations have been concerned only with *electrically* anisotropic bodies. But there are also *magnetic* crystals. We mentioned the most important ferromagnetic ones in Vol. III at the beginning of Sec. 12. These are, however, of no interest in optics because the magnetization cannot follow the rapid optical oscillations, but dies out in the far infrared. For this reason we can henceforth set $\mu = \mu_0$, that is, treat only magnetically isotropic media. Correspondingly, our earlier considerations in Sec. 3 C in which the distinction between μ and μ_0 was important, referred not to optical but to centimeter waves.

25. The Structure of the Plane Wave and its Polarization

As we know (Vol. III, Sec. 4), Maxwell's equations are valid in crystals as well as in isotropic media. Since we can set $\mu = \mu_0$, these equations contain the three quantities E, D, and H where D and E are connected by eq. (24.1). Thus, if we assume the crystal to be non-conducting, we have

$$\mu_0 \frac{\partial \mathsf{H}}{\partial t} = -\,\mathrm{curl}\,\mathsf{E}, \qquad \frac{\partial \mathsf{D}}{\partial t} = \mathrm{curl}\,\mathsf{H}. \tag{1}$$

From $\mathrm{div\,curl} = 0$, it follows that $\mathrm{div}\,\mathsf{H}$ and $\mathrm{div}\,\mathsf{D}$ are constant with respect to time. Both of these constants are to be set equal to zero; the first one because the magnetic force lines are free of sources, the second because we can assume the crystal to be free of charges and because the charge density is generally to be defined by $\mathrm{div}\,\mathsf{D}$. Hence

$$\mathrm{div}\,\mathsf{H} = 0, \qquad \mathrm{div}\,\mathsf{D} = 0. \tag{2}$$

The condition

$$\mathrm{div}\,\mathsf{D} = \frac{\partial D_1}{\partial x_1} + \frac{\partial D_2}{\partial x_2} + \frac{\partial D_3}{\partial x_3} = 0 \tag{2 a}$$

holds in every cartesian coordinate system, not merely in the principal axis system of the dielectric. If one were to replace D by E by means of (24.1), a rather unwieldy formula would result. Only in the coordinate system of the principal axes does this substitution have the relatively simple form

$$\varepsilon_1 \frac{\partial E_1}{\partial x_1} + \varepsilon_2 \frac{\partial E_2}{\partial x_2} + \varepsilon_3 \frac{\partial E_3}{\partial x_3} = 0. \tag{2 b}$$

We mention this mainly in order to show that

$$\mathrm{div}\,\mathsf{E} = \frac{\partial E_1}{\partial x_1} + \frac{\partial E_2}{\partial x_2} + \frac{\partial E_3}{\partial x_3} \tag{2 c}$$

cannot vanish simultaneously with $\mathrm{div}\,\mathsf{D}$; this occurs neither in the principal axis system nor in a general cartesian coordinate system.

Limiting ourselves to the case of the plane wave, we will make the following consistent assumptions as to the forms of D and E:

$$\mathsf{D} = \mathsf{A} \exp i\,\{\mathsf{k}\cdot\mathsf{r} - \omega t\}, \qquad \mathsf{E} = \mathsf{B} \exp i\,\{\mathsf{k}\cdot\mathsf{r} - \omega t\}. \tag{3}$$

These are valid in any cartesian coordinate system. In this way we express the fact that the space and time dependences of both these vectors are the same, but that their amplitudes and directions will differ. As in the isotropic case of Sec. 2, these assumptions express an *ideal* state which is completely monochromatic (single frequency ω) and directed completely parallel (single wave vector k). We have already discussed in Sec. 2 how such a state can be approximated with natural light by using a monochromator and collimator.

The **D**-wave is *transverse*, i. e. the vector **D** is *perpendicular* to the wave vector **k**. This follows from (2 a). For, according to (3),

$$\text{div}\,\mathbf{D} = i\,(A_1 k_1 + A_2 k_2 + A_3 k_3) \exp i\,\{\mathbf{k}\cdot\mathbf{r} - \omega t\} = i\,\mathbf{k}\cdot\mathbf{D} = 0. \tag{4}$$

Hence, **D** has no component in the direction of **k**. As a result of the remarks accompanying (2 b, c), this is *not* true of the field vector **E**.

We now inquire into the connection between ω, **k**, and the phase velocity u of our plane wave. In the isotropic case this connection was given by

$$u = \frac{\omega}{k} = \frac{1}{\sqrt{\varepsilon\mu}}. \tag{5}$$

The first of these equalities is valid also for crystals. To prove this we need only differentiate the phase $\varphi = \mathbf{k}\cdot\mathbf{r} - \omega t$ with respect to t and follow the progress of a certain phase value by setting $d\varphi/dt$ equal to zero:

$$\frac{d\varphi}{dt} = \mathbf{k}\cdot\dot{\mathbf{r}} - \omega = 0 \tag{5 a}$$

here $\dot{\mathbf{r}}$ is nothing else but the vector **u** which has the same direction as the wave vector **k** so that $(\mathbf{k}\cdot\dot{\mathbf{r}}) = |\mathbf{k}|\,u = k\,u$ and

$$\omega = k\,u, \qquad u = \frac{\omega}{k}. \tag{5 b}$$

The second equality in (5) for isotropic media resulted from the wave equation which, written in terms of **D** instead of **E** and for $\mu = \mu_0$, reads

$$\varepsilon\,\mu_0 \frac{\partial^2 \mathbf{D}}{\partial t^2} = \Delta\,\mathbf{D}. \tag{6}$$

We must now see what takes the place of this equation in the anisotropic case. For this purpose we eliminate **H** from the two eqs. (1) by applying the curl operator to the first and the operator $\mu_0\,\partial/\partial t$ to the second. Thus we obtain

$$\mu_0 \frac{\partial^2 \mathbf{D}}{\partial t^2} = -\,\mathbf{curl}\,\mathbf{curl}\,\mathbf{E}, \tag{6 a}$$

or using a well-known, actually only symbolic, vector relation (see Vol. III, eq. (6.2)

$$\mu_0 \frac{\partial^2 \mathbf{D}}{\partial t^2} = \Delta\,\mathbf{E} - \mathbf{grad}\,\text{div}\,\mathbf{E}. \tag{6 b}$$

This present wave equation is considerably more complicated than eq. (6). The last term on the right-hand side does not vanish, as was remarked in connection with (2 c); nor can $\Delta\,\mathbf{E}$ be written in vector form as a sum of derivatives of $\mathbf{D}$. Therefore, we shall not discuss (6 b) further but shall return to eq. (6 a). Performing the indicated differentiations on the expressions (3) we can write

$$\frac{\partial^2 \mathbf{D}}{\partial t^2} = -\omega^2 \mathbf{D}, \qquad \operatorname{curl} \mathbf{E} = i\,[\mathbf{k} \times \mathbf{E}], \qquad \operatorname{curl}\operatorname{curl} \mathbf{E} = -[\mathbf{k} \times [\mathbf{k} \times \mathbf{E}]].$$

Equation (6 a) yields then

$$-\mu_0\,\omega^2 \mathbf{D} = [\mathbf{k} \times [\mathbf{k} \times \mathbf{E}]] = \mathbf{k}\,(\mathbf{k} \cdot \mathbf{E}) - k^2\,\mathbf{E}. \tag{7}$$

Using (5 b) to express ω in terms of u and dividing by k^2, this becomes

$$-\mu_0\,u^2\,\mathbf{D} = \frac{\mathbf{k}}{k^2}(\mathbf{k} \cdot \mathbf{E}) - \mathbf{E}. \tag{8}$$

We now decompose this vector equation into its components along the three principal axis directions. In the principal axis system we get from (3) and the connection (24.5) between $\mathbf{D}$ and $\mathbf{E}$

$$B_j = \frac{A_j}{\varepsilon_j}, \qquad j = 1, 2, 3.$$

Using the principal light velocities u_j defined in (24.6 b) and cancelling the common exponential factor, (8) can be rewritten

$$(u^2 - u_j^2)\,A_j = k_j\,K \tag{9}$$

with the abbreviation

$$K = -\frac{1}{k^2}\sum_i u_i^2\,k_i\,A_i. \tag{9 a}$$

Formula (9) is a system of homogeneous linear equations for the A's which is solvable only if its determinant vanishes. Instead of setting up this determinant, it is simpler to do the following: we multiply (9) by $\dfrac{k_j}{(u^2 - u_j^2)}$ and sum over j. This gives

$$\sum_j k_j\,A_j = K \sum \frac{k_j^2}{u^2 - u_j^2}. \tag{9 b}$$

The left-hand side of this equation vanishes because by (4)

$$\mathbf{k} \cdot \mathbf{A} = \sum_j k_j\,A_j = 0.$$

The factor K on the right-hand side of (9 b) generally does not vanish (the principal dielectric axes with their special values of A_j and k_j form an exception). Therefore we conclude from (9 b) that

$$\frac{k_1^2}{u^2 - u_1^2} + \frac{k_2^2}{u^2 - u_2^2} + \frac{k_3^2}{u^2 - u_3^2} = 0. \tag{10}$$

This is a quadratic equation for u^2, as can be seen by multiplying it by the product of the denominators. Therefore, to every direction $\mathbf{k}$ there correspond two, generally different, values of u^2. The fact that each of these still yields the two u-values $\pm u$ means, of course, that the same value of $|u|$ corresponds to the two directions $\pm \mathbf{k}$.

We denote the two roots belonging to any k by u'^2 and u''^2. We shall call the corresponding dielectric displacements $\mathbf{D}'$, $\mathbf{D}''$ and their amplitude coefficients A_j', A_j''. We now assert that $\mathbf{D}'$ and $\mathbf{D}''$ are *mutually perpendicular*, that is, that

$$\mathbf{D}' \cdot \mathbf{D}'' = 0. \tag{11}$$

This follows from the two equations

$$A_j' = K' \frac{k_j}{u'^2 - u_j^2}, \qquad A_j'' = K'' \frac{k_j}{u''^2 - u_j^2}$$

which are contained in (9). Multiplying these and summing gives

$$\sum_j A_j' A_j'' = K' K'' \sum_j \frac{k_j^2}{(u'^2 - u_j^2)(u''^2 - u_j^2)} = \tag{11 a}$$

$$= \frac{K' K''}{u''^2 - u'^2} \left\{ \sum_j \frac{k_j^2}{u'^2 - u_j^2} - \sum_j \frac{k_j^2}{u''^2 - u_j^2} \right\}.$$

Because of (10), the last two summations over j vanish so that $\sum_j A_j' A_j''$ and $\mathbf{D}' \cdot \mathbf{D}''$ also vanish.

These calculations of u', u'' and the resulting facts concerning $\mathbf{D}'$, $\mathbf{D}''$ can be illustrated and made more definite by means of the geometrical construction in fig. 33. We begin with the *index ellipsoid* (24.11). If we replace the n_i by the principal light velocities u_i, the equation of that ellipsoid reads

$$u_1^2 x_1^2 + u_2^2 x_2^2 + u_3^2 x_3^2 = C, \qquad C = 2 W_e \varepsilon_0 c^2 = \frac{2 W_e}{\mu_0}. \tag{12}$$

We place a plane perpendicular to $\mathbf{k}$ through the center of the ellipsoid. Its equation is

$$k_1 x_1 + k_2 x_2 + k_3 x_3 = 0. \tag{13}$$

We now construct the ellipse which forms the intersection between the plane and the ellipsoid. We assert that the principal axes of this ellipse are (aside from a common factor) equal to the reciprocal values of u', u'', and that their directions coincide with the directions of $\mathbf{D}'$, $\mathbf{D}''$.

We find these principal axes by computing the extrema of $x_1^2+x_2^2+x_3^2$, subject to the subsidiary conditions (12) and (13). Using the Lagrangian multipliers λ_1 and λ_2, we write

(14) $$\delta\{x_1^2+x_2^2+x_3^2+\lambda_1(u_1^2x_1^2+u_2^2x_2^2+u_3^2x_3^2)+ \\ +\lambda_2(k_1x_1+k_2x_2+k_3x_3)\}=0.$$

After introducing λ_1 and λ_2, the variations δx_j of the coordinates x_j belonging to the vertices can be considered as independent of one another. Hence, the coefficients of δx_j resulting from (14) must individually be equal to zero. Thus we obtain three conditions for the x_j:

(14 a) $$2x_j(1+\lambda_1 u_j^2)+\lambda_2 k_j=0.$$

To determine λ_1 [1] we multiply (14 a) by x_j and sum over j. Applying the conditions (12) and (13), we obtain

$$\sum x_j^2+\lambda_1 C=0$$

where Σx_j^2 means a^2 or b^2 (a and b are the major and minor axes of the ellipse). If we introduce the noncommittal abbreviation C/u^2 to cover both of these possibilities, then

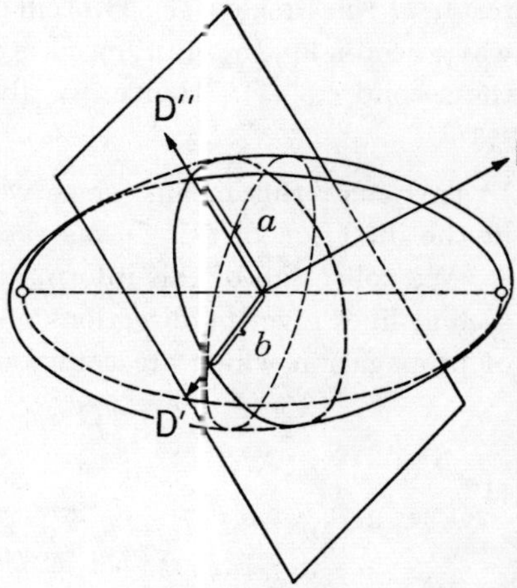

Fig. 33.
Index ellipsoid and construction of the D-vectors belonging to the wave number vector **k**.

(15) $$\lambda_1=-\frac{1}{u^2}.$$

Equation (14 a) reads then

(15 a) $$\frac{2x_j}{u^2}(u^2-u_j^2)=-\lambda_2 k_j \quad \text{or} \quad \frac{k_j}{u^2-u_j^2}=-\frac{2x_j}{\lambda_2 u^2}.$$

If we multiply the last equation by k_j and sum over j, then the right-hand side vanishes because of (13), and we obtain as in (10)

(16) $$\sum\frac{k_j^2}{u^2-u_j^2}=0.$$

Here u has the same meaning as in (10), and *our two velocities of propagation u', u'' are,* aside from the factor C as defined in (12), *equal to the reciprocals of the two principal axes a and b* as was claimed.

[1] We shall not need the value of λ_2. But we could determine it also from (14 a) by multiplying by k_j and summing over j.

To determine the *directions* of the two principal axes, we form from (15 a) the ratios

$$x_1 : x_2 : x_3 = \frac{k_1}{u^2 - u_1{}^2} : \frac{k_2}{u^2 - u_2{}^2} : \frac{k_3}{u^2 - u_3{}^2}. \tag{17}$$

According to (9), these same ratios also hold for the coefficients $A_1 : A_2 : A_3$ of our **D**-vectors. *Thus the directions of oscillation of the two* **D**-*vectors coincide with the directions of the principal axes of our intersectional ellipse.*

The directions of oscillation of the **H**-vectors are now also fixed. As a result of the first eq. (1), **H** oscillates *transversely*, i. e., perpendicularly to the wave vector **k**. In addition, **H** is perpendicular to **D** as is easily proved from the second eq. (1). Hence, we always have

$$\mathbf{H} \cdot \mathbf{k} = 0, \qquad \mathbf{H} \cdot \mathbf{D} = 0. \tag{18}$$

In particular this means, since we know the position of **D**, that **H** oscillates in the direction of b if **D** has the direction of a, and vice versa.

We collect all of this information in the following two tables in which the second lines give the directions of oscillation and the third lines the velocities of propagation which are common to the pair of vectors **D**, **H**:

$$\begin{array}{cc|cc} \mathbf{D}' & \mathbf{H}' & \mathbf{D}'' & \mathbf{H}'' \\ a & b & b & a \\ \multicolumn{2}{c}{u' = \sqrt{C}/a} & \multicolumn{2}{c}{u'' = \sqrt{C}/b} \end{array} \tag{19}$$

In connection with (19) we must still discuss the physical meaning of the principal light velocities u_1, u_2, u_3 which were introduced only formally in (24.6 b). We consider, for example, a wave vector **k** in the direction of the first principal axis. To it belong two sets of vectors **D**, **H** whose respective velocities u', u'' are, according to (19), the reciprocals of the principal axes of the intersection ellipse formed by the plane perpendicular to **k**. Hence u', u'' are equal to u_2, u_3, respectively.

Therefore, the principal light velocities u_2 and u_3 are the velocities of the two waves which propagate in the direction of the first principal axis of the index ellipsoid. A corresponding statement holds, with cyclic interchange of the indices, for the other two principal axes.

The most important result of this section is that *all monochromatic plane waves propagating in a crystal are completely linearly polarized in directions which are determined by the crystalline structure.*

How does this result compare with the optical behavior of isotropic media? To make a valid comparison we must, of course, recall the properties of *monochromatic* and *parallel*, i. e. perfectly collimated, light in an isotropic medium, and not those of completely unpolarized *natural* light. In Sec. 2 we saw that such light must necessarily be *elliptically polarized.* Therefore the distinction between isotropic and anisotropic media does not lie in the fact of polarization as such; the distinction lies in the type of polarization of the light. The crystal structure of an anisotropic medium permits two waves with different linear polarizations and different velocities to propagate in any given direction. In an isotropic medium this completely specified type of polarization is smeared out[1] into the oscillatory pattern of an ellipse whose orientation and size of axes remain unspecified. This is quite understandable, because in an isotropic medium all waves have the same velocities and all directions are equivalent and, as we know, the elliptic polarization can be considered as a superposition of two perpendicular plane oscillations with different phases. In contrast to the waves in crystals, any two such oscillations propagate with equal velocities in isotropic media and are therefore indistinguishable. On the other hand, it is just the fact that these two oscillations are distinguishable in anisotropic media which gives crystals their importance as principal components of polarization apparatus (calcite, mica, etc., see Sec. 29).

So far we have characterized the state of polarization of the light wave by the state of the excitation $\mathbf{D}$, whereas previously we have usually considered the field strength as the actual light vector. But since $\mathbf{D}$ and $\mathbf{E}$ are connected by the *unique* linear relationship of Sec. 24, it is clear that if $\mathbf{D}$ is linearly polarized, so is $\mathbf{E}$. The next section will deal with the directions of oscillation imposed on $\mathbf{E}$ by the crystal structure.

26. Dual Relations[2], Ray Surface and Normal Surface, Optic Axes

The calculations of Sec. 25 can immediately be transposed from the index ellipsoid to Fresnel's ellipsoid. They then yield information about the *field vector* $\mathbf{E}$ and the propagation of the *ray*

$$\mathbf{S} = \mathbf{E} \times \mathbf{H}.$$

---

[1] Using an expression common in wave mechanics we could say: the two linearly polarized oscillations in a crystal degenerate into elliptically polarized oscillations in isotropic media.

[2] For a complete treatment of this dual relationship we refer to the excellent textbook by T. Liebisch, Physikalische Kristallographie, Leipzig 1891. What is involved here is the same duality which exists in projective geometry between the coordinate spaces of points and planes. If one considers the components of $\mathbf{E}$ as point coordinates, then the components of $\mathbf{D}$ are plane coordinates. From this point of view the Fresnel and index ellipsoids represent the same surface, one in point coordinates, the other in plane coordinates. The elementary-geometric method of exercises IV.1 and IV.2 adheres closely to Liebisch's textbook.

In fig. 34 $\mathbf{s} = s_1, s_2, s_3$ is the unit vector in the direction of $\mathbf{S}$ and F is the intersection of the plane of the drawing with Fresnel's ellipsoid. The plane of the drawing is perpendicular to $\mathbf{H}$; hence $\mathbf{H}$ projects into the center O of the ellipse F. $\mathbf{E}$ is in the plane of the drawing and so is $\mathbf{S}$ by virtue of the last equation, but because of the conditions (25.18) $\mathbf{D}$ and $\mathbf{k}$ are also in the plane of the drawing. The diameter $W W'$ which is perpendicular to $\mathbf{k}$ indicates the trace of the plane of the wave or, in other words, the trace of the plane of the intersectional ellipse of Sec. 25. The diameter $S S$ is the trace of the plane which passes through O and is perpendicular to $\mathbf{S}$. The tangents drawn through the points S (traces of plane tangential to Fresnel's ellipsoid) are perpendicular to $\mathbf{D}$, and hence parallel to $\mathbf{k}$ according to the polar construction of Sec. 24.

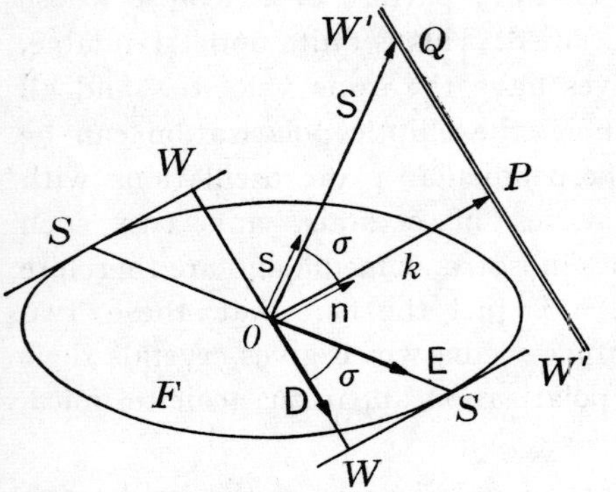

Fig. 34.
Intersection of Fresnel's ellipsoid with the plane of the drawing ($\perp \mathbf{H}$). Construction of the wave and ray velocities.

In the upper right-hand portion of fig. 34 the plane of the wave $W' W'$ (plane of constant phase) which is perpendicular to $\mathbf{k}$ is indicated by means of a double line. If the wave propagates in the direction of $\mathbf{k}$ through the distance OP with a velocity u, the ray $\mathbf{S}$ must, in order to stay in phase with the wave, cover the longer distance $O Q$ at the larger velocity v. From the right triangle OPQ it follows that

$$\cos\sigma = \frac{u}{v}. \tag{1}$$

As shown in fig. 34 the directions of $\mathbf{D}$ and $\mathbf{E}$ intersect at this same angle σ. Hence, also

$$\cos\sigma = \frac{\mathbf{E}\cdot\mathbf{D}}{|\mathbf{E}|\,|\mathbf{D}|}. \tag{1 a}$$

The coplanar position of $\mathbf{D}$, $\mathbf{E}$, and $\mathbf{k}$ has already been expressed in eq. (25.8). Introducing, in analogy to the unit vector $\mathbf{s}$ of the ray, the unit "wave normal" vector

$$\mathbf{n} = \frac{\mathbf{k}}{k} \tag{2}$$

we rewrite eq. (25.8) as

$$\mu_0 u^2 \mathbf{D} = \mathbf{E} - (\mathbf{n}\cdot\mathbf{E})\,\mathbf{n}. \tag{3}$$

We want to derive the equation which is the dual of this one; that is, the equation which expresses the coplanar position of E, D and s. First we write, using two undetermined coefficients p and q,

$$p\,\mathsf{E} = \mathsf{D} - q\,\mathsf{s}. \tag{4}$$

Because $\mathsf{s}\cdot\mathsf{E} = 0$ and $\mathsf{s}\cdot\mathsf{s} = 1$, it follows that $q = \mathsf{s}\cdot\mathsf{D}$ and hence

$$p\,\mathsf{E} = \mathsf{D} - (\mathsf{s}\cdot\mathsf{D})\,\mathsf{s}. \tag{4 a}$$

Now we form the scalar products of (3) with s and of (4 a) with n. Because $\mathsf{s}\cdot\mathsf{E} = 0$ and $\mathsf{n}\cdot\mathsf{D} = 0$, these result in

$$\mu_0\,u^2\,(\mathsf{s}\cdot\mathsf{D}) = -(\mathsf{n}\cdot\mathsf{E})\,(\mathsf{s}\cdot\mathsf{n}),$$
$$p\,(\mathsf{n}\cdot\mathsf{E}) = -(\mathsf{s}\cdot\mathsf{D})\,(\mathsf{s}\cdot\mathsf{n}).$$

If we multiply the right- and left-hand sides of these two equations and remember that $\mathsf{s}\cdot\mathsf{D}$ and $\mathsf{n}\cdot\mathsf{E}$ do not vanish, then

$$\mu_0\,u^2\,p = (\mathsf{s}\cdot\mathsf{n})^2. \tag{5}$$

According to fig. 34 and eq. (1)

$$\mathsf{s}\cdot\mathsf{n} = \cos\sigma = \frac{u}{v}.$$

Hence, according to (5)

$$p = \frac{1}{\mu_0\,v^2}. \tag{6}$$

Thus eq. (4 a) becomes

$$\frac{1}{\mu_0\,v^2}\,\mathsf{E} = \mathsf{D} - (\mathsf{s}\cdot\mathsf{D})\,\mathsf{s}. \tag{7}$$

This equation has precisely the same form as eq. (3); it is its "dual".

We now recall the plane wave expression (25.3) with the coefficients A_j and B_j. By writing this in the principal dielectric axis system and setting $B_j = A_j/\varepsilon_j$, we were able to derive in eq. (25.9) a linear system of equations for the coefficients A_j. We shall now similarly compute the coefficients B_j from (7) by setting $A_j = \varepsilon_j\,B_j$. First, we get from (7)

$$\frac{1}{\mu_0\,v^2}\,B_j = \varepsilon_j\,B_j - s_j \sum_i s_i\,\varepsilon_i\,B_i; \tag{8}$$

then, multiplying by μ_0 and rearranging, we obtain

$$\left(\varepsilon_j\,\mu_0 - \frac{1}{v^2}\right) B_j = \mu_0\,s_j \sum_i s_i\,\varepsilon_i\,B_i.$$

Making use of the principal light velocities $u_j = (\varepsilon_j \mu_0)^{-1/2}$, we have

$$\left(\frac{1}{u_j^2} - \frac{1}{v^2}\right) B_j = s_j K', \tag{9}$$

$$K' = \sum_i s_i \frac{B_i}{u_i^2}. \tag{9 a}$$

These equations correspond precisely to eq. (9) and (9 a) of Sec. 25. The same is true of the following equation which is derived from the two previous ones just as in Sec. 25:

$$\sum_j s_j B_j = K' \sum_j \frac{s_j^2}{\dfrac{1}{u_j^2} - \dfrac{1}{v^2}}. \tag{9 b}$$

The left-hand side is zero because $\mathbf{s} \perp \mathbf{E}$. Since $K' \neq 0$, it follows, as in (25.10), that

$$\frac{s_1^2}{\dfrac{1}{v^2} - \dfrac{1}{u_1^2}} + \frac{s_2^2}{\dfrac{1}{v^2} - \dfrac{1}{u_2^2}} + \frac{s_3^2}{\dfrac{1}{v^2} - \dfrac{1}{u_3^2}} = 0. \tag{10}$$

Equation (10) is quadratic in v^2 just as eq. (25.10) was quadratic in u^2. Therefore, to every ray direction $\mathbf{s}$ there correspond two values v' and v'' (if we leave the $\pm$ sign out of consideration). From a construction on Fresnel's ellipsoid analogous to that of fig. 33, one sees that the corresponding field vectors $\mathbf{E}'$ and $\mathbf{E}''$ are mutually perpendicular.

We now summarize the transformation from $\mathbf{D}$, $\mathbf{n}$, u to $\mathbf{E}$, $\mathbf{s}$, v which we have thus developed in the form of the useful "transformation rule":

$$\mathbf{D},\ \mathbf{E},\ \mathbf{n},\ \frac{u_i}{c},\ \frac{u}{c} \rightleftharpoons \varepsilon_0\ \mathbf{E},\ \frac{\mathbf{D}}{\varepsilon_0},\ \mathbf{s},\ \frac{c}{v_i},\ \frac{c}{v}. \tag{11}$$

The reader may convince himself that this rule does indeed transform eqs. (3) and (7) into each other, both as far as their general form and their coefficients are concerned. The same is also true of the expressions (10) and (25.10). We have departed from the usual formulation of this rule only insofar as we have throughout related to each other only quantities with the same dimensions.

A. Discussion of the ray surface

We shall now construct a complete picture of the distribution of the ray velocities v', v'' for all possible spatial directions of $\mathbf{s}$. For this purpose we plot these velocities as radius vectors in the direction of $\mathbf{s}$ from the origin of an orthogonal coordinate system ξ_1, ξ_2, ξ_3. In this way we obtain a two-sheeted surface in our ξ_1, ξ_2, ξ_3 space, one sheet of which corresponds to v', the other one to v''. The points on this surface have the coordinates

$$\xi_i = s_i\, v. \tag{12}$$

Equation (10) can be written in terms of these as

$$\sum_i \frac{\xi_i^2\, u_i^2}{u_i^2 - v^2} = 0. \tag{13}$$

Because $v^2 = \Sigma \xi_i^2$, this equation appears to be of the sixth order in the ξ_i; but upon multiplication by the product of the denominators, one finds that it reduces to the following fourth order equation:

$$\begin{aligned} &\xi_1^2\, u_1^2\, (u_2^2 - v^2)\, (u_3^2 - v^2) + \xi_2^2\, u_2^2\, (u_3^2 - v^2)\, (u_1^2 - v^2) + \\ &\quad + \xi_3^2\, u_3^2\, (u_1^2 - v^2)\, (u_2^2 - v^2) = 0 \end{aligned} \tag{13 a}$$

or, grouped according to powers of v:

$$\begin{aligned} &v^4\, (u_1^2\, \xi_1^2 + u_2^2\, \xi_2^2 + u_3^2\, \xi_3^2) - v^2\, \{\xi_1^2\, u_1^2\, (u_2^2 + u_3^2) + \xi_2^2\, u_2^2\, (u_3^2 + u_1^2) + \\ &\quad + \xi_3^2\, u_3^2\, (u_1^2 + u_2^2)\} + u_1^2\, u_2^2\, u_3^2\, (\xi_1^2 + \xi_2^2 + \xi_3^2) = 0. \end{aligned} \tag{13 b}$$

Since the last term contains the factor $v^2 = \Sigma \xi_i^2$, a factor v^2 can be cancelled, and the equation indeed represents a surface of only the fourth order.

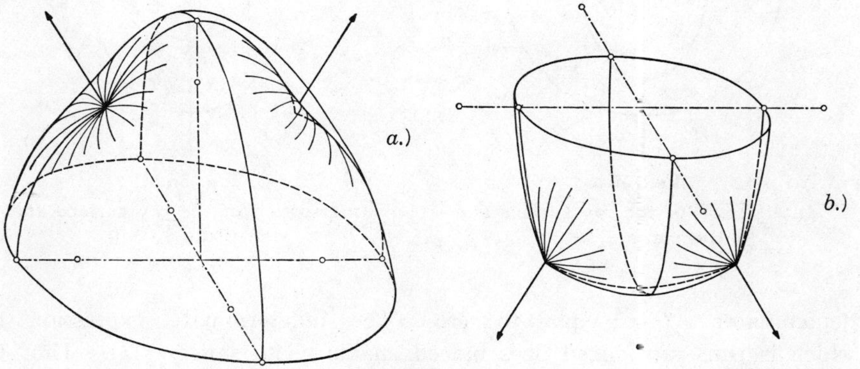

Fig. 35 a, b.

Ray surface: a) upper half of the outer sheet, b) lower half of the inner sheet. The directions of the arrows indicate the two optic axes.

We shall call this surface the *ray surface*. It used to be commonly called "Fresnel's wave surface". Our name indicates the origin of the surface from the ray velocity v. There are beautiful plaster models of the ray surface which can be taken apart and so reveal the way in which the two sheets are connected. Figure 35 represents the upper half of the outer sheet and the lower half of the inner sheet. The missing half of each surface is the mirror image of the half which is shown. We shall study only the principal sections of the ray surface in greater detail, that is, its traces in the planes $\xi_1 = 0$, $\xi_2 = 0$, $\xi_3 = 0$. We may assume for this purpose that

$$u_1 > u_2 > u_3. \tag{14}$$

For $\xi_1 = 0$, we obtain from (13), by multiplying by the product of the two remaining denominators and by cancelling the factor $(\xi_2^2 + \xi_3^2)$,

$$\frac{\xi_2^2}{u_3^2} + \frac{\xi_3^2}{u_2^2} = 1. \tag{15}$$

This is an ellipse with the principal axes u_3 and u_2. There is, however, yet another solution of (13) which is obtained by setting both ξ_1 and the

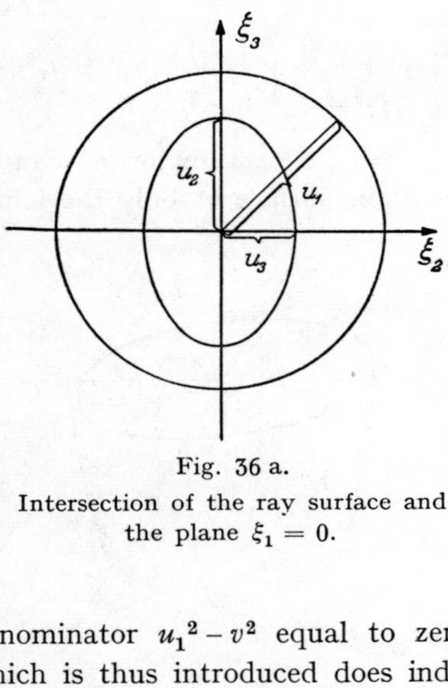

Fig. 36 a.
Intersection of the ray surface and the plane $\xi_1 = 0$.

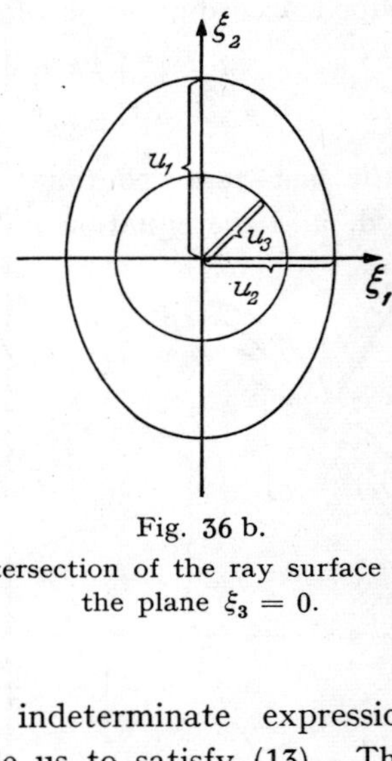

Fig. 36 b.
Intersection of the ray surface and the plane $\xi_3 = 0$.

denominator $u_1^2 - v^2$ equal to zero. The indeterminate expression 0/0 which is thus introduced does indeed enable us to satisfy (13). Thus the second solution for $\xi_1 = 0$ becomes

$$\xi_2^2 + \xi_3^2 = u_1^2. \tag{15 a}$$

This is a circle of radius u_1. Because of (14) this circle encloses the ellipse (15), see fig. 36 a.

These two solutions (15), (15 a) can obviously be obtained also from the complete expression (13 a) for the ray surface. For, if one sets $\xi_1 = 0$, then each of the remaining terms contains the factor $(u_1^2 - v^2)$. If this factor is taken out, an expression equivalent to (15) remains.

Next we consider the principal section $\xi_3 = 0$ which is in the plane perpendicular to the smallest axis of Fresnel's ellipsoid. Again the section consists of a circle and an ellipse, but now the circle lies inside the ellipse. For, from (13) we find

$$(16) \qquad \frac{\xi_1^2}{u_2^2} + \frac{\xi_2^2}{u_1^2} = 1 \qquad \text{and} \qquad \xi_1^2 + \xi_2^2 = u_3^2, \text{ respectively;}$$

(see fig. 36 b.)

The principal section $\xi_2 = 0$ is more interesting. It yields

$$(17) \qquad \frac{\xi_1^2}{u_3^2} + \frac{\xi_3^2}{u_1^2} = 1, \qquad \xi_1^2 + \xi_3^2 = u_2^2.$$

(see fig. 36 c.) Now the radius u_2 of the circle is smaller than the major axis u_1 but larger than the minor axis u_3 of the ellipse. The circle, therefore, intersects the ellipse. At the points of intersection the two branches of the ray surface interpenetrate. What is the significance of the two axes which join the diametrically opposite points of intersection?

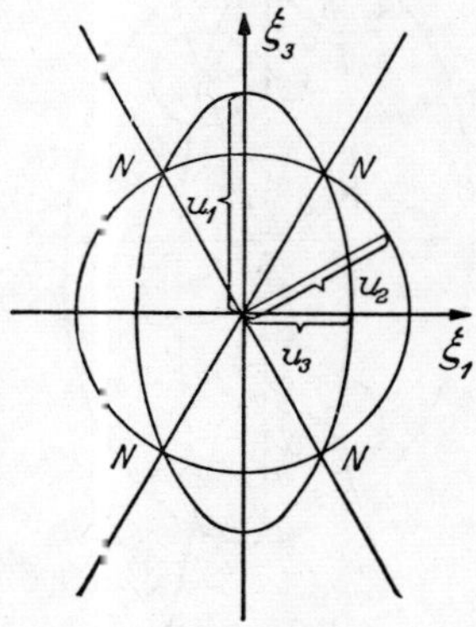

Fig. 36 c.
Intersection of the ray surface and the plane $\xi_2 = 0$.

B. The optic axes

In these axes the two ray velocities v' and v'' are identical just as in isotropic media. They are, therefore, called the *axes of isotropy* or the *optic axes.* The latter name indicates that these axes are even more important in crystal optics than are the principal axes of the Fresnel, or index ellipsoid which were called "principal dielectric axes".

As we have seen, the ray velocities v', v'' may be determined from the principal axes of an elliptic section of Fresnel's ellipsoid, and in the special case $v' = v''$ this elliptic section degenerates into a circle. Therefore, we see that the optic axes are perpendicular to the planes which intersect the ellipsoid in circles. There are well-known cardboard models of triaxial ellipsoids which consist of two sets of parallel circular discs which are fitted into each other and still have a certain degree of mobility. These models provide an interesting and complete representation of the surface of a triaxial ellipsoid. The points at which the normals to these circular discs intersect the surface of the ellipsoid are known as the *umbilical points* (German: "Nabelpunkte", hence the notation *NN* in fig. 36 c.) Fig. 36 d shows the positions of both pairs of umbilical points on Fresnel's ellipsoid and their connecting lines which are the optic axes; the relation of these lines to the principal axes 1 and 3 is also shown. If we call the lengths of the principal axes of Fresnel's ellipsoid u_1, u_2, u_3, as before, and denote the angle between the two optic axes by 2 δ_s, then

$$\tan \delta_s = \frac{u_1}{u_3} \sqrt{\frac{u_2^2 - u_3^2}{u_1^2 - u_2^2}}. \tag{18}$$

This expression agrees with the value of ξ_3/ξ_1 which is obtained from the two eqs. (17) for the intersection points of the circle and ellipse.

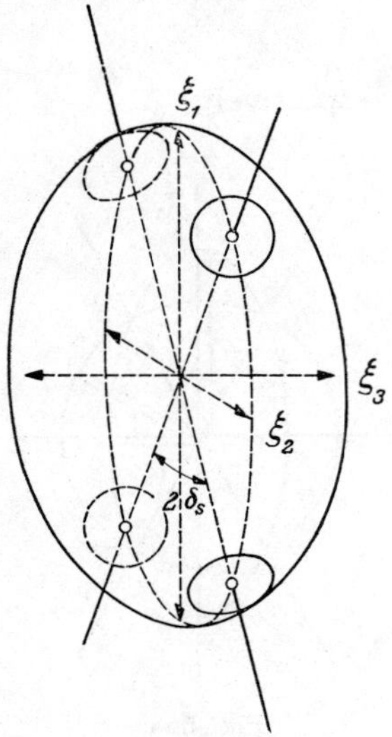

Fig. 36 d.
Construction of the optic axes as perpendiculars to the centers of the circular sections of Fresnel's ellipsoid.

If we define the *polarization of the ray* by the direction of the **E**-vector (in Sec. 25 we correspondingly defined the *polarization of the wave* by the direction of the **D**-vector), we can say: the polarization is linear for all ray directions. The planes of polarization of the two rays propagating in any one direction are perpendicular to one another. The optic axes form the only exception. Because of the circular shape of the sections belonging to these axes, no direction of polarization is preferred over any other. This gives further motivation to the name "axes of isotropy".

C. The normal surface

We generate this surface by plotting in every wave number direction **k** the two phase velocities u', u'' of the waves propagating in that direction. If we describe this locus again by means of the rectangular coordinates ξ_1, ξ_2, ξ_3, then we must write in place of (12)

$$\xi_i = \frac{k_i}{k} u, \qquad \sum \xi_i^2 = u^2 \tag{19}$$

and instead of (13)

$$\sum \frac{\xi_i^2}{u^2 - u_i^2} = 0 \tag{19 a}$$

because of eq. (25.10). Multiplying by the product of the denominators we obtain instead of (13 b)

$$(19\text{ b})\qquad u^6 - u^2\,[\xi_1^{\,2}\,(u_2^{\,2} + u_3^{\,2}) + \xi_2^{\,2}\,(u_3^{\,2} + u_1^{\,2}) + \xi_3^{\,2}\,(u_1^{\,2} + u_2^{\,2})] + \\ + \xi_1^{\,2}\,u_2^{\,2}\,u_3^{\,2} + \xi_2^{\,2}\,u_3^{\,2}\,u_1^{\,2} + \xi_3^{\,2}\,u_1^{\,2}\,u_2^{\,2} = 0.$$

Since the last set of terms does not contain a factor u^2, this equation represents a sixth order surface. By (19 a) the principal section $\xi_1 = 0$ is found to consist of a circle

$$\xi_2^{\,2} + \xi_3^{\,2} = u_1^{\,2}$$

and an "oval"

$$(\xi_2^{\,2} + \xi_3^{\,2})^2 - u_3^{\,2}\,\xi_2^{\,2} - u_2^{\,2}\,\xi_3^{\,2} = 0,$$

which is a fourth order curve (containing also its center $\xi_2 = \xi_3 = 0$ as an isolated point of the curve). The other two principal sections consist of similar curves. These sections can again be illustrated by figs. 36 a, b, c, where, however, the ellipses must be replaced by slightly differently shaped ovals.

As in fig. 36 c there are two pairs of intersection points in the principal section $\xi_2 = 0$. They correspond to the umbilical points of the index ellipsoid and the lines connecting them define the "optic normal axes". The angle between these axes, which we shall call $2\,\delta_n$, is only slightly larger than the angle $2\,\delta_s$ defined in (18). Its magnitude is determined by

$$(20)\qquad \tan\delta_n = \sqrt{\frac{u_2^{\,2} - u_3^{\,2}}{u_1^{\,2} - u_2^{\,2}}}\,.$$

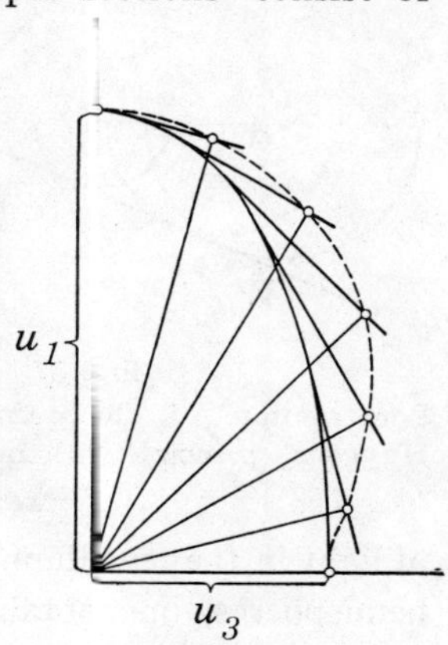

Fig. 36 e.

The normal surface as the pedal surface of the ray surface and the ray surface as the envelope of the normal surface.

In view of the connection of these axes with the circular sections of the index ellipsoid, they are the "isotropy axes" of the **D**-vector. This vector is not necessarily linearly polarized for propagation along these and only these axes.

A simple geometrical connection exists between the normal surface and the ray surface: the normal surface is the pedal surface to the ray surface. The section in the 1 – 3 plane yields, for example, the picture shown in fig. 26 e. This figure also demonstrates that the oval of the normal surface section (dotted) differs only slightly from the ellipse of the ray surface section (full line). It has already been shown in fig. 34 that the ray surface is the *envelope* of the wave planes (planes of equal phase).

27. The Problem of Double Refraction

If we place a slab of calcite over a sample of writing on a piece of paper, then on looking through the calcite the writing appears double. One image is displaced parallel with respect to the other. If we consider, for simplicity, a negative picture, that is white writing on a black background, then we can say the following: the waves emitted by the writing reach the eye by two paths which have different directions in the calcite and after the refraction have also different directions in the air. The eye extrapolates the perceived directions wrongly by projecting them in straight lines to the bottom of the calcite and therefore the impression is created that the writing is double.

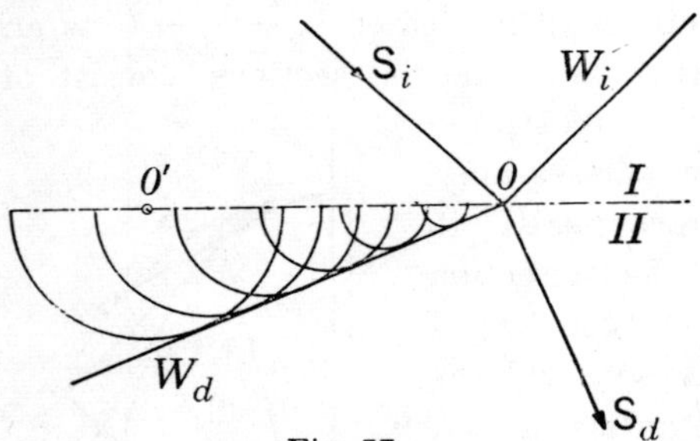

Fig. 37.
Construction of the refracted wave by Huygens' principle in the isotropic case.

The beautiful construction based on Huygens' principle for the refracted wave in an isotropic medium (see fig. 37) is well known. A plane wave propagating in the direction $\mathbf{S}_i$ is incident upon another medium. At the instant that the wave front W_i reaches the point O on the boundary plane, all previous positions O' of the wave have already radiated waves (which are drawn as hemispheres in fig. 37) into the second medium with the velocity of light in that medium. Upon drawing from O the envelope of this system of hemispheres, one obtains a straight line which is a wave front W_d of the refracted ray. The normal $\mathbf{S}_d$ to W_d is the direction of propagation of the refracted light.

A. Double refraction according to Huygens' principle

Huygens himself[1] extended this construction with ingenious foresight to the case of the (optically uniaxial) calcite crystal. He assumed the surface of propagation of the light in the crystal to be not a sphere but a certain combination of a sphere and an oblate ellipsoid of rotation (compare with fig. 39 b below). Thus he obtained two envelopes and thereby two wave fronts, one for the system of spheres and another for the system of ellipsoids. Here we shall describe the general (optically biaxial) case by replacing the combina-

[1] The complete title of Huygens' book is: Traité de la lumière, où sont expliquées les causes de ce qui lui arrive dans la réflexion et dans la réfraction et particulièrement dans l'étrange réfraction du cristal d'Islande. Leiden 1690.

tion sphere + ellipsoid by our two-sheeted ray surface (26.13); see fig. 37 a. The envelopes of the two sheets again yield two wave surfaces W_d' and W_d'' of the refracted light and the lines connecting the center O' of the ray surface with the points of contact of the envelope give two ray directions $\mathbf{S}_d'$ and $\mathbf{S}_d''$. The perpendiculars from O' upon the two envelopes (in the figure these perpendiculars are erected at O) represent the propagation vectors (the "wave normals") $\mathbf{k}_d'$, $\mathbf{k}_d''$.

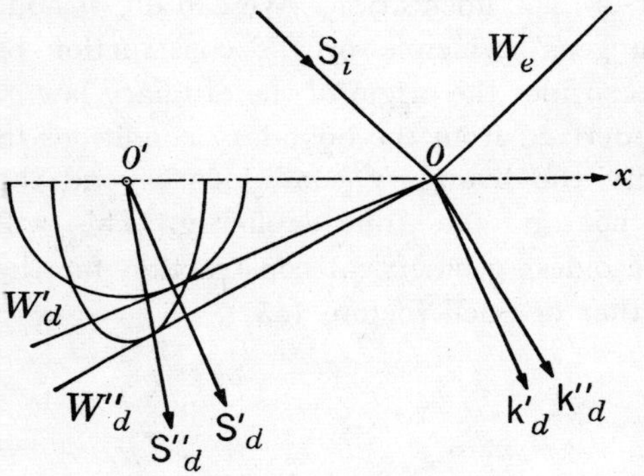

Fig. 37 a.

Construction of the two refracted waves by Huygens' principle in the anisotropic case.

This construction illustrates well the origin of double refraction, and it is therefore widely used in the literature; however it gives an incomplete description in several respects.

1. The construction presupposes that a diverging bundle of rays which originates from a point source behaves in the same way as a system of mutually independent plane waves. It is the latter type of wave which we have considered in Sec. 26 and whose velocities of propagation we symbolized purely geometrically by means of our ray surface. Lamé[1] was the first to recognize that this presents a mathematical problem which is by no means simple. He posed the problem of representing the complex of waves sent out by a concussion center (which is analogous to the spherical wave in an isotropic medium) in a precise mathematical manner by computing the three displacement components. He was, indeed, led (upon excluding the longitudinal waves) to the form of the ray surface. These results were criticized and extended by V. Volterra[2].

2. Figure 37a as it stands gives no information about the polarization of either the $\mathbf{D}$ or the $\mathbf{E}$ wave. In this respect our description would have to be completed by means of the results obtained in Secs. 25 and 26.

3. The construction leaves the question of the amplitude ratios between the different waves unanswered.

[1] In his leçons sur la théorie mathématique de l'élasticité, Paris 1852. The differential equations integrated by Lamé agree with the differential equations for the *magnetic* field components in electromagnetic optics.

[2] Acta Mathematica, Vol. **16**, p. 153, 1892.

B. The law of refraction as a boundary value problem

The preceding remarks indicate that a complete and quantitative theory of double refraction can be obtained only by considering the boundary conditions, following the method of Chap. I. The reasoning remains the same; we operate again with infinitely extended plane waves as the only known strict solutions of the optical differential equations, and not with the limited rays which usually form the objects of experimentation. We can do without additional hypotheses, such as Huygens' principle or the construction of envelopes. Let us, however, first reconsider the origin of the ordinary law of refraction and reflection which was derived from the boundary conditions in Sec. 3. This time we will not specify the boundary conditions beyond the fact that they exist and do not contain the time explicitly. This will lead us to a reinterpretation of the oldest geometrical construction for the angles of reflection and refraction, that of Snell (before 1637).

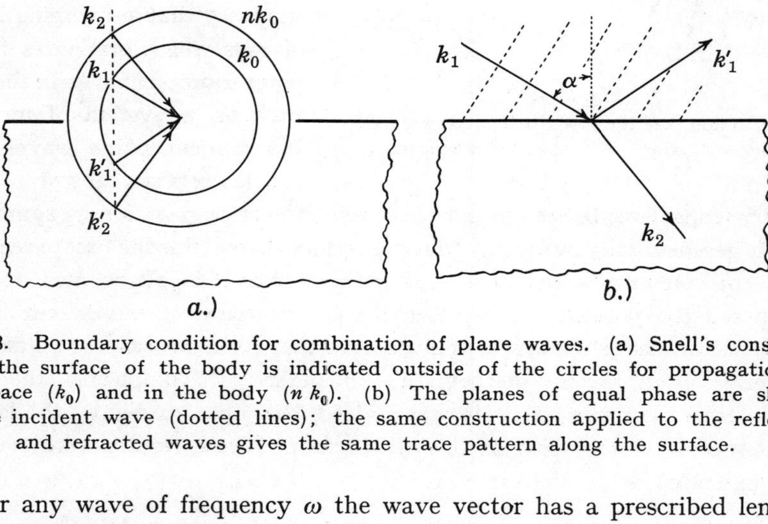

Fig. 38. Boundary condition for combination of plane waves. (a) Snell's construction; the surface of the body is indicated outside of the circles for propagation in free space (k_0) and in the body ($n\, k_0$). (b) The planes of equal phase are shown for the incident wave (dotted lines); the same construction applied to the reflected and refracted waves gives the same trace pattern along the surface.

For any wave of frequency ω the wave vector has a prescribed length, namely $|k_0| = k_0 = \omega/c$ in free space, and $|k| = \omega/u = n\, k_0$ in the body of refractive index n (assumed > 1 in the figures). This fact can be expressed geometrically by drawing two concentric spheres of radii k_0 and $n\, k_0$; any vector drawn from the surface of one of the spheres to its center is then a possible wave vector of a plane wave in the corresponding medium. Outside of the spheres we indicate in the figure the position of the body and its surface. We assume an incident wave to fall on the surface and show its wave vector k_1 of length k_0 in fig. 38 a. In fig. 38 b the incident wave itself is drawn by indicating its planes of equal phase. These planes travel along the arrow with velocity c; their trace forms a wave pattern on the surface which travels along the surface from left to right with velocity $c/\sin\alpha$ to which corresponds a wave vector tangential to the surface and of length $|k_1|\sin\alpha$.

We may then say that the surface wave obtained as the intersection of the incident wave with the surface is described by a wave vector which is the tangential component of k_1. The same will hold for any other wave, in free space or in the medium: the field produced at the surface is a surface wave the wave vector of which is the component of the spatial wave vector along the surface.

Now a time-independent relation on the surface can exist only between waves which give surface traces of the same mode of propagation. Otherwise wave amplitudes fulfilling the boundary conditions at selected points or times would fall out of step between these points or times.

Thus, going back to fig. 38 a, only three other wave vectors can be combined with k_1, namely those having the same tangential components as k_1. They are easily obtained by drawing the normal to the surface through the tail point of k_1. Of these vectors k_1' is the wave vector of the reflected wave (in free space), k_2 that of the refracted wave, and its mirror image k_2' would be a second internal wave compatible with the boundary conditions. This last wave exists in a plate with parallel surfaces: it is the internally reflected refracted wave which is essential for obtaining the interference pattern of the Lummer-Gehrke plate and for calculating the corresponding optical field directly from the boundary conditions instead of by the method of summing repeated reflexions and refractions as in section 7 F. If we wish to deal with a single surface and a single wave incident on it — and this is the assumption which leads to the Fresnel formulae — we have to omit the wave of wave vector k_2'. It is easily seen from the construction that the directions of k_1, k_1' and k_2 give the geometrical laws of reflection and refraction, in fact fig. 38 a is Snell's construction.

We can now generalize this construction to the case of a doubly refracting crystal. It was pointed out in the discussion following eq. (25.10) that to each direction of k belong two values of u, viz. u', u''. Since $|k| = \omega/u$ there are thus in each direction (except those of the optic axes) two vectors k of different length. These correspond, according to fig. 33, to two linearly polarized waves of which the D-vectors are in fixed, mutually orthogonal directions. In Snell's construction this means that the outer sphere has to be replaced by a double surface. This surface is actually the dual counterpart of the ray-surface of paragraph 26 A and fig. 35, a, b and it can be obtained from it by the translation rule 26 (11). Without going into more detail, however, we see that Snell's construction now yields two refracted waves progressing at different angles of refraction β' and β'' and polarized at right angles to one another. Each of the directions satisfies the law of refraction

$$\frac{\sin\alpha}{\sin\beta'} = \frac{c}{u'}; \qquad \frac{\sin\alpha}{\sin\beta''} = \frac{c}{u''}.$$

It is worth noting, however, that this law gives the angles β', β'' implicitly only, because u', u'' vary with these angles.

C. The Amplitudes of Reflected and Refracted Rays

Let us limit the calculation to the case that the incident ray lies in the plane containing the greatest and the smallest axis of the index ellipsoid (eq. (24.11)), i. e. that, as in Sec. 26 B, it also lies in the plane of the optic axes. Then we know, by symmetry, that either **E** and **D** both lie in the plane of incidence and **H** at right angles to it, or, for the other case of polarization, **E** and **D** are both normal to the plane (and thus parallel to one another) and **H** is in the plane. In the latter case the boundary conditions are simpler, and we restrict the derivation to it.

a) *Plane of Polarization Parallel to Plane of Incidence*

In medium I (air, $y > 0$), we assume, as in eq. (3.1), cf. fig. 3 a,

$$\mathbf{E} = E_z = A\, e^{i k_0 (x \sin\alpha - y\cos\alpha)} + C\, e^{i k_0 (x\sin\alpha' + y\cos\alpha')} \tag{1}$$

and in the crystal

$$\mathbf{E} = E_z = B\, e^{i(\mathbf{k}\cdot\mathbf{r})}, \qquad B_x = B_y = 0; \qquad B_z \equiv B. \tag{2}$$

Together with the (suppressed) time factor $\exp(-i\,\omega\,t)$, these terms represent the incident, reflected, and refracted waves, respectively. Snell's construction gives

$$\alpha' = \alpha \qquad \text{and} \qquad k\sin\beta = k_0\sin\alpha. \tag{3}$$

Continuity of the tangential component of **E** at the surface $y = 0$ is expressed in the equation

$$A + C = B. \tag{4}$$

Since differences of permeability are neglected, the second boundary condition may be expressed as: **H** continuous. Now in the first medium at the boundary $(y = 0)$

$$\begin{aligned} H_x &= -\frac{1}{i\,\omega}\frac{\partial E_z}{\partial y} = -\frac{k_0}{\omega}(-A\cos\alpha + C\cos\alpha')\, e^{i k_0 x \sin\alpha}, \\ H_y &= +\frac{1}{i\,\omega}\frac{\partial E_z}{\partial x} = \frac{k_0}{\omega}(A\sin\alpha + C\sin\alpha')\, e^{i k_0 x\sin\alpha}, \end{aligned} \tag{5}$$

and in the second medium

$$(6)\qquad \begin{aligned} H_x &= -\frac{1}{i\,\omega}\frac{\partial E_z}{\partial y} = -\frac{k}{\omega}(-B\cos\beta)\,e^{i k x \sin\beta}, \\ H_y &= +\frac{1}{i\,\omega}\frac{\partial E_z}{\partial x} = \frac{k}{\omega}(B\sin\beta)\,e^{i k x \sin\beta}. \end{aligned}$$

The boundary condition thus is, using (27.3),

$$(7)\qquad \begin{aligned} A - C &= \frac{\tan\alpha}{\tan\beta} B, \\ A + C &= \frac{\sin\alpha}{\sin\beta} B \frac{\sin\beta}{\sin\alpha} = B. \end{aligned}$$

The second of these conditions is the same as that for E. In terms of the incident amplitude we obtain the refracted and reflected waves

$$(8)\qquad \begin{aligned} B/A &= \frac{2}{1+\dfrac{\tan\alpha}{\tan\beta}}, & C/A &= \frac{\tan\beta-\tan\alpha}{\tan\beta+\tan\alpha}, \\ &= \frac{2\sin\beta\cos\alpha}{\sin(\beta+\alpha)}, & &= \frac{\sin(\beta-\alpha)}{\sin(\beta+\alpha)}. \end{aligned}$$

This result is, of course, the same as that obtained in Sec. 3 A for the reflected and refracted waves of the same polarization in the case of an isotropic body (3.12).

b) Plane of Polarization Normal to Plane of Incidence

In this case we cannot expect to obtain the same result as for an isotropic body, which is expressed in (3.16). Whereas in the isotropic body the field vectors **E** and **D** (electric force and dielectric displacement) have the same direction, shown by the amplitude vector **B** in fig. 3 b, their directions differ in an anisotropic body. The boundary conditions are: continuity of the tangential component of **E** and of the normal component of **D**, as well as continuity of the magnetic vector **H**. Now the decomposition of **E** and **D** into tangential and normal components means their decomposition according to the axial system of fig. 3 b (x parallel, y normal to surface), but the anisotropic relation between the two vectors finds its simple expression if they are decomposed according to the principal axes of the dielectric tensor or index ellipsoid (conf. Sec. 24). This double decomposition complicates the derivation and expression of the amplitude ratios for this case of polarization.

28. The Optical Symmetry of Crystals

So far we have concerned ourselves only with the general form of the dielectric tensor ε_{ik}. This tensor is defined by 6 parameters (because $\varepsilon_{ik} = \varepsilon_{ki}$). We can find the number of parameters not only from the tensor scheme but also from its geometrical interpretation, namely Fresnel's ellipsoid. The ellipsoid is defined by its 3 principal axes and by their positions in space which in turn are defined by 3 angle parameters. A crystal without symmetry properties is called *triclinic*. Such a crystal is described by the general (ε_{ik})-tensor.

If the crystal has symmetry, the number of independent parameters is reduced. By *symmetry direction* let us understand either an axis of rotational symmetry or the direction normal to a mirror plane, as the case may be. Then if a crystal has a single symmetry direction it must coincide with one of the three symmetry directions of the Fresnel ellipsoid. If the crystal has two symmetry directions at right angles to one another, there exists a third one which is orthogonal to both, and the symmetry directions of the Fresnel ellipsoid must coincide with those of the crystal. Furthermore, the existence of a symmetry axis[1] of higher order than 2 necessitates equal magnitude of the axes of the Fresnel ellipsoid at right angles to its direction; thus the ellipsoid will be one of revolution. Again, if there are, as in the cubic system, four threefold axes of symmetry (along the body diagonals of a cube), then the Fresnel ellipsoid degenerates into a sphere, and a single constant is left over from the ε_{ik} scheme.

Let us consider some cases in more detail. If a crystal has only one symmetry direction, it is called *monoclinic*. One of the principal axes of the Fresnel ellipsoid being fixed by this direction, there remain four parameters. These can be found from the scheme of the ε_{ik}. Assume, for instance, that there is a mirror plane normal to the direction of symmetry (index 2), so that $+x_2$ and $-x_2$ are symmetrically equivalent directions. If, for convenience, we write x_i, y_i as the variables instead of E_i, D_i, the general scheme

$$\begin{aligned} y_1 &= \varepsilon_{11} x_1 + \varepsilon_{12} x_2 + \varepsilon_{13} x_3, \\ y_2 &= \varepsilon_{21} x_1 + \varepsilon_{22} x_2 + \varepsilon_{23} x_3, \\ y_3 &= \varepsilon_{31} x_1 + \varepsilon_{32} x_2 + \varepsilon_{33} x_3 \end{aligned} \tag{1}$$

must remain unchanged if we substitute $-x_2$ for x_2 and simultaneously $-y_2$ for y_2. Thus we have also (reversing all signs in the second line),

[1]A rotational symmetry axis of order n is one about which rotation of the crystal by $2\pi/n$ will bring it to the nearest covering position.

$$
\begin{aligned}
y_1 &= \varepsilon_{11} x_1 - \varepsilon_{12} x_2 + \varepsilon_{13} x_3, \\
y_2 &= -\varepsilon_{21} x_1 + \varepsilon_{22} x_2 - \varepsilon_{23} x_3, \\
y_3 &= \varepsilon_{31} x_1 - \varepsilon_{32} x_2 + \varepsilon_{33} x_3.
\end{aligned} \tag{1 a}
$$

Eq^s. (1) and (1 a) are consistent only if $\varepsilon_{12} = \varepsilon_{21} = 0$ and $\varepsilon_{23} = \varepsilon_{32} = 0$. The tensor scheme for this crystal therefore reduces to the four parameters

$$
\begin{pmatrix} \varepsilon_{11} & 0 & \varepsilon_{13} \\ 0 & \varepsilon_{22} & 0 \\ \varepsilon_{13} & 0 & \varepsilon_{33} \end{pmatrix}. \tag{2}
$$

Next consider a crystal with two symmetry directions. There follows automatically a third such direction, but we assume that these three directions are not further related to one another. This is the case

(i) in the *orthorhombic* crystal system where the directions are either twofold axes of rotation or normals of mirror planes,

(ii) in the *rhombohedral* crystal system where one direction is that of threefold rotation symmetry,

(iii) in the *hexagonal* system, where one direction is a sixfold axis.

Since now three angular parameters of the ellipsoid are fixed, the number of free parameters is reduced to 3. Assume, for instance, that direction 3 is that of a twofold rotation axis, and direction 2, as before, corresponds to a mirror plane. The rotation about 3 means that no change is brought about by the simultaneous substitutions $(x_1, x_2) \to (-x_1, -x_2)$, $(y_1, y_2) \to (-y_1, -y_2)$. Applying this to the equations condensed in the scheme eq. (2), we see that the addition of the twofold axis to the mirror plane makes necessary the vanishing of ε_{13}. The tensor scheme thus reduces to the terms of the principal diagonal.

If we assume a fourfold rotation axis in direction 3 (together with the mirror plane normal to 2), then also a rotation by 90° should produce no change; this is given by the substitution

$$(x_1, x_2) \to (-x_2, x_1), \qquad (y_1, y_2) \to (-y_2, y_1).$$

Since the repetition of the 90° rotation produces a rotation by 180°, which is that of a twofold axis, we may use the previous simplified scheme for reducing it further by this substitution. In this way we obtain the following sets of relations

$$
\begin{aligned}
y_1 &= \varepsilon_{11} x_1 & \qquad -y_2 &= -\varepsilon_{11} x_2 \\
y_2 &= \varepsilon_{22} x_2 & y_1 &= \varepsilon_{22} x_1 \\
y_3 &= \varepsilon_{33} x_3 & y_3 &= \varepsilon_{33} x_3.
\end{aligned}
$$

From these follows $\varepsilon_{11} = \varepsilon_{22}$, and this leaves only two constants undetermined.

Crystals which have three symmetry directions of which two are symmetrically equivalent are called *tetragonal.* If all three directions become equivalent, they are called *cubic.* If one of the symmetry directions contains a threefold or sixfold axis of rotation, then the Fresnel ellipsoid has full rotational symmetry.

Summarizing we have the following optical properties in the seven crystal systems:

Cubic crystals are optically isotropic, i. e. optically they do not differ from amorphous bodies, glasses or liquids.

Tetragonal, hexagonal and rhombohedral crystals have an ellipsoid of rotation. This implies that they are optically uniaxial and have two principal indices of refraction.

Orthorhombic, monoclinic, and triclinic crystals have three principal indices of refraction and are optically biaxial. The directions of the principal axes of refraction are fixed (for all wavelengths and temperatures) in the orthorhombic crystals, but the angle of the optic axes may vary, since it depends on the values of the principal refractive indices. In the monoclinic system only one, and in the triclinic system none, of the directions of principal index of refraction is fixed.

The reason why optically only three groups, the biaxial, uniaxial and isotropic groups, are distinct lies in the fact that eqs. (1) relate two vector quantities (field strength and excitation) to one another. In the theory of elasticity two tensor quantities, stress and strain, are related, and this leads to the distinction of many more groups than by the crystal-optical behavior. Electrostriction and piezoelectricity connect a vector (field strength) with a tensor (deformation) and give again a different classification.

The fact that in optics we are only concerned with the Fresnel or index ellipsoids, i. e. with surfaces of a very restricted type, makes it understandable that the more elaborate symmetry properties need not be discussed in this connection. They are systematically enumerated in the 32 *crystal classes* which are unequally distributed over the seven crystal systems. These classes contain a complete description of all geometrically possible symmetry relations between directions in space which pass through one point. According to the modern conception of crystal structure, crystals are bodies with an internal three-dimensional periodicity of atomic arrangement; the complete enumeration of the symmetry-types compatible with this periodicity led Barlow, Schoenflies and Fedorow to 230 different *space groups.* Crystal optics cannot probe into these; only X-ray analysis is able to do this (see Sec. 32).

In the uniaxial case the familiar constructions which we have used in the biaxial case are considerably simplified. Letting $\varepsilon_1 = \varepsilon_2 \neq \varepsilon_3$, we introduce the notation

(3)
$$u_1 = u_2 = u_o \quad \text{ordinary wave velocity}$$
$$u_3 = u_e \quad \text{extraordinary wave velocity,}$$

and correspondingly, v_o the ordinary and v_e the extraordinary ray velocity. Then one of the principal axes a, b of the elliptic section in fig. 33 lies in the equatorial plane of the index ellipsoid; the other principal axis lies in the meridian plane through the optic axis. The planes of polarization of the respective waves are that meridian plane and the plane perpendicular to it which passes through the direction of propagation but not through the optic axis. The same is true for the construction of the ray directions from Fresnel's ellipsoid.

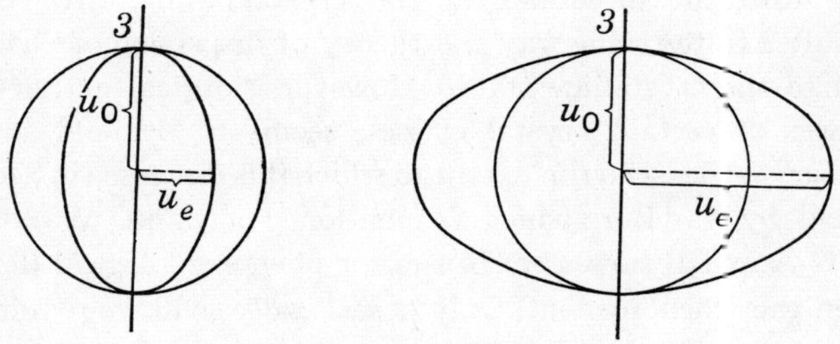

Fig. 39.

Ray surfaces of optically uniaxial crystals
a) $u_o > u_e$ uniaxial positive; example, quartz.
b) $u_o < u_e$ uniaxial negative; example, calcite.

The shape of the *ray surface* is now particularly simple. Not only can the factor v^2 be removed in eq. (26.13 b) for this surface, but since $u_1 = u_2 = u_o$, also a factor $u_o^2 - v^2$ cancels out. Thus the equation becomes

$$\{\xi_1^2 + \xi_2^2 + \xi_3^2 - u_o^2\}\{(\xi_1^2 + \xi_2^2)\, u_o^2 + \xi_3^2 u_e^2 - u_o^2 u_e^2\} = 0$$

which separates into a sphere of radius u_o and an ellipsoid of revolution

(4)
$$\frac{\xi_1^2 + \xi_2^2}{u_e^2} + \frac{\xi_3^2}{u_o^2} = 1.$$

These two surfaces touch at the points $\xi_3 = \pm u_o$. This verifies the form of the radiation surface which Huygens predicted for calcite.

Depending on the way in which the sphere and ellipsoid touch, one distinguishes positive and negative optically uniaxial crystals, see fig. 39 a, b.

The wave surface (normal surface) looks less simple. It separates into a sphere of radius u and a fourth order surface of revolution which is called an "ovaloid", namely

(5)
$$(\xi_1^2 + \xi_2^2 + \xi_3^2)(\xi_1^2 + \xi_2^2 + \xi_3^2 - u_e^2) = \xi_3^2 (u_o^2 - u_e^2).$$

In the limiting case of isotropy $u_o = u_e$, (4) becomes a sphere of radius u_o and hence the ray surface consists of this sphere, counted twice. At the same time (5) degenerates into a sphere of radius zero (isolated double point) and a sphere of radius u_o.

29. Optically Active Crystals and Fluids

By means of the structure theory of crystals it is possible to confirm the purely phenomenological discussion of this chapter, beginning with the relation (24.1). For, in the summation over the lattice structure of elementary constituents, different directions in the crystal would give dielectrically different results. In the same way the theory of dispersion of Chap. III could be extended to the crystalline state. However, "optical activity", i. e. the rotating power of certain crystal classes, seems to demand some use of structure theory, at least to the extent to which it is discussed in Secs. 75 and 84 of the textbook by Max Born which we mentioned on p. 40. We will, however, show in parts A and B how a much shorter phenomenological discussion can describe even this phenomenon. Only occasionally shall we require structural considerations in order to illustrate our results.

In subsection C we shall give only a very cursory presentation of optical activity in fluids and optically isotropic crystals even though this subject is of enormous theoretical importance for stereochemistry and of considerable practical importance in industry. A more precise treatment would require a more profound discussion of molecular structure than is compatible with the scope of these lectures. For such treatments we refer the reader to Secs. 84 and 99 of Born's book and to the articles by Born, C. W. Oseen, and W. Kuhn which are cited there.

A. The gyration vector of solenoidal crystal structures

We can write our linear vector function (24.1) in the abbreviated form

(1) $$D_j = \varepsilon_{jh} E_h$$

where the ε_{jh} form a real *symmetric tensor*. We now discard the reality condition for the ε and replace

$$\varepsilon_{jh} \text{ by } \varepsilon_{jh} + i\gamma_{jh}.$$

The additional terms γ, which are supposed to be *small*, are not to have an ohmic, dissipative character as in the complex dielectric constant of metal optics. Rather, they shall be conservative, which means that the γ_{jh} must not contribute to the electric energy density $W_e = 1/2\,(\mathbf{D}\cdot\mathbf{E})$ as computed from Maxwell's equations. This latter condition is fulfilled if they form an *antisymmetric tensor* as do the conservative gyroscopic terms in mechanics (see Vol. I, Sec. 30).

In order to explain this in greater detail, we write instead of (1)

$$D_j = \varepsilon_{jh} E_h + i\gamma_{jh} E_h \tag{2}$$

and find

$$\mathbf{D}\cdot\mathbf{E} = \sum_j \sum_h \varepsilon_{jh} E_j E_h + i \sum_j \sum_h \gamma_{jh} E_j E_h.$$

In order that the contribution of the γ term vanish for arbitrary values of the $\mathbf{E}$ components, every γ need not be identically zero, but it suffices if the γ fulfill the following conditions:

$$\gamma_{ii} = 0, \qquad \gamma_{jh} + \gamma_{hj} = 0.$$

These are precisely the conditions for the antisymmetry of the γ-tensor. We see now also why the γ-tensor must be purely imaginary[1]. For if it had a real part, this would add to the ε-tensor and disturb its symmetry, but we know that the ε-tensor must be symmetric for general reasons of energy.

An antisymmetric tensor (γ_{jh}) can always be replaced by a vector $\boldsymbol{\gamma}$ with the components

$$\gamma_1 = \gamma_{23} = -\gamma_{32}, \qquad \gamma_2 = \gamma_{31} = -\gamma_{13}, \qquad \gamma_3 = \gamma_{12} = -\gamma_{21}.$$

Thus we obtain

$$\sum_i \gamma_{1h} E_h = \gamma_3 E_2 - \gamma_2 E_3 = -[\boldsymbol{\gamma}\times\mathbf{E}]_1 \text{ etc.}$$

Then (2) becomes

$$D_j = \sum_h \varepsilon_{jh} E_h - i[\boldsymbol{\gamma}\times\mathbf{E}]_j. \tag{3}$$

We call $\boldsymbol{\gamma}$ the *gyration vector*. It is not a polar but an *axial* vector like the angular velocity $\boldsymbol{\omega}$ in Vol. I, Sec. 22. We quote from there:

"Axial vectors are properly represented by an axis provided with a sense of rotation and a magnitude of rotation." "The signs of their components do not change under inversion of the coordinate system (interchange of

---

[1]Physically, the factor i is due to the fact that the value of $\mathbf{D}$ at any given point depends not only on the value of $\mathbf{E}$ at that particular point but also on the behavior of $\mathbf{E}$ in the vicinity of this point; that is to say, $\mathbf{D}$ depends also on the local derivatives of $\mathbf{E}$. Owing to the wave character of $\mathbf{E}$, these derivatives contain the factor i. From an atomistic viewpoint this is due to the influence of neighboring ions which are in a field differing from that at the point under consideration. In order that these effects shall not cancel one another, the lattice has to have a certain amount of asymmetry which will be determined presently.

x, y, z with $-x, -y, -z$)." "The vector product of an axial and a polar vector is a polar vector." "Under inversion a right-handed coordinate system becomes a left-handed coordinate system." *But thereby also the sense of rotation* $+i$ *of the complex plane is replaced by* $-i$.

Let us consider a crystal which has a symmetry center. For such a crystal eq. (3) must be invariant under inversion. Upon inversion the components of the polar vectors **D** and **E** change sign as do the components of $\boldsymbol{\gamma} \times \mathbf{E}$. But since i also changes its sign, the sign of the last term in (3) remains unchanged. Hence after an inversion (3) reads

$$-D_j = -\sum_h \varepsilon_{jh} E_h - i\,[\boldsymbol{\gamma} \times \mathbf{E}]_j. \tag{3 a}$$

This is consistent with (3) only if $\boldsymbol{\gamma} \times \mathbf{E} = 0$. Therefore our invariance condition for a centrally symmetric crystal requires that $\boldsymbol{\gamma} = 0$. *Only a crystal without a center of symmetry can possesss a gyration vector.* There are examples of such acentric crystals among each of the seven crystal systems. The most common example is *quartz* (silicon dioxide, SiO_2).

Writing eq. (3) in the principal axes of the dielectric, 1, 2, 3, we obtain

$$\begin{aligned} D_1 &= \varepsilon_1 E_1 \qquad\qquad + i\gamma_3 E_2 - i\gamma_2 E_3, \\ D_2 &= \varepsilon_2 E_2 - i\gamma_3 E_1 \qquad\qquad + i\gamma_1 E_3, \\ D_3 &= \varepsilon_3 E_3 + i\gamma_2 E_1 - i\gamma_1 E_2. \end{aligned} \tag{4}$$

We see here that in general, in a triclinic acentric crystal, for instance, the $\boldsymbol{\gamma}$-direction is by no means determined by the principal dielectric axes. However, in the case of quartz with its rotational dielectric symmetry, the gyration vector must also submit to this symmetry and must be parallel to the principal axis. Hence $\gamma_1 = \gamma_2 = 0$, $\gamma_3 = \gamma$, and (4) becomes

$$\begin{aligned} D_1 &= \varepsilon_1 E_1 + i\gamma E_2, \\ D_2 &= \varepsilon_2 E_2 - i\gamma E_1, \\ D_3 &= \varepsilon_3 E_3. \end{aligned} \tag{4 a}$$

B. The rotation of the plane of polarization in quartz.

We proceed exactly as in Sec. 26 except that the above eq. (4 a) is used in place of $D_j = \varepsilon_j E_j$. As a result, a number of correction terms which differ for the different directions $j = 1, 2, 3$ have to be added in eq. (26.8). For $j = 1$, we obtain in place of (26.8)

$$\frac{1}{\mu_0 v^2} B_1 = \varepsilon_1 B_1 + i\gamma B_2 - s_1 \left\{ \sum_j s_j \varepsilon_j B_j + i\gamma\, (s_1 B_2 - s_2 B_1) \right\}. \tag{5}$$

Introducing the abbreviation,

(5 a) $$K = \sum s_j \varepsilon_j B_j + i\gamma (s_1 B_2 - s_2 B_1),$$

multiplying (5) by μ_0, and introducing the ordinary and extraordinary principal light velocities $u_o = (\varepsilon_1 \mu_0)^{-1/2} = (\varepsilon_2 \mu_0)^{-1/2}$, $u_e = (\varepsilon_3 \mu_0)^{-1/2}$, one obtains in place of (26.9)

(6) $$\left(\frac{1}{u_o^2} - \frac{1}{v^2}\right) B_1 + i\mu_0 \gamma B_2 = \mu_0 s_1 K.$$

Correspondingly, we find for $j = 2$ and 3

(7) $$\left(\frac{1}{u_o^2} - \frac{1}{v^2}\right) B_2 - i\mu_0 \gamma B_1 = \mu_0 s_2 K,$$

(8) $$\left(\frac{1}{u_e^2} - \frac{1}{v^2}\right) B_3 = \mu_0 s_3 K.$$

We now multiply eqs. (6), (7), and (8) by the factors

$$\frac{s_1}{\left(\frac{1}{u_o^2} - \frac{1}{v^2}\right)}, \quad \frac{s_2}{\left(\frac{1}{u_o^2} - \frac{1}{v^2}\right)}, \quad \frac{s_3}{\left(\frac{1}{u_e^2} - \frac{1}{v^2}\right)}$$

respectively, and add the three resulting equations, obtaining

(9) $$\sum s_j B_j + i\mu_0 \gamma \frac{s_1 B_2 - s_2 B_1}{\frac{1}{u_o^2} - \frac{1}{v^2}} = \mu_0 K \left\{ \frac{s_1^2 + s_2^2}{\frac{1}{u_o^2} - \frac{1}{v^2}} + \frac{s_3^2}{\frac{1}{u_e^2} - \frac{1}{v^2}} \right\}.$$

The first term on the left-hand side vanishes because $\mathbf{s} \perp \mathbf{E}$. But the second term also vanishes to an order higher than the first, for the presence of the factor γ allows us to approximate B_1 and B_2 by means of eq. (26.9). Since in this equation B_1 and B_2 are proportional to s_1 and s_2 respectively, $s_1 B_2 - s_2 B_1$ is at least of the first order in γ. Hence the right-hand side of (9) must vanish at least like γ^2. Since $K \neq 0$, we find that to the same degree of accuracy

(9 a) $$\frac{s_1^2 + s_2^2}{\frac{1}{u_o^2} - \frac{1}{v^2}} + \frac{s_3^2}{\frac{1}{u_e^2} - \frac{1}{v^2}} = 0.$$

This is our former eq. (26.10) specialized to the case of an uniaxial crystal. Therefore, *except for differences of the second order in γ the ray surface of an optically active crystal agrees with that of an inactive crystal.*

But this is true only "generally", namely only so long as the denominator of the second term on the left side of eq. (9) does not itself become as small as γ; but for a ray in the approximate direction of the optic axis one has $v^2 \sim u_o^2$.

For such a ray s_1 and s_2 are small of the first order, and so is B_3, because of the transversality condition. Therefore, according to (5), K also becomes small of the first order. This means that the right-hand sides of eqs. (6) and (7) are second order quantities in s_1 and s_2 and can be neglected, while eq. (8) is automatically satisfied in the first order. Equations (6) and (7) thus become

$$(10)\qquad \begin{aligned} &\left(\frac{1}{u_o^2}-\frac{1}{v^2}\right)B_1+i\mu_0\gamma\,B_2=0 &&\Big|\ 1\\ &\left(\frac{1}{u_o^2}-\frac{1}{v^2}\right)B_2-i\mu_0\gamma\,B_1=0 &&\Big|\ \pm i \end{aligned}$$

Multiplication by the factors indicated on the right and addition gives rise to two equations which must be fulfilled simultaneously:

$$(11)\qquad \begin{aligned} &\left(\frac{1}{u_o^2}-\frac{1}{v^2}+\mu_0\gamma\right)(B_1+i\,B_2)=0,\\ &\left(\frac{1}{u_o^2}-\frac{1}{v^2}-\mu_0\gamma\right)(B_1-i\,B_2)=0. \end{aligned}$$

If we satisfy the first equation by choosing v^2 so as to make the first factor zero, then we must satisfy the second equation by setting its second factor equal to zero, and vice versa. Thus there are two solutions for v^2 which correspond to the two branches of the ray surface. As before, we shall denote the two solutions by v' and v'', and the corresponding values of B will also be distinguished by primes and double primes. Then our two solutions read

$$(11\text{ a})\qquad \frac{1}{u_o^2}-\frac{1}{v'^2}+\mu_0\gamma=0,\qquad B_1'-i\,B_2'=0,$$

$$(11\text{ b})\qquad \frac{1}{u_o^2}-\frac{1}{v''^2}-\mu_0\gamma=0,\qquad B_1''+i\,B_2''=0.$$

These expressions represent two opposite circularly polarized waves which propagate with the respective velocities

$$(12)\qquad \left.\begin{matrix} v'\\ v'' \end{matrix}\right\}=u_o\left(1\mp\frac{g}{2}\right),\qquad g=\mu_0\gamma\,u_o^2$$

and whose directions of rotation are determined by the two non-vanishing complex quantities $B_1'+i\,B_2'$ and $B_1''-i\,B_2''$.

This situation is illustrated in fig. 40. However, in contrast to fig. 39 a, this figure refers not to linearly but to circularly polarized waves. Where the optic axis passes through them, the two branches of the ray surface are now separated by the small distance

$$(13)\qquad v''-v'=u_o\,g$$

instead of being tangent as before. For all other ray directions for which the two branches were separated anyway in fig. 39 a, the additional separation is a second order correction and can be neglected. In these other directions, in particular normally to the optic axis, ordinary double refraction of linearly polarized waves takes place.

As in Sec. 20 we interpret the difference between v' and v'' in terms of a *rotation of the plane of polarization.* Let us consider a linearly polarized wave which is normally incident on a quartz plate whose surfaces are cut perpendicular to the optic axis. We decompose the linear polarization into two equal and oppositely directed circular polarizations. In passing through the plate of thickness l the phase of one of these waves lags behind that of the other. When they emerge from the crystal, we again combine the two waves into one linearly polarized wave. Its plane of polarization differs from that of the incident wave. The plane of polarization has been turned through an angle χ which is proportional to the thickness l and to the difference $k_+ - k_-$. Hence χ is also proportional to the difference $v'' - v'$ which along the optic axis is identical with the difference $u'' - u'$ between the wave velocities. The difference between magnetic and "natural" rotation of the plane of polarization, which we discussed at the end of Sec. 20, is due to the fact that if we reverse the direction of the ray, our present gyration vector, which depends on the structure of the quartz, does not change sign, while in Sec. 20 the magnetic field strength does change its sign under reversal of the direction of propagation.

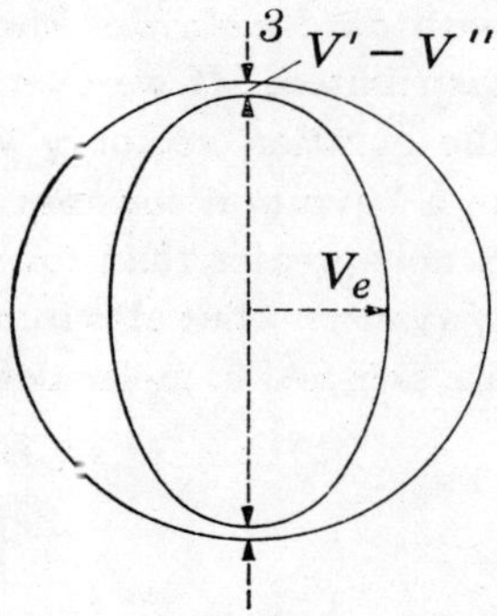

Fig. 40.
Ray surface of an optically active uniaxial crystal. (In contrast to fig. 39 a, this refers to a circularly polarized wave.)

The absence of a center of symmetry as a *"conditio sine qua non"* is indicated by the outward shape of quartz. There are "right-handed and left-handed quartzes" which are distinguished by the enantiomorphic trapezoidal faces which truncate their hexagonal prisms to the right or to the left. The rotating power of cinnabar HgS is several times stronger than that of quartz. Optical activity has also been found to exist in the axis directions of optically biaxial crystals (cane sugar, Rochelle salt) (Voigt, Pocklington). If cubic crystals, in which every direction is a principal axis and an optic axis, are optically active at all, they are optically active in every direction, e. g. $NaClO_3$. Optical activity is not due to the crystal lattice, i. e., to the mere internal periodicity of the crystal, but to its structure, that is, to the symmetry of arrangement of the elements of structure, the atoms, within the unit of periodic repeat. In molecular crystals, like cane sugar, the part of the rotatory power that resides in the molecule is retained when the crystal is dissolved in a liquid; in atomic crystals, like $NaClO_3$, rotation resides entirely in the crystal structure, and the solution, in which the molecule has dissociated into ions, is optically inactive.

C. Optically active fluids.

Here we are not concerned with a rigid structure as in crystals but rather with *fluid molecules* whose spatial positions and orientations are statistically distributed. If we average over all possible orientations of these molecules, the gyration vector $\boldsymbol{\gamma}$ which was introduced in the assumption (3) reduces to a "gyration constant" γ. The degree of asymmetry necessary for activity is now greater than for crystals. Not only must the molecule have no *center of symmetry* but also it must not have any *plane of symmetry*. These conditions are satisfied in molecules which contain an asymmetrical carbon atom, i. e. one

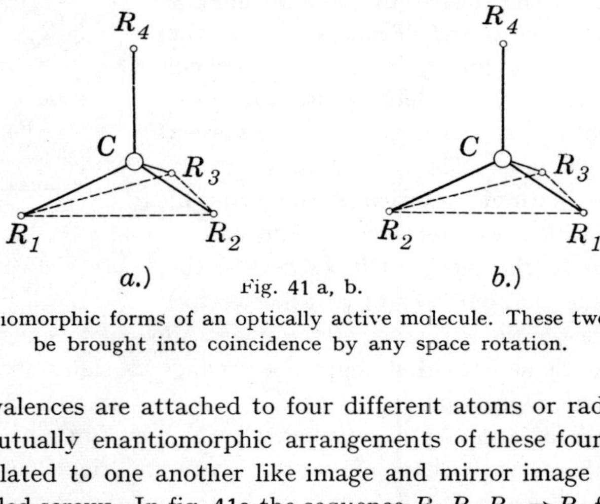

Fig. 41 a, b.

The two enantiomorphic forms of an optically active molecule. These two forms cannot be brought into coincidence by any space rotation.

whose four valences are attached to four different atoms or radicals. There exist two mutually enantiomorphic arrangements of these four substituents which are related to one another like image and mirror image or like right- and left-handed screws. In fig. 41a the sequence $R_1\ R_2\ R_3 \rightarrow R_4$ forms a right-handed screw; in fig. 41b this sequence forms a left-handed screw. These two forms cannot be brought into coincidence by any motion in three-dimensional space. Examples of this kind of molecule are the two sugar types, grape sugar, dextrose, and fruit sugar, levulose. The rotating power of solutions and mixtures of these sugars can be determined with extreme accuracy. A balanced mixture of right- and left-rotating molecules is called a "racemic" state. We have already mentioned at the end of Sec. 20 the great importance of activity measurements in the sugar industry.

Lindman[1] devised a macroscopic model which serves to illustrate the molecular process responsible for optical activity. The arrangement is the following: a cardboard box contains several hundred wire spirals which are about 2 cm. in diameter and consist of about two windings, all having the same rotational sense. The spirals are wrapped individually in tissue paper and the box is shaken so that the spirals assume arbitrary positions. Linearly polarized electromagnetic dipole radiation of about 10 cm. wavelength falls on the box. This radiation can be thought of as decomposed into two waves, right- and left-circularly polarized. One of these two waves is accelerated in its propagation by the metal spirals while the other wave is retarded. Behind the box the two circularly polarized waves again combine into one linearly polarized wave whose direction of oscillation is, however, rotated with respect to that of the incoming wave. This rotation of the plane of polarization can be detected by rotating a linear antenna which is tuned to the primary radiation and is connected to a receiver. The rotation of the plane of polarization can be cancelled by inserting a second box which is identical to the first and contains an equal number of spirals which are, however, twisted in the opposite rotational sense. The two boxes together constitute a racemic mixture.

The description of this attractive model experiment will have to serve as a substitute for the treatment of the actual molecular theory of optical activity which, unfortunately, cannot be taken up here.

30. Nicol's Prism, Quarter Wave Plate, Tourmaline Tongs, and Dichroism

A. Nicol's prism

Looking at a model of the structure of calcite $CaCO_3$, one gains the impression that the constituents Ca^{++} and CO_3^{--} would assume a cubic arrangement (as do the constituents Na^+ and Cl^- of rock salt) if it were not for the fact that the plane triangular radical CO_3 (in contrast to the spherical Cl-ion) imposes special lateral conditions on the spatial arrangement. These requirements force the cubic structure which characterizes rock salt into a rhombohedral structure. The transition between these two types of structures can be imagined as a lateral stretching or, alternatively, as a longitudinal compression of the cubic model. In this process one of the threefold symmetry axes formed by the diagonals of the cubic structure becomes the threefold principal axis and at the same time the optic axis of the rhombo-

[1] K. F. Lindman, Ann. d. Phys. **63**, p. 621, 1920 and **69**, p. 270, 1922.

hedral crystal. The fundamental cell of such a lattice is a rhombohedron bounded by 3 + 3 rhombuses. This is easily confirmed because the crystals can be split along the face planes of the cells. The double refraction of the famous crystal clear "Iceland Spar" was discovered by Bartholinus and studied by Huygens.

A Nicol prism (which is really not a prism but a parallelepipedon) is made from a cleavage rhombohedron which is about three times as long as it is wide. The end surfaces AB, CD are then cut so that they form an angle of 68° with the long edge; see fig. 42 (the natural surfaces AB', CD' which are drawn dotted in the figure make an angle of 70°52′ with the long edges). Finally, the resulting parallelepiped is cut in two along a plane perpendicular to the end surfaces AB, CD, and the two halves I and II are glued together with Canada balsam. The index of refraction of Canada balsam is 1.55. The two principal indices of refraction of calcite are

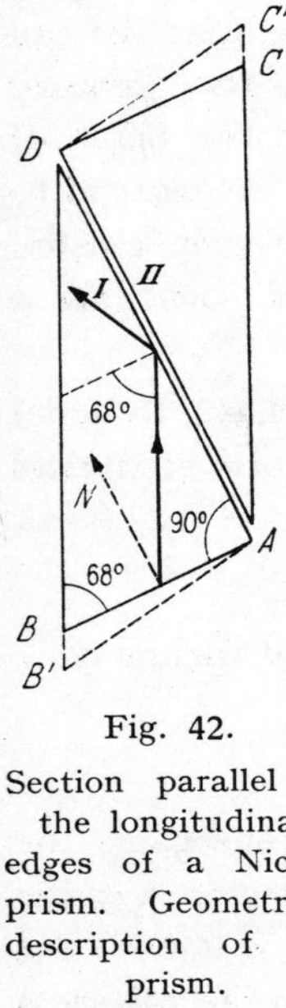

Fig. 42.
Section parallel to the longitudinal edges of a Nicol's prism. Geometrical description of the prism.

$$n_o = 1.66, \qquad n_e = 1.49. \tag{1}$$

For the ordinary ray the Canada balsam is the rarer medium; for the extraordinary ray it is the denser medium. The ordinary ray can enter the balsam only if its angle of incidence is less than the limiting angle of total reflection. According to Sec. 5 the latter is given by

$$\sin \alpha_{\text{tot}} = \frac{1.55}{1.66}, \qquad \alpha_{\text{tot}} = 69°10'. \tag{2}$$

If a ray parallel to the longitudinal edges is incident on one of the end surfaces, the ordinary ray is refracted at that surface so that it falls on the balsam at an angle of incidence of about 77°. This ray is therefore totally reflected at the balsam layer and does not enter the second half II of the crystal. It is deflected towards the side face BD. The same is also true for a certain easily calculated interval of neighboring directions of the incident ray. The face BD is blackened so that it will absorb these ordinary rays.

The extraordinary ray for which the balsam is the denser medium cannot be totally reflected (see fig. 42a). Furthermore, because $n_e < n_o$, this ray is less strongly refracted toward the normal N upon entering the calcite. After passing through the balsam, the ray traverses the crystal II in a direction parallel to that in I and it emerges from the Nicol prism with a direction parallel to that of the incoming ray. In the figure this direction is parallel to

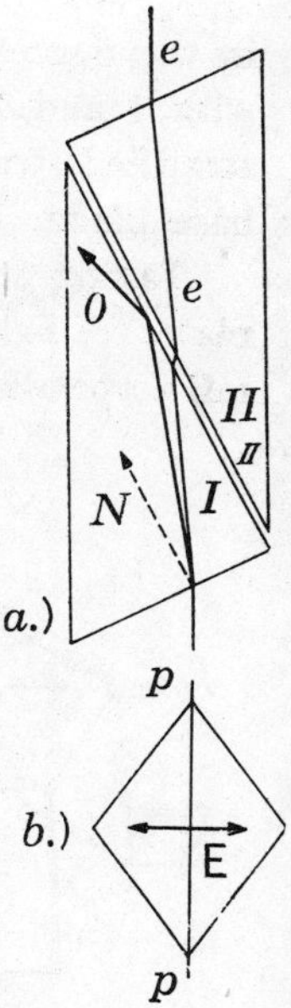

Fig. 42 a, b.
a) Ray paths in the Nicol prism. The section is the same as that drawn in fig. 42. e is the extraordinary ray passing through the prism. o is the totally reflected ordinary ray.
b) The Nicol prism viewed end on. The position of the plane of polarization is indicated by pp. The direction of oscillation is shown by **E**.

the longitudinal edges of the prism. Its plane of polarization is that of the extraordinary ray in calcite, namely, see Sec. 28, parallel to the optic axis. This plane of polarization is indicated by the longer diagonal of the end surface of the prism as shown in fig. 42 b.

Therefore the Nicol prism produces *linearly polarized light whose direction of oscillation is known.* Because the light oscillating perpendicularly to that direction is suppressed by total reflection, the polarization is *complete.*

When two Nicol prisms which can be rotated about their longitudinal axes are placed one behind the other in the path of a light ray, then the first prism is called a *polarizer* and the second an *analyzer.* If the analyzer is oriented perpendicular to the polarizer and there is no birefringent or optically active material between the two, then no light leaves the analyzer. If the analyzer is now rotated, a gradually increasing amount of light passes through it which attains its maximum intensity when the analyzer is oriented parallel to the polarizer. In Sec. 31 we shall discuss the interesting intensity and color patterns which result if a double refracting crystal plate is placed between the polarizer and the analyzer and if the incident light is parallel. We shall also treat the still more interesting patterns produced by *converging* light.

B. The Quarter-Wave Plate and the Babinet Compensator

Mica (alkali-aluminum silicate) is a monoclinic crystal with an extraordinarily pronounced cleavage parallel to the base plane. For optical purposes the transparent potassium mica, $KH_2Al_3\ (SiO_4)_3$, called *muscovite* is of particular interest. The twofold crystallographic symmetry axis is identical with our dielectric principal axis 2; the base plane is identical with the principal axis plane 12, see fig. 43. The plane perpendicular to 2 is the crystallographic symmetry plane. This plane contains the dielectric principal axis 3 and the two optic axes as well as the crystallographic[1] axis 3′ (which is drawn dotted

[1] The angle between the two crystallographic axes 1′ and 3′ is $\beta = 95°5'$ and hence differs little from $\pi/2$. For this reason mica used to be thought of as orthorhombic or hexagonal because of the frequently occuring hexagonal shape of the base plane which is shown in fig. 43.

in the figure). The other two crystallographic axes 1′ and 2′ are identical with 1 and 2, respectively. The structural model[2] of mica illustrates the stratified structure of the crystal and its prominent cleavage parallel to the base plane.

Taking advantage of this cleavage property we make a very thin mica plate. A light ray which has been polarized by a Nicol prism is allowed to fall perpendicularly, that is in the direction 3, on this plate. The Nicol prism is oriented so that the trace of the polarization plane on the mica surface bisects the angle between the axes 1 and 2 (drawn as a dot-dash line in the figure). For convenience the directions of these axes can be marked on the frame holding the mica plate. In the crystal the ray is decomposed into two linearly polarized waves of equal amplitude and direction of propagation (neither wave is refracted, see Sec. 27 B), which oscillate in the directions 2 and 1, respectively. These two waves propagate with the principal light velocities u_1 and u_2. In the yellow part of the spectrum (D-line) the corresponding indices of refraction are

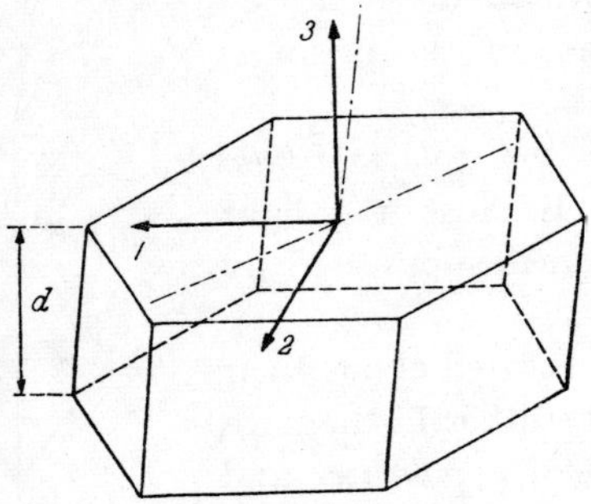

Fig. 43.
Crystallographic model of mica.

$$n_1 = 1.5941, \qquad n_2 = 1.5997. \tag{3}$$

Since one wave propagates faster than the other, a phase difference develops between them which at a depth x amounts to

$$(k_2 - k_1)\, x = k\,(n_2 - n_1)\, x = \frac{2\pi}{\lambda}(n_2 - n_1)\, x; \tag{4}$$

k and λ are the wave number and wavelength in air. At the rear surface of the plate, $x = d$, the two waves emerge again without refraction, so that they still propagate in the same direction and are polarized in mutually perpendicular directions. While originally their phases were the same, they now differ. Combination of the two waves yields *elliptically polarized light.*

If we make the phase difference (4) for $x = d$ equal to $\pi/2$, that is, if we set

$$\frac{\pi}{2} = \frac{2\pi}{\lambda}(n_2 - n_1)\, d; \qquad d = \frac{\lambda/4}{n_2 - n_1}, \tag{5}$$

[2] This was first designed by Lawrence Bragg during an extended visit in Munich and was constructed by Karl Selmayr, the skillful mechanic of the Institute of Theoretical Physics.

then we obtain circular polarization. From (3) and (5) we find for $\lambda = 5.9 \times 10^{-4}$ mm.

$$d = \frac{5.9 \times 10^{-4}}{4 \times 0.0056} \text{ mm.} \sim 0.026 \text{ mm.} \tag{5 a}$$

The name "quarter-wave plate" is somewhat misleading (especially for examination candidates!): its thickness is not $\lambda/4$, but fortunately it is larger than $\lambda/4$ by a factor of one over the small quantity $n_2 - n_1$. We could, of course, use any odd multiple of $\pi/2$ instead of $\pi/2$ in (5). Then d would be three times, five times, . . . as large as the thickness given by (5 a). However, because of their stronger dispersion, these thicker plates are optically less advantageous than the actual quarter-wave plate.

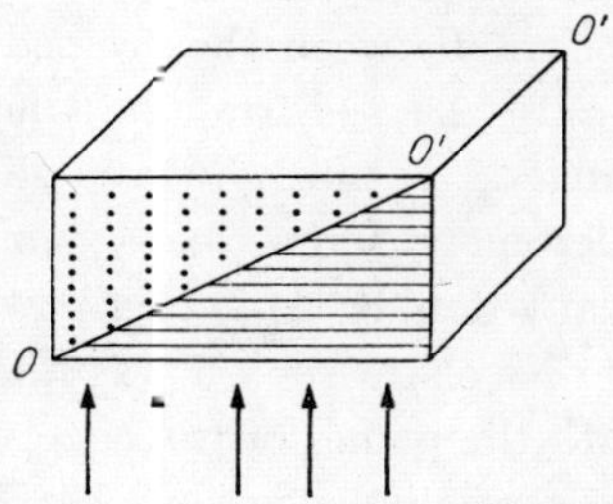

Fig. 44.
Babinet Compensator

The same is true for every other birefringent crystal. Mica is used only because of its extremely pronounced cleavage. Formula (4) for the phase difference can be applied to any arbitrary thickness. If we cut a crystal such as quartz in the shape of a wedge and view the light passing through the varying thickness of the wedge, we observe a whole range of phase differences emerging from the crystal. Since we are here interested in double refraction and not in rotation of the plane of polarization, the light must pass through the quartz wedge in a direction *perpendicular to the optic axis.* The horizontal hatched lines in the lower part of the front surface in fig. 44 indicate the position of the optic axis. The latter is therefore perpendicular to the edge of the wedge which passes through O.

We now cut a second quartz wedge which is outwardly congruent to the first one, but in which the optic axis is *parallel* to the edge $O'\,O'$ of the wedge. We place the two wedges against each other so that together they form a plane parallel plate. The position of the optic axis in the upper wedge is indicated by the dots in fig. 44. Let us consider a point on the crystal where the upper wedge has a thickness x_2 and the lower wedge a thickness x_1. Then by (4) phase differences for a ray passing through the two wedges at that point are

$$\frac{2\pi}{\lambda}(n_e - n_o)\, x_1 \quad \text{and} \quad -\frac{2\pi}{\lambda}(n_e - n_o)\, x_2.$$

We have written here n_e and n_o (since quartz is uniaxial) instead of n_1 and n_2 (as in the case of the biaxial mica). The negative sign is due to the reversed positions of the two wedges. The total phase difference is therefore

$$\Delta = \frac{2\pi}{\lambda}(n_e - n_o)(x_1 - x_2). \tag{6}$$

This combination of quartz wedges when placed between crossed Nicol prisms is called a *Babinet Compensator.* (In the figure the polarizer will be below the plate and the analyzer above it. The polarization planes of the Nicols must be placed at an angle of 45° with respect to the wedge edges.) If such a compensator is illuminated with monochromatic light, then for $x_2 = x_1$ a dark fringe appears because there $\Delta = 0$. Thus in the center of the plate we observe complete extinction just as though no birefringent medium were present between the crossed Nicols. The same is true for the points where $\Delta = \pm 2\pi, \pm 4\pi, \ldots$ One obtains therefore a system of equidistant dark fringes. If one wedge is shifted with respect to the other by means of a micrometer screw, the system of fringes moves also. The same effect is achieved if any doubly refracting plate is inserted between the wedge plate and one of the Nicols in such a way that the new optic axis also makes an angle of 45° with the planes of the Nicols. The resulting shifted fringes can be returned to their original positions by displacing the wedges with respect to each other (hence the name compensator). From this shift, which is read on the micrometer, the amount of double refraction of the inserted sample (its $n_e - n_o$ or $n_1 - n_2$) can be determined by means of (6).

If the compensator is illuminated with white light, it again produces a dark center fringe at $x_2 = x_1$. To the right and left of this line the Newtonian colors of thin plates appear.

We need not go into the many existing modifications of this apparatus.

C. Tourmaline and the Polarization Filter

Tourmalines are boron silicates of various chemical compositions. Their crystalline structure belongs to a class of the hexagonal system without symmetry center and with a "polar principal axis". The latter is the reason for the pyroelectric property[1] of tourmaline. A plate made from suitable material and cut parallel to the principal axis has a transparent green appearance, while a plate cut perpendicularly to the principal axis looks

[1] See Vol. III, 11 E. The permanent electric moment of tourmaline, which is ordinarily compensated by a surface charge, becomes apparent if the temperature is changed.

almost black. This property of tourmaline is called *dichroism.* It is a special form of *pleochroism* (multicoloredness).

Thus the absorptivity of tourmaline depends on direction, and this dependence must, of course, conform to the symmetry of the crystal structure. This is true for all absorbing crystals. In tourmaline the ordinary ray is almost completely absorbed, while the extraordinary ray is absorbed only weakly. The light emerging from a plate which is cut parallel to the principal axis consists almost entirely of the extraordinary ray. Therefore this light is almost completely linearly polarized. The long familiar instrument known as the "tourmaline tongs" is based on this fact.

Modern commercial "*polarization filters*" are made from impregnated plastic materials which are subjected to a strong tension. The absorbing pigment is thereby given an anisotropic arrangement which causes *complete polarization* of any light passing through the material. The same effect can be obtained with strongly dichroic dyes (methylene blue) which are crystallized on glass in thin layers like "frost flowers".

In Sec. 6 we described the absorption in isotropic metals by adding to the dielectric constant the conduction term $i\,\sigma/\omega$. In this way we obtained the complex dielectric constant of (6.1). For a crystal we are led to the complex tensor

$$\varepsilon'_{jh} = \varepsilon_{jh} + i\,\frac{\sigma_{jh}}{\omega} \tag{7}$$

which consists of the dielectric and the conductivity tensors.

The nature of the imaginary part of (7) clearly differs from that in (29.2). The tensor γ_{jh} of Sec. 29 was non-dissipative and therefore had to be antisymmetric. Our present imaginary tensor, on the other hand, is dissipative, as in the case of metallic reflection, and it can therefore be assumed to be symmetric. Its principal axes need not agree with those of the ε_{jh} tensor. However, the symmetry rule which we used in Sec. 29 still holds: if the principal axes of one of the tensors are completely determined by the crystallographic structure, then the axes of the other tensor must also be completely determined, and the two principal axes systems must be identical. By this rule which applies to tourmaline because of its hexagonal structure, the calculations of Sec. 24 ff. can be formally extended without change to the absorbing crystals. This leads to complex principal dielectric constants and therefore the principal light velocities defined in (24.6 b) also become complex. The wave velocities corresponding to a given direction of the wave number vector are again defined by the quadratic equation (25.10). Since the roots u', u'' of this equation are now complex, the components of the corresponding electric vectors D', D'' are also complex. This means that D', D'' now describe not linearly, but *elliptically* polarized oscillations. Thus it is seen that the quantitative theory of anisotropic absorption requires, at least for sufficiently symmetric crystals, no essentially new mathematical development. Since absorption, like double refraction, depends in general on the wavelength, we have thus obtained a general scheme for the explanation of the *pleochroism* of crystals.

31. Interference Phenomena Due to Crystal Plates in Parallel and Converging Polarized Light

Consider a thin crystal plate of the type customarily used in petrography which is placed between two, usually crossed, Nicol prisms. We shall assume the light to be monochromatic except where we expressly state it to be white. For observations with parallel light the rays shall fall perpendicularly on the plate. By "converging light" we mean an arrangement of converging lenses in front of and behind the plate (see below) which enables us to observe simultaneously all bundles of parallel rays which pass through the plate in arbitrary directions which, however, do not differ very much from the normal to the plate.

There are two "principal directions of oscillation" in the crystal plate. These are the principal axes of the ellipse formed by the intersection of the plate surface (which is also a wave surface) with Fresnel's ellipsoid (or the index ellipsoid). The plate is assumed to be in the "diagonal position" between the Nicols; this means that the two principal directions of oscillation are bisected by the plane of polarization of the polarizer (and also by that of the perpendicularly placed analyzer). In this position the amplitudes of the two components of the *incident* light along the principal directions are equal, and so are their phases, since they arise from one linearly polarized oscillation in the polarizer. Since we assume the crystal to be transparent (not dichroic), the amplitudes of these components of the light *emerging* from the plate are also equal. But, as we shall presently show, their *phases differ*. Therefore when the two components are recombined, the resultant emerging intensity differs from the incident intensity. The value of the former varies between maximum brightness and complete darkness as the crystal is rotated out of its diagonal position. We can neglect the small intensity changes which take place when the light enters and emerges from the crystal.

A. Parallel light

The wave velocities u', u'' of the two principal oscillations H_1, H_2 (fig. 45) determine the two indices of refraction $n_1 = c/u'$, $n_2 = c/u''$. These in turn determine the phase difference which arises between the two component waves as the light passes through a crystal plate of thickness d. As in (30.4) this phase difference is given by

$$(1) \qquad \Delta = \frac{2\pi}{\lambda}(n_2 - n_1)\, d.$$

If we call a the amplitude of the wave incident from the polarizer (whose polarization plane is indicated by PP in the figure), then the initial amplitudes of the principal components of this wave are $a/\sqrt{2}$ if the crystal plate is in the diagonal position. After passage through the plate, these oscillations are represented by

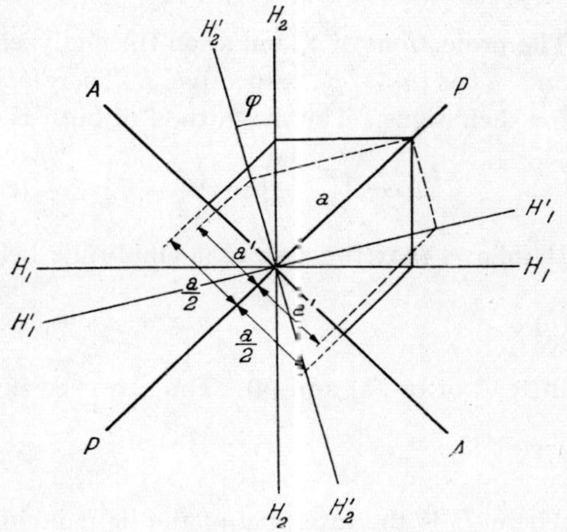

Fig. 45.

Crystal plate in normally incident parallel light for different positions H_1, H_2 and H_1', H_2' of the principal oscillation directions. The plate is between crossed Nicols whose oscillation directions are PP and AA, respectively.

$$\frac{a}{\sqrt{2}}\begin{Bmatrix} e^{\frac{2\pi i}{\lambda} n_1 d} \\ e^{\frac{2\pi i}{\lambda} n_2 d} \end{Bmatrix} e^{-i\omega t}$$

which we shall write in the form

$$(2) \qquad \frac{a}{\sqrt{2}}\begin{Bmatrix} 1 \\ e^{i\Delta} \end{Bmatrix} \exp\left(\frac{2\pi i}{\lambda} n_1 d - i\omega t\right).$$

We now project this oscillation onto the polarization plane of the analyzer (AA in fig. 45). The signs of these projections are determined from the solid lines in the figure, and the amplitude of the resultant oscillation behind the analyzer is found to be

$$(3) \qquad \frac{a}{2}\,|1 - e^{i\Delta}|.$$

Now we use

$$|1 - e^{i\Delta}|^2 = (1 - e^{i\Delta})(1 - e^{-i\Delta}) = 2 - 2\cos\Delta = 4\sin^2\frac{\Delta}{2}$$

which gives for (3) simply

$$(4) \qquad a \sin\frac{\Delta}{2}.$$

If the plate is rotated into the position H_1', H_2' making an angle φ with the diagonal position (see the dotted lines in fig. 45), then the initial amplitudes are, instead of $a/\sqrt{2}$

$$a_1 = a \cos\left(\frac{\pi}{4} - \varphi\right), \qquad a_2 = a \cos\left(\frac{\pi}{4} + \varphi\right). \tag{5}$$

The projections of a_1 and a_2 on the analyzer plane are given by $a \cos(\pi/4 + \varphi)$ and $a \cos(\pi/4 - \varphi)$, respectively. Therefore both projections are equal except for their signs. The magnitude of both is

$$a \cos\left(\frac{\pi}{4} - \varphi\right) \cos\left(\frac{\pi}{4} + \varphi\right) = \frac{a}{2}(\cos^2\varphi - \sin^2\varphi) = \frac{a}{2}\cos 2\varphi.$$

It follows that the resultant amplitude behind the analyzer is given by

$$a' = a\,|\cos 2\varphi|\,\sin\frac{\Delta}{2} \tag{6}$$

instead of by (3) and (4). The observed intensity is therefore

$$J = J_0 \cos^2 2\varphi \sin^2\frac{\Delta}{2} \tag{7}$$

where J_0 is the intensity of the light incident on the plate.

According to (7), for one complete revolution of the plate the intensity observed behind the analyzer changes four times between maximum brightness at the diagonal positions

$$\varphi = 0, \quad \frac{\pi}{2}, \quad \pi, \quad \frac{3\pi}{2} \tag{8}$$

and complete darkness whenever H_1', H_2' coincide with P or A, that is, when

$$\varphi = \frac{\pi}{4}, \quad \frac{3\pi}{4}, \quad \frac{5\pi}{4}, \quad \frac{7\pi}{4}. \tag{8 a}$$

If the illumination is monochromatic, the plate appears of varying but uniform brightness in its entirety.

If white light is used, the positions (8 a) again yield darkness. In the intermediate positions the entire plate is uniformly colored with a *mixed color*.

Only for very thin or very thick plates does the color remain white. For very thin plates this is true because there is no wavelength for which $\Delta/2$ attains the value π. For very thick plates there are very many points distributed over the whole spectrum for which $\Delta/2$ is a multiple of π. In this case the spectrum (not to be confused with the appearance of the plate!) has a large number of *dark lines* but retains its white character. For moderately thin or moderately thick plates there are only *one* or *a few such dark lines*. The missing wavelengths and the intensity variations in the remaining portion of the spectrum cause the deviation of the color from white and determine the character of the mixed color which is seen by the eye. If the observation is made with parallel instead of crossed Nicols, precisely the complementary mixed color is seen, and full illumination takes the place of darkness at the positions (8 a).

The pattern becomes much more interesting if, instead of a single crystal, a mosaic of crystal fragments is used; for instance, granite which consists of feldspar, quartz, mica, hornblende, etc. In that case each constituent yields under white illumination a different color, which is determined by the material and its orientation with respect to the surface of the plate. The principal directions H_1, H_2 belonging to the individual crystal fragments are distributed at random in the plate. Therefore, when a thin plate of this type is rotated, its various components extinguish the light at different angular positions. By the same token the various pieces exhibit different intensities at different positions of the plate. Petrographic investigations depend to a large extent upon such observations.

We shall not investigate the appearance of the plate when the polarizer and analyzer are in an intermediate position, i. e. neither crossed nor parallel.

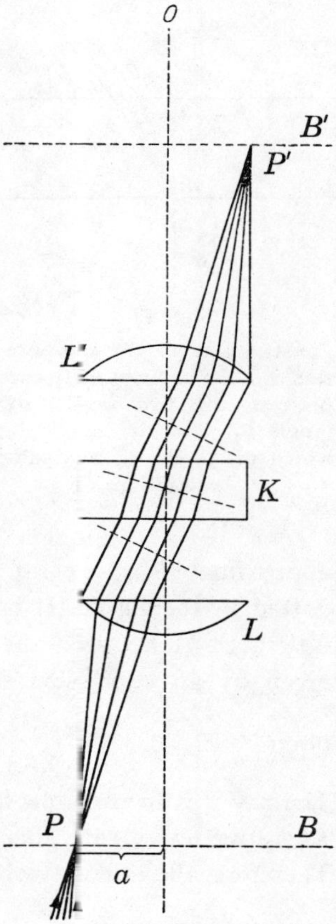

Fig. 46.
Crystal plate illuminated by "converging" light, that is by parallel light bundles which form a finite solid angle around the normal to the plate.

B. Converging Light

As was stated at the beginning of this section we are in fact again considering bundles of parallel rays, which now, however, in passing through the plate, assume all possible directions in the neighborhood of the normal to the plate; they are then *simultaneously* focused at the eye (at O) by means of the converging lens L' in fig. 46, B' being the focal plane of L' on which the eye is focused (with the help of a lens or microscope). B is the focal plane of the converging lens L. The light source is an *extended* luminous surface which is placed below B. We need follow the rays originating from that surface only after they have passed through B. L converts the originally divergent rays, e. g. the rays emerging from P, into parallel light. L' converts the parallel rays emerging from the crystal plate K into converging light, e. g. into light focused at P'. The polarizer is below B and the analyzer is between B' and O.

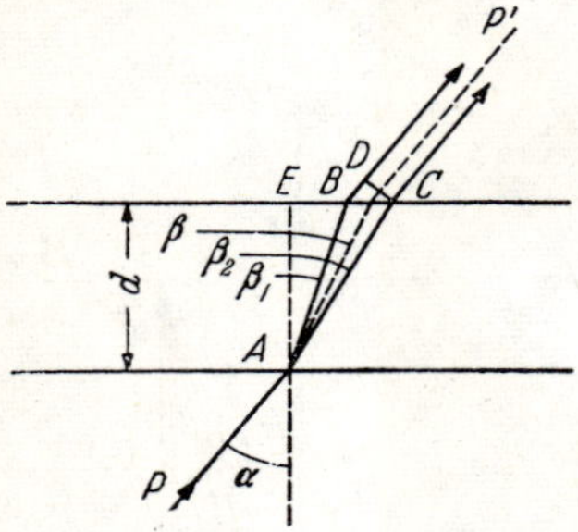

Fig. 47.

Construction for the computation of the phase difference between the two rays ABD (angle β_1) and AC (angle β_2) which are produced by double refraction.

The rays drawn in fig. 46 which are parallel inside K cannot interfere because they originate from different points on the luminous surface. Therefore, the *intensities* of these rays add. The phenomena at P' are made quantitatively observable by the fact that we have parallel ray bundles of considerable width, as is indicated in fig. 46.

Interference takes place only between any two rays which are created by double refraction in K and originate from the same ray coming from P. Each such pair of interfering rays is parallel as it leaves the plate K. Figure 47 indicates that the directions of the interfering ray pairs which have angles of refraction β_1, β_2 in the plate can (for β_1, β_2 not too large) be approximated by one ray with the average direction β. This ray is drawn dotted in the figure. It is easily verified, see exercise IV. 3, that the phase difference Δ between the two waves which propagate in this direction is given by an expression similar to eq. (1):

$$\Delta = \frac{2\pi}{\lambda} \frac{n_2 - n_1}{\cos\beta} d. \tag{9}$$

The ray β is focused on the same point P' in the focal plane B' as the rays β_1, β_2 (see figure), and this point is characterized by the value of Δ given by (9). Therefore, the intensity observed from O at the point P' is, in analogy to (7),

$$J = J_0 \cos^2 2\varphi \sin^2 \frac{\Delta}{2} \tag{10}$$

where J_0 is the intensity incident at A and the angle φ depends on the orientations of the Nicols with respect to the principal oscillation directions of the plate and also on the ray direction β.

According to (10) we have

$$J = 0, \text{ extinction, if } \Delta/2 = g\pi \ (g = \text{integer}). \tag{11}$$

By (9) this condition means that

$$\cos\beta = \frac{d}{\lambda g}(n_2 - n_1). \tag{12}$$

Fig. 48.

Calcite plate, cut perpendicularly to the optic axis, in sodium light between crossed Nicols.

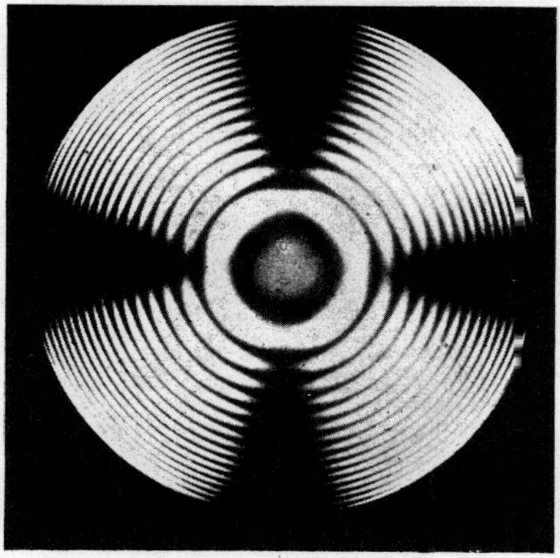

Fig. 49.

Quartz plate, cut perpendicularly to the optic axis, in sodium light between crossed Nicols. Note the bright center.

Fig. 50.

Calcite plate, cut parallel to the optic axis, in white light. Diagonal position.

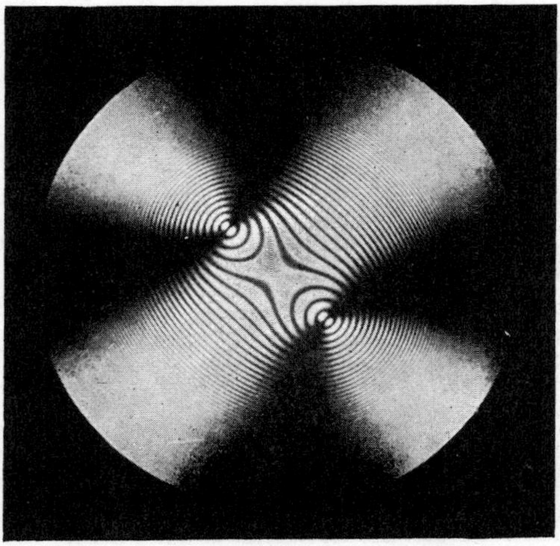

Fig. 51.

Cerussite, biaxial, cut perpendicularly to the bisector of the angle between the optic axes. Diagonal position.

Now

$$n_1 = \frac{c}{u'}, \qquad n_2 = \frac{c}{u''}$$

where u', u'' are the two wave velocities belonging to the direction β. Therefore, we have also

$$\cos\beta = \frac{d}{\lambda g}\left(\frac{c}{u''} - \frac{c}{u'}\right) = \frac{c\,d}{\lambda g}\frac{u' - u''}{u'\,u''}. \tag{13}$$

This formula suggests that we find those directions $\beta = \beta_g$ on the Fresnel wave surface for which the above equation is satisfied. In the uniaxial case in which one branch of the wave surface is the sphere $u = u_0$ and the other the ovaloid of rotation about the optic axis (28.5), this becomes a relatively simple algebraic calculation which, however, we shall not carry out here.

Rather, we shall immediately turn to the completely symmetric situation:

crystal plate $\perp$ optic axis.

In this case the angle of refraction lies in a meridian plane of the ovaloid, and every cone $\beta =$ constant intersects the ovaloid in a *circle*. Because of the condition (13) one obtains, therefore, in the focal plane B' a family of concentric circles

$$\beta = \beta_1,\ \beta_2, \ldots \beta_g, \ldots \tag{14}$$

at which the intensity vanishes. By (13) their radii depend on the ratio d/λ. The difference between successive radii decreases steadily with increasing g.

But according to eq. (10), we have extinction not only for $\sin \Delta/2 = 0$ which led to (14) but also for $\cos 2\varphi = 0$. The latter condition indicates extinction along the two mutually perpendicular directions

$$\varphi = \pm\frac{\pi}{4}. \tag{15}$$

The figures 48 to 51 are reproduced from the famous collection of photographic plates "Interferenzerscheinungen im polarisierten Licht" by H. Hauswaldt, Magdeburg, 1902 and 1904. Figure 48 is obtained with calcite (1/2 mm thick) in sodium light between crossed Nicols. The system of concentric circles represents the β-values (14). The dark cross which coincides with the polarization planes of the Nicols represents the φ-values (15). With white light illumination the pattern is colored and fewer interference circles are discernible. The curves $\Delta =$ constant are called *isochromatics* since each of them is characterized by its own mixed color.

Figure 49 is obtained with a quartz plate (7 mm thick because the double refraction of quartz is less than that of calcite) which is photographed in the same manner. The pattern differs from that in fig. 48 in that the center is bright. This indicates the rotation of the plane of polarization for rays parallel to the optic axis of quartz. The center bright spot illustrates the gap between the two branches of the ray surface shown in fig. 40.

Figure 50 shows calcite which is cut parallel to the optic axis and is placed diagonally between the Nicols. The black cross is missing. The isochromatics are rectangular hyperbolas. It can be proved generally that for arbitrarily cut slabs of uniaxial crystals the circles in fig. 48 become conic sections. Indeed, these conic sections are formed by the intersections of the circular cones $\beta =$ constant with the plate surface.

Figure 51 is obtained with a biaxial orthorhombic crystal, namely cerussite, $PbCO_3$, cut perpendicularly to the bisector of the angle between the two optic axes and placed in diagonal position. The cross of extinction of figures 48 and 49 is here pulled apart so that two branches meet at each of the optic axes. The isochromatics are not conic sections as for uniaxial crystals but are fourth degree curves (lemniscates). It is wonderful to observe with a polarization microscope how nature traces these varied patterns with geometric prescision and colors them so brilliantly.

CHAPTER V

THE THEORY OF DIFFRACTION

Any deviation of light rays from rectilinear paths which cannot be interpreted as reflection or refraction is called diffraction. Reflection and refraction, clearly, occur only when the bodies causing the deviations of the rays from straight lines have surfaces whose radii of curvature are everywhere large compared to the wavelength of the light.

The phenomenon of the shadow, which seemed to pose difficulties for the elementary wave theory, is explainable only by the theory of diffraction. According to the latter, the border of a shadow is shown to be diffuse and composed of diffraction bands. The conflict between geometrical and wave optics is resolved by the theory of diffraction.

Geometrical optics is the limiting case of wave optics as $\lambda \to 0$. In this limiting case there is no diffraction. Hence, in contrast to ordinary refraction, rays of greater wavelength are diffracted more strongly than rays of shorter wavelength. Diffraction, then, generally deflects the red end of the spectrum more strongly away from the geometrical direction of the rays than the violet end, which is just the opposite of prismatic refraction. The *coronae* around the sun and the moon are *diffraction phenomena* which are caused by water droplets randomly distributed in a layer of haze and are especially strong when the droplets are of approximately uniform size. Their outer rims are colored red. The *halos* around the sun and the moon, on the other hand, are caused by *refraction* in the ice crystals of thin cirrus clouds. The sequence of colors, when visible, is the opposite in this phenomenon: red inside and violet outside. It is well known that Descartes had already explained the principal features of the *rainbow* by refraction and reflection in raindrops. The complete treatment of the rainbow, however, involves also a difficult diffraction problem.

Because of their low intensities and small dimensions, diffraction phenomena are, in daily life, generally not noticeable to the naked eye, but there are exceptions. If we view a distant light source through a fine fabric (an opened umbrella, for instance) we see the beautiful colored figures of a Fraunhofer cross grating. When we squint with nearly closed eyes at a distant candle, the eyelashes act like a (very distorted) line grating and decompose the candle light into its natural spectrum.

In Secs. 32 and 33 we shall treat those phenomena in which the difficulty of low intensity is overcome by the use of a large number of diffracting elements. Such devices are: the regular gratings in which the *amplitudes* of the oscillations add by interference, and random distributions of diffraction centers in which case the *intensities* add. In these devices the diffraction of single grating elements or particles plays as yet no essential rôle. This latter problem will be studied in various degrees of approximation in Sec. 34 ff. For the visual observation or photographic recording of this diffraction of single elements, a telescope or a lens is required.

While up to this point it has been sufficient to operate with a *plane* wave, the *spherical* wave will now come into its own. Only in the two following paragraphs, and later on in the treatment of Fraunhofer diffraction, shall we continue to use plane waves. The classical theory of diffraction, which is based upon *Huygens' principle*, operates essentially with the scalar spherical wave.

32. Theory of Gratings

A. Line gratings

The first diffraction gratings, made by Fraunhofer, consisted of parallel, stretched thin wires. Later, he used a glass plate covered with lampblack which he scored with a ruling engine in such a way that a system of equidistant transparent lines appeared on the glass. Fraunhofer's original gratings are preserved in the "Deutsches Museum" in Munich.

Famous, and hardly excelled even today, are *Rowland's reflection gratings*. They consist of up to 1800 ruled lines per millimeter on a metallic mirror surface; altogether some 100,000 lines. Faultless uniformity of the spacing[1] of the lines over the whole length of the grating is important.

We take the y-axis along the direction of the grating lines and assume that the lines are spaced at intervals of length d measured along the x-axis. Let the total number of lines be N. Let the plane of incidence be the xz-plane; and let $z = 0$ be the plane of the grating. Let the incident light be white and its rays be made parallel by an ideal collimator. The vectorial wave numbers $\mathbf{k}$ of the monochromatic components of the white light have the direction cosine α_0 with the positive x-axis. (α_0 is now not the cosine of the angle of incidence, but rather the cosine of the so-called "glancing angle" which is the complement of the angle of incidence.) We now assume a cylindrical

[1] Periodically repeated irregularities of the ruling engine produce "ghosts", i. e. false lines in the diffraction spectrum.

wave sent out from each grating line. This process we call, as is commonly done, "diffraction", although we could instead use the more general term "scattering". The waves emerging from different grating lines are capable of interference because they originate from the same incident wave.

We now look at fig. 52 and recall the similar figs. 9 and 47 in which we were also concerned with the phase differences between neighboring rays. The ray emerging from O having direction cosine α with the x-axis covers a distance which exceeds that covered by the ray emerging from P by the path difference

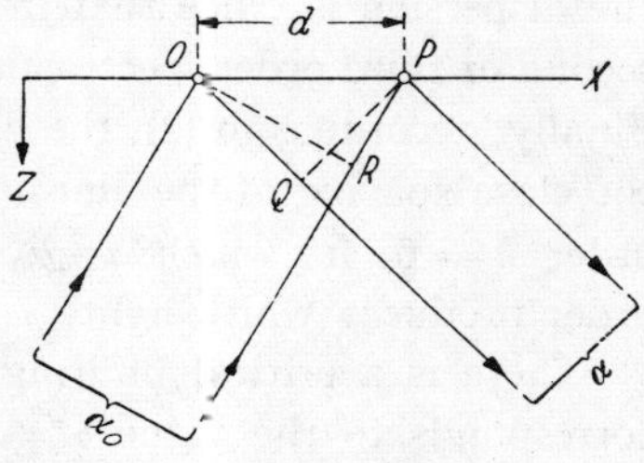

Fig. 52.
Determination of the phase difference in a reflection grating $OP = d =$ grating constant.

$$OQ - RP = \alpha d - \alpha_0 d,$$
$$d = OP = \text{"grating constant"}.$$

The phase difference between these two rays is, therefore,

$$(1) \qquad \Delta = kd\,(\alpha - \alpha_0), \quad k = \frac{2\pi}{\lambda}.$$

This must be an even multiple of π in order that, sufficiently far from the grating and in the direction indicated by the rays, these two waves shall show maximum reinforcement owing to interference. The condition for this is

$$(2) \qquad \alpha - \alpha_0 = h\,\frac{\lambda}{d}, \qquad h = \text{positive or negative integer.}$$

$h = 0$ corresponds to ordinary reflection $\alpha = \alpha_0$; $h = \pm 1$ corresponds to the first order spectrum on the right or left of the regularly reflected light; $h = \pm 2$ corresponds to the second order spectrum, etc.

If we collect the cylindrical waves emerging from all the different grating lines with the same phase difference, we obtain, at a distance large compared to the grating constant, a *plane* wave. In the case given by (2) this wave has its maximum amplitude; its amplitude is zero when Δ is an odd multiple of π. For a given λ the dependence of the amplitude of this wave upon Δ is shown by the intensity curve in fig. 53 which will be calculated below. For its observation one employs, following Fraunhofer, a telescope focused at infinity.

That the grating actually does produce spectra in this way, i. e. that it separates the colors, follows from the fact that, according to (2), $\alpha - \alpha_0$ depends on the wavelength. Hence, the different colors are diffracted in different directions. Since $\alpha - \alpha_0$ increases with λ, red is diffracted more strongly than violet, as mentioned in the introduction to this chapter. The dispersion, that is, the separation of the various colors, is directly proportional to λ. Thus, the grating yields a spectrum of the incident white light, whose scale of wavelengths is normal and quantitatively correct. Furthermore, the dispersion is directly proportional to the order number h. In the second order spectrum the dispersion is twice that in the first order spectrum. For this reason the second or third order spectra are preferred for exact wavelength measurements. Finally, according to (2), the dispersion is inversely proportional to d; hence, the close spacing of the lines of the Rowland gratings. An exception is the order $h = 0$, for which $\alpha - \alpha_0$ is independent of color. The spectrum of zero order produces white light.

There is a critical limiting value, not necessarily integer, $h = h_{cr}$ which corresponds to the value $\alpha = 1$. If h_{cr} happens to be integer the diffracted light ray is parallel to the plane of the grating, just like the reflected wave in the limiting case of total reflection[1]. In any case, even for $h > h_{cr}$ a diffracted wave runs parallel to the plane of the grating, however, not as a regular wave, but as an inhomogeneous wave. This is also analogous to total reflection.

We now wish to show that the grating produces practically pure spectral colors. In order to do this, we must estimate the width of the maximum computed in (2). We do this by considering monochromatic light with a given λ instead of the white light employed until now.

We define the radiation emitted by an arbitrary grating groove in the direction α by means of an amplitude factor $f(\alpha)$, which is the same for all grooves and can be considered as a function which varies slowly as α varies between the limits $\alpha = \pm 1$ (positive and negative x-axis). Consecutive grating grooves shall have the abscissas

$$x_0, \ldots x_n, \ldots x_{N-1}, \qquad x_n = x_0 + n\, d.$$

Omitting the time factor, we write for the oscillations emitted by the nth groove in the direction α the expression

$$u_n = f(\alpha) \exp\{i\, k\, (\alpha\, x + \gamma\, z) + i\, n\, \Delta\} \tag{3}$$

regardless of whether we are concerned with the vectors **E**, **D**, or **H**. For the significance of $f(\alpha)$ and a systematic development of (3), we refer the reader to Sec. 36. Δ is the phase difference defined in (1). Superposition of the effects of all the lines gives

$$u = \sum u_n = f(\alpha)\, S \exp\{i\, k\, (\alpha\, x + \gamma\, z)\} \tag{4}$$

[1] Lord Rayleigh, Phil. Mag. **14**, 60, 1907 and Proc. Roy. Soc. **79**, 399, 1907; W. Voigt, Göttinger Nachr. **40** (1911); also U. Fano, Ann. d. Phys. **32**, 393, 1938, Phys. Rev. **38**, 921, 1948.

with

(4 a)
$$S = \sum_{n=0}^{N-1} e^{in\Delta} = \frac{1-e^{iN\Delta}}{1-e^{i\Delta}} = \frac{e^{iN\frac{\Delta}{2}}}{e^{i\frac{\Delta}{2}}} \frac{\sin N\frac{\Delta}{2}}{\sin\frac{\Delta}{2}}.$$

Changing from the amplitude to the intensity, we get

(5)
$$J = |u|^2 = f^2(\alpha)\,|S|^2 = f^2(\alpha)\,\frac{\sin^2 N\frac{\Delta}{2}}{\sin^2\frac{\Delta}{2}}.$$

This expression is composed of two parts. The first factor corresponds to the intensity arising from a single groove and varies slowly with α. The second factor arises from the sequence of many grooves, and is a very rapidly varying function of $\alpha - \alpha_0$.

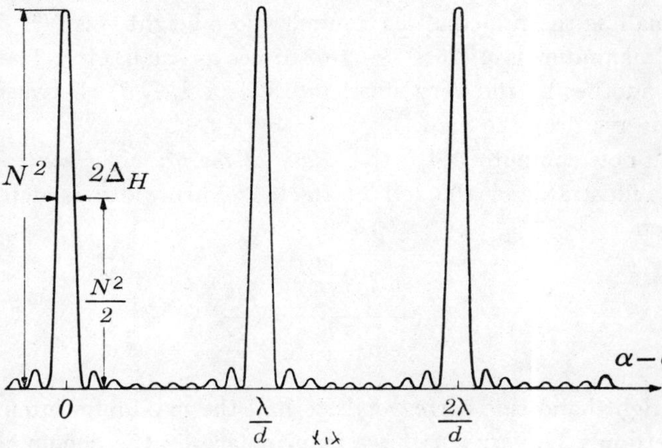

Fig. 53.

The function $\dfrac{\sin^2 N\Delta/2}{\sin^2 \Delta/2}$ plotted against $\alpha - \alpha_0 = \dfrac{\lambda\Delta}{2\pi d}$.

The incident primary intensity is contained in the first factor of (5). Figure 53 shows the dependence of the second factor on $\alpha - \alpha_0$. Its *principal maxima* lie, in accordance with (1) and (2), at

(6)
$$\frac{\Delta}{2} = \pi h.$$

These maxima are given by (5) as 0/0. The value in the limit, as computed in the well-known manner (de l'Hôpital's rule), is the same for all h, namely N^2.

In addition, there are *subsidiary maxima* corresponding to the rapid fluctuations of the numerator. Since the denominator varies slowly, the positions of these maxima are given sufficiently accurately by the maxima of the numerator. Thus, in the vicinity of the hth principal maximum, the positions of the subsidiary maxima are given by

$$\frac{\Delta}{2} = \pi\left(h + \frac{\nu}{2N}\right), \qquad \nu = (1), 3, 5. \tag{6 a}$$

The value $\nu = 1$ is put in parentheses because the corresponding peak is masked by the flank of the principal maximum. The heights of the subsidiary maxima are

$$\frac{1}{\sin^2 \dfrac{\nu\pi}{2N}} \sim \frac{4N^2}{\pi^2 \nu^2}.$$

Hence, the first subsidiary maximum to be taken into account is $4/(9\pi^2) \approx 1/22$ times as small as the principal maximum, whose height was N^2. The second subsidiary maximum is $4/(25\pi^2) \approx 1/62$ times as small, etc. These maxima follow one another at the very small interval of $\lambda/(N d)$; between them the intensity always drops to zero.

We shall now compute $2\Delta_H$, the *width of the principal maximum at half intensity*, as illustrated on the left of fig. 53. This width is determined by the equation

$$\frac{\sin^2 N \dfrac{\Delta_H}{2}}{\sin^2 \dfrac{\Delta_H}{2}} = \frac{N^2}{2} \tag{6 b}$$

where the right-hand side corresponds to half the maximum intensity. Since Δ_H must certainly be very small, we may replace, in the denominator on the left-hand side, the sine function by its argument. Thereby, we get

$$\sin^2 x = \frac{x^2}{2}, \qquad \text{where} \qquad x = N\frac{\Delta_H}{2}. \tag{6 c}$$

The solution of the equation $\sin x = x/\sqrt{2}$ is found from the table of sines to be

$$x \sim 80° = 1.38, \qquad \text{hence} \qquad \Delta_H = \frac{2 \times 1.38}{N}.$$

Since N is a very large number the width at half intensity which is twice this Δ_H, namely $5.5/N$, is extremely small. It follows from this that the principal maxima belonging to distinct colors of our spectrum fall next to one another, so that no appreciable overlapping of colors occurs.

That does not, of course, exclude the possibility that the ends of spectra of *different* orders may overlap. Since the dispersion increases with h, mixed colors can be produced in this way. In fact, one sees immediately that the red end of the second order spectrum, for instance, will overlap the violet end of the third order spectrum since

$$2\,\lambda_{\text{red}} > 3\,\lambda_{\text{violet}} \qquad \text{because} \qquad \lambda_{\text{red}} \sim 2\,\lambda_{\text{violet}}.$$

Finally, we must multiply the second part of the intensity expression (5), as illustrated in fig. 53, by the factor f^2. In general this factor will decrease steadily with increasing $|\alpha|$ and will, therefore, attenuate the spectra with larger values of h in comparison with those which have smaller h values. However, this is true only "in general". In each particular case the form of f depends entirely on the shape of the groove (i. e. on the shape of the diamond of the ruling machine). Nor need f be an even function of α. The spectra with $h > 0$ can, for instance, be enhanced over those with $h < 0$. It can even happen that most of the intensity is thrown into a single spectrum. Under certain circumstances this may be especially desirable. More details are given in Sec. 36 D.

B. Cross gratings

Two systems of grating lines which intersect at right (or oblique) angles are called a cross grating. We have, then, on the plane of the grating a set of dark rectangles (or parallelograms) extending in two directions. Or we could, instead, consider a two-dimensional system of bright rectangles (as in the above-mentioned example of the opened umbrella) or of arbitrarily shaped bright spots (circles for instance) on a dark background. These also will be called cross gratings.

As in the case of the line grating, let the grating plane be the xy-plane. For convenience we will consider our cross grating to be oriented along the x- and y-axes, that is, to be rectangular. Here we must replace the summation over n in eq. (4) by a double sum over n_1 and n_2:

$$S = \sum_{n_1=0}^{N_1-1} \sum_{n_2=0}^{N_2-1} \exp\{i\,n_1 \Delta_1 + i\,n_2 \Delta_2\}, \tag{7}$$

$$\Delta_1 = 2\pi\, d_1 \frac{(\alpha-\alpha_0)}{\lambda},$$

$$\Delta_2 = 2\pi\, d_2 \frac{(\beta-\beta_0)}{\lambda}.$$

Performing the summation and computing the intensity, we obtain instead of (5)

$$J = f^2(\alpha, \beta) \frac{\sin^2 N_1 \frac{\Delta_1}{2}}{\sin^2 \frac{\Delta_1}{2}} \frac{\sin^2 N_2 \frac{\Delta_2}{2}}{\sin^2 \frac{\Delta_2}{2}}. \tag{8}$$

According to eqs. (1) and (2), principal maxima will now occur if

$$\Delta_1 = 2\pi h_1 \text{ and simultaneously } \Delta_2 = 2\pi h_2 \tag{8 a}$$

where h_1 and h_2 are arbitrary positive or negative integers. The directions α, β of the deflected rays corresponding to these principal maxima are given, according to (7), by

$$\alpha - \alpha_0 = h_1 \frac{\lambda}{d_1}, \qquad \beta - \beta_0 = h_2 \frac{\lambda}{d_2}. \tag{9}$$

The corresponding intensity is proportional to $N_1^2 N_2^2$. If only one of the two conditions (8 a) is fulfilled, then the intensity is only proportional to N_1^2, or N_2^2, and is, therefore, imperceptibly small compared to the intensity of the principal maxima. Also the subsidiary maxima, described by (6), can be disregarded when compared to the principal maxima. Since, according to (9), a definite pair of values α, β is associated with every λ, each of the cases characterized by (9) constitutes a complete color spectrum. The spectra extend parallel to the direction of the x-axis when $h_2 = 0$, and in the direction of the y-axis when $h_1 = 0$. In the general case $h_1 \neq 0$ and $h_2 \neq 0$, the spectra are directed radially, i. e. towards the central point α_0, β_0. Only at this latter point is the light not spectrally decomposed but white. Again, owing to the factor $f^2(\alpha, \beta)$ in (8) the outer spectra of this manifold and colorful display are generally strongly attenuated.

C. Space gratings

We ask next how a three-dimensional grating could be produced. Neither a ruling machine nor stacked layers of the sheerest fabric produce an optically useful grating. Max von Laue had the ingenious idea that nature herself offers us an ideal space grating in the form of a flawless, non-absorbing crystal. Though useless in the field of optics, such crystals find application in the much more interesting spectral range of X-rays. In this connection it is to be noted that this range was not even known in 1912 but was determined quantitatively only by means of Laue's discovery. For optical purposes the mesh of a crystal lattice is far too fine, but for the analysis of X-rays its order of magnitude is just right. In fact, the spacing between atoms in a crystal is approximately the same as the wavelength of soft X-rays (several Å units, 1 Å $= 10^{-8}$ cm), just as the spacing of the lines in the Rowland grating agrees approximately with the wavelength of red light ($1/2\,\mu$, $1\,\mu = 10^{-4}$ cm).

In order to make the formulae as clear as possible, we shall restrict ourselves here to the special case of an orthorhombic crystal, but it is to be emphasized that if an oblique coordinate system is used, the general triclinic crystal presents no difficulties. Let the sides of the fundamental orthorhombic cell have lengths a, b and c (this would correspond to d_1, d_2 and d_3 in our previous notation). Rewriting eq. (9) in three dimensions, we obtain immediately *Laue's Fundamental Equations*:

$$\alpha - \alpha_0 = h_1 \frac{\lambda}{a}, \qquad \beta - \beta_0 = h_2 \frac{\lambda}{b}, \qquad \gamma - \gamma_0 = h_3 \frac{\lambda}{c}. \tag{10}$$

The special cases of the tetragonal and cubic systems ($b = a$, $c = b = a$), are, of course, contained in (10).

In Laue's experimental arrangement the rays are made to pass through thin crystal slabs. In particular, for the first photographs by Friedrich and Knipping, made in the spring of 1912, thin slabs of zinc blende, ZnS, sliced perpendicularly to the fourfold or threefold axis of symmetry, were used. The crystal acts here not as a reflection grating, but as a transmission grating. The rays emerging from the crystal produce the surprisingly beautiful "Laue Diagrams" on a photographic plate placed beyond the crystal. The original pictures are preserved in the "Deutsches Museum" in Munich and have been reproduced in countless textbooks.

The intensity is computed by the properly extended formula (8). The number N of the contributing lattice elements is determined by the thickness of the crystal slab and by the cross-section of the incident X-ray beam. The "atomic form factor", much discussed in the theory of crystal analysis, takes the place of $f(\alpha, \beta)$ in eq. (8).

The difference between this theory and the theory of cross gratings arises from the fact that, because of the condition $\alpha^2 + \beta^2 + \gamma^2 = 1$, the three eqs. (10) are not compatible for any arbitrarily given values of λ. While the cross grating produces complete spectra containing all λ, the space grating is *selective*. To every Laue spot corresponds a characteristic λ. (However, for reasons of symmetry several spots may correspond to the same λ; as, for instance, in the fourfold symmetrical picture of zinc blende in which each λ occurs, in general, in eight spots.) The polychromatic character of the cross grating spectra occurs again in the Laue diagram in the sense that every Laue spot selects its own special "color" from the incident "X-ray light".

We confirm this analytically by squaring and adding the values of α, β, γ as determined by (10), and by taking into account the condition $\alpha_0^2 + \beta_0^2 + \gamma_0^2 = 1$. Cancelling a common factor λ, we thus obtain:

$$\lambda = -2 \frac{\alpha_0 \frac{h_1}{a} + \beta_0 \frac{h_2}{b} + \gamma_0 \frac{h_3}{c}}{\frac{h_1^2}{a^2} + \frac{h_2^2}{b^2} + \frac{h_3^2}{c^2}}. \tag{11}$$

Therefore, once the order numbers h_1, h_2, h_3 of the interference are determined for every Laue spot, the wavelength producing each such spot is (for a known crystal lattice) also determined. At the same time we now see that in contrast to Bragg's method, Laue's arrangement uses the continuous X-ray spectrum (the so-called "white X-ray light" or "Bremsstrahlung").

Before discussing Bragg's experiment, we draw yet another conclusion from eq. (10). We form the sum of the squares of the left-hand sides of (10)

$$\begin{aligned}(\alpha - \alpha_0)^2 + (\beta - \beta_0)^2 + (\gamma - \gamma_0)^2 &= 1 - 2(\alpha\alpha_0 + \beta\beta_0 + \gamma\gamma_0) + 1 \\ &= 2 - 2\cos 2\vartheta = 4\sin^2\vartheta.\end{aligned} \tag{12}$$

2ϑ is here the angle between the incident ray $\alpha_0, \beta_0, \gamma_0$ and the diffracted ray α, β, γ (see fig. 54). The plane E bisects the angle 2ϑ between these two rays. Next, we compute the sum of the squares of the right-hand sides of (10), namely,

$$\lambda^2 \left\{ \left(\frac{h_1}{a}\right)^2 + \left(\frac{h_2}{b}\right)^2 + \left(\frac{h_3}{c}\right)^2 \right\} = \frac{\lambda^2}{D^2} \tag{13}$$

where D is a length of the order of magnitude of the sides a, b, c. We can make the definition of D, which is implicit in (13), more precise if we rid the integers h of a possible common factor and write

$$h_1 = n\, h_1^*, \qquad h_2 = n\, h_2^*, \qquad h_3 = n\, h_3^*, \tag{14}$$

$$D = \frac{d}{n}, \qquad d = \left\{ \left(\frac{h_1^*}{a}\right)^2 + \left(\frac{h_2^*}{b}\right)^2 + \left(\frac{h_3^*}{c}\right)^2 \right\}^{-\frac{1}{2}}. \tag{14 a}$$

By equating (12) and (13) and applying (14) and (14 a), there results *Bragg's equation*:

$$2\, d \sin\vartheta = n\,\lambda. \tag{15}$$

We have encountered this equation before as eq. (8.6) in connection with Wiener's standing light waves. There the length d denoted the distance between two neighboring layers of the "screen" which is produced by standing light waves and is utilized in Lippmann's color photography. We must now investigate the significance of d in the case of the crystal lattice.

For this purpose we construct the equation of the plane E in fig. 54. We use a system of rectangular coordinates x, y, z whose axes are parallel to the crystal axes a, b, c and whose origin lies in the plane E at the lattice point O. This is the same point which in fig. 54 was considered as the point of origin of the diffracted ray. From O we mark off the segments OP along the extension of the incoming ray direction and OQ along the direction of the diffracted ray and we let

$$OP = OQ = 1.$$

The coordinates of the points P and Q are then

$$\alpha_0, \beta_0, \gamma_0 \qquad \text{and} \qquad \alpha, \beta, \gamma.$$

respectively. We can now define the plane E as the locus of all points equidistant from P and Q:

$$(x-\alpha_0)^2 + (y-\beta_0)^2 + (z-\gamma_0)^2 = \\ = (x-\alpha)^2 + (y-\beta)^2 + (z-\gamma)^2.$$

This simplifies to

$$(\alpha-\alpha_0)\, x + (\beta-\beta_0)\, y + (\gamma-\gamma_0)\, z = 0.$$

Substituting (10) and (14) into this yields

$$\frac{h_1^*}{a} x + \frac{h_2^*}{b} y + \frac{h_3^*}{c} z = 0. \tag{16}$$

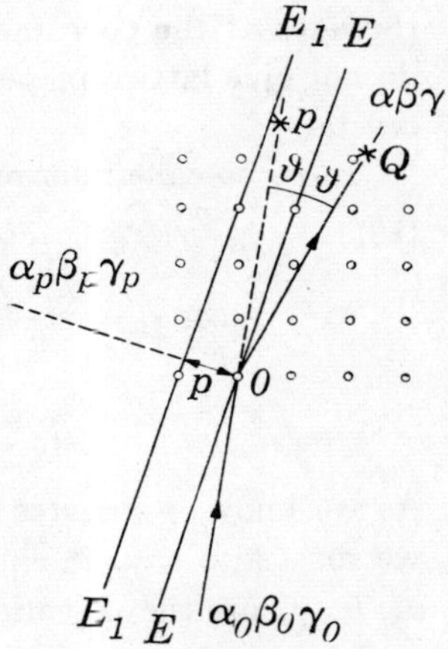

Fig. 54.
Diffraction of X-rays in a space lattice (description in the text).

This plane E is a *lattice plane* of the crystal, which means that in an unbounded crystal the plane E contains an infinite number of lattice points. (If the plane contains three lattice points, then because of the periodicity of the lattice it must contain an infinite number of lattice points.) The numbers h^* are called the *Miller indices* of the lattice plane. (Their magnitudes determine the density of lattice points on the plane; small values of the h^* imply a large density; large values of the h^* imply a small density. Only planes with small h^* values occur as natural boundary planes of a crystal.) A plane parallel to our lattice plane intersects the crystal axes a, b, c at coordinates which have the ratios

$$\frac{a}{h_1^*} : \frac{b}{h_2^*} : \frac{c}{h_3^*} . \tag{17}$$

This was the original definition of the Miller indices as used in macroscopic crystallography where one did not speak of lattice planes but only of the natural crystal faces. The a, b, c were then defined only as relative lengths (for instance, setting $b = 1$). Since we are here dealing with the microscopic theory of structure, we can introduce the lengths of the edges of the orthorhombic unit cell as absolute values of a, b, c. Therefore the quantities given in (17) also become the absolute lengths of the axis segments, and the lattice plane E which is parallel and nearest to (16) has the equation

$$\frac{h_1^*}{a}x + \frac{h_2^*}{b}y + \frac{h_3^*}{c}z = 1. \tag{18}$$

If we replace the 1 in (18) by any integer n, we obtain another lattice plane which is parallel to E. This will be the nth lattice plane E_n which intersects the axes at the coordinates $n\,a/h_1^*$, $n\,b/h_2^*$, $n\,c/h_3^*$. Non-integer values of n do not give lattice planes; they are in contradiction to the periodicity of the crystal.

In its so-called normal form eq. (18) reads

$$\cos\alpha_p\,x + \cos\beta_p\,y + \cos\gamma_p\,z = p, \tag{19}$$

$$\cos\alpha_p = \frac{h_1^*\,p}{a}, \qquad \cos\beta_p = \frac{h_2^*\,p}{b}, \qquad \cos\gamma_p = \frac{h_3^*\,p}{c},$$

$$p = \left\{\left(\frac{h_1^*}{a}\right)^2 + \left(\frac{h_2^*}{b}\right)^2 + \left(\frac{h_3^*}{c}\right)^2\right\}^{-\frac{1}{2}}.$$

As we know, p denotes the length of the perpendicular from O to E_1 or, as we may also say, it equals the distance between the planes E and E_1. The $\alpha_p, \beta_p, \gamma_p$ are the direction cosines of p (see fig. 54). But according to (19), p is identical with the length d which was introduced in (14 a). *Hence our former quantity d is the spacing of that system of parallel lattice planes whose Miller indices are equal to our interference numbers h* (divided by any common divisor which they may contain).

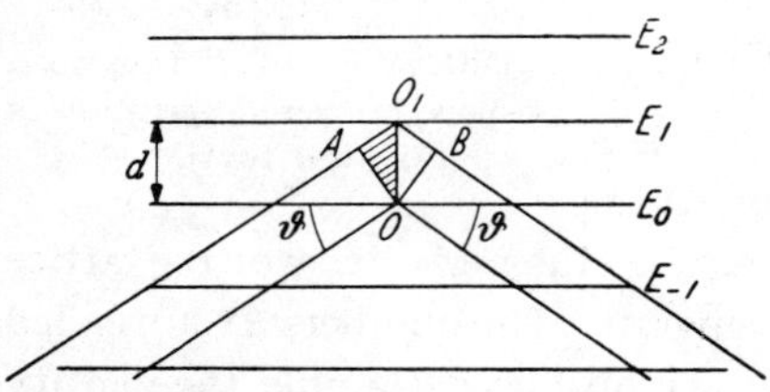

Fig. 55.
Direct derivation of Bragg's equation. Reflection at the lattice planes E.

Bragg's eq. (15) may be visualized as "reflection on the lattice planes", and not merely on a single such plane but on the entire system of parallel lattice planes. This is seen immediately and independently from Laue's theory by means of fig. 55. In order that the wave reflected at E_0 be reinforced by the wave reflected at E_1, that is, in order that the amplitude be doubled, the path difference of the two rays must be a multiple of λ. This path difference is $A\,O_1 + O_1\,B$. From the shaded triangle with the hypotenuse $O\,O_1 = d$ in the figure it follows that

$$A\,O_1 = d \sin\vartheta = O_1\,B.$$

Hence, the above condition becomes, in agreement with (15),

$$2\,d \sin\vartheta = n\,\lambda.$$

This same condition also guarantees the enhancement of the reflections from the lattice planes $E_2, E_3, \ldots E_{-1}, E_{-2}$ and leads to Laue's amplification factor N^2 for the intensities.

This derivation shows that in the present case, only the regularity of the sequence of lattice planes E is important and not the periodic arrangement of the crystal atoms within the planes E. Even if these atoms were distributed completely at random, as are the silver grains in the Wiener layers, the interference effect would not be disturbed. In that case we would have, so to say, a one-dimensional crystal. The three-dimensional crystal lattice differs from it by producing simultaneous interference on many sets of parallel lattice planes.

Immediately after Laue, William Bragg and his son, Lawrence Bragg, (successor of Rutherford at the Cavendish Laboratory) were the first to determine most of the now known simpler crystal structures (rock salt, diamond, fluorite, pyrite, etc.). Later they also determined the structures of some highly complicated organic and inorganic crystals (see, for instance, the remarks on mica on p. 167). They observed the "glancing angles" ϑ for various crystal faces and from them determined the lattice plane spacings d by means of eq. (15). Instead of a continuous X-ray spectrum they used for this purpose a known characteristic line such as the Cu K_α-line $\lambda = 1.537$ Å. In addition to the directions of the reflected rays, their intensities were of essential importance in these structure determinations. Particularly significant were observations on the extinction of even or odd orders of possible reflections.

Debye gave a general explanation of the extent to which thermal motion in crystals influences the intensities of reflections. C. G. Darwin studied the effect of disorientations found in most crystals (the so-called mosaic structure of crystals). P. P. Ewald in his "dynamic theory" of X-ray interference accomplished a profound extension of Laue's original theory. This theory takes into account the attenuation of the primary radiation during its passage through the lattice and the mutual radiation passing from lattice point to lattice point. The valuable notion of the "reciprocal lattice" is also due to Ewald.

33. Diffraction Arising from Many Randomly Distributed Particles

We shall here consider a glass plate which is covered with condensed fog droplets or has been dusted with lycopodium powder. The light source shall be as small as possible and very distant, and we shall observe this source through the glass plate with the eye focused at infinity. We shall assume that the droplets or spore grains are of uniform size and circular in shape. By means of a filter we select a small spectral range of wave number k from the source.

The glass plate shall lie in the xy-plane whose origin $x = 0$, $y = 0$ will be taken to lie on the straight line connecting the eye with the source. The centers of the small circular diffracting discs will be described by the coordinates x_n, y_n. We write the formula for the radiation reaching the eye from each particle and having the direction of propagation α, β, γ, in the form

$$u_n = f(\alpha, \beta, \gamma) \exp\{i k [\alpha (x - x_n) + \beta (y - y_n) + \gamma z]\}. \tag{1}$$

The factor $f(\alpha, \beta, \gamma)$ will be further explained in Sec. 36. As in (32.4) we find for the total amplitude

$$\left| \sum u_n \right| = f(\alpha, \beta, \gamma)\, S, \tag{2}$$

$$S = \left| \sum_{n=1}^{N} \exp[-i k (\alpha x_n + \beta y_n)] \right|, \tag{3}$$

where N is the total number of particles. We have here put the factors containing x, y, and z in front of the summation sign and have taken into account the fact that they disappear when the absolute value is taken.

Because the particle coordinates x_n, y_n are unknown, the summation cannot, of course, be performed algebraically as was done in Sec. 32. Therefore, we must apply a statistical procedure. The value of k in eq. (3) is given; α, β are arbitrary but are to be chosen and then held fixed; the x_n, y_n in the summation assume completely random values. Equation (3) tells us to add N unit vectors of random directions in the complex plane and to determine the length of the resultant vector. A theorem in the theory of probability says: if all directions have *equal probabilities, the length of the resultant vector is* $\sqrt{N}$. This theorem is used, for instance, in the theory of Brownian motion where, as in our diffraction problem one is interested, in the addition of a large number of, on the average, equal impulses which the observed colloidal particles receive by collisions with the molecules of the surrounding fluid.

To prove this theorem we set the exponents in (3) (reduced modulo 2π) equal to $i\varphi_n$ and obtain

$$S = \left| \sum_{n=1}^{N} e^{i\varphi_n} \right|, \qquad S^2 = \sum_n e^{i\varphi_n} \sum_m e^{-i\varphi_m}.$$

We now find the statistical mean value $\overline{S}$ of S, which we define as the square root of the average value of S^2:

$$\overline{S} = \sqrt{\overline{S^2}}, \qquad \overline{S^2} = \frac{1}{2\pi}\int_0^{2\pi} d\varphi_1 \frac{1}{2\pi}\int_0^{2\pi} d\varphi_2 \ldots \frac{1}{2\pi}\int_0^{2\pi} d\varphi_N\, S^2. \tag{4}$$

Thus we average S with respect to each φ_n over its whole range of values from 0 to 2π. The assumption of equal probabilities of all angles for each φ_n and the mutual independence of all φ_n which is implied in (4) is motivated by our complete lack of knowledge of the values of x_n and y_n in eq. (3) which defines the φ_n.

First we compute only the integral with respect to φ_1 occurring in (4):

$$\text{(4 a)} \qquad \frac{1}{2\pi}\int_0^{2\pi} d\varphi_1 \left(e^{i\varphi_1} + \sum_{n=2}^{N} e^{i\varphi_n}\right)\left(e^{-i\varphi_1} + \sum_{m=2}^{N} e^{-i\varphi_m}\right).$$

The product of the two parentheses in this integral gives

$$1 + \ldots + \ldots + S_1{}^2, \qquad S_1{}^2 = \sum_{n=2}^{N} e^{i\varphi_n} \sum_{m=2}^{N} e^{-i\varphi_m}.$$

The two middle terms which have not been written down contain the factors $\exp(i\varphi_1)$ and $\exp(-i\varphi_1)$ and therefore vanish on integrating over φ_1. The two other terms are independent of φ_1 and therefore integrate to $1 + S_1{}^2$.

Next, we calculate

$$\text{(4 b)} \qquad \frac{1}{2\pi}\int d\varphi_2\,(1 + S_1{}^2) = 1 + 1 + S_2{}^2, \qquad S_2{}^2 = \sum_{n=3}^{N} e^{i\varphi_n} \sum_{m=3}^{N} e^{-i\varphi_m}.$$

Continuation of this procedure leads to

$$\text{(5)} \qquad \overline{S^2} = 1 + 1 + 1 + \ldots = N, \qquad \overline{S} = \sqrt{N}.$$

This proves our probability theorem. As was to be expected from the symmetry of the arrangement, $\overline{S}$ is independent of α, β.

Returning now to eq. (2), we find for the intensity of the diffraction pattern

$$\text{(6)} \qquad J = N J_0, \qquad J_0 = f^2(\alpha, \beta, \gamma).$$

J_0 is the intensity arising from one individual diffracting disc. *For random distributions of the diffracting elements, the intensities add and not the amplitudes as was the case in the grating theory.* Instead of the amplification factor N^2 of fig. 53, eq. (6) contains the factor N.

In the case of our circular diffracting discs J_0 does not, of course, depend on α, β and γ individually but only on the radial angular distance $s = (\alpha^2 + \beta^2)^{1/2} = (1 - \gamma^2)^{1/2}$. As we shall see in Sec. 36, J_0 has a flat maximum in the center of the diffraction pattern and vanishes for the first time at a sharply defined position $s = s_1$. A considerably weaker maximum follows and then a less sharp zero, etc. According to Sec. 36, s_1 is given by

$$s_1 = 0.61 \frac{\lambda}{a} \tag{7}$$

where a is the radius of the diffracting discs. Hence, the radial extension of the diffraction pattern increases as the ratio a over λ decreases.

If we use *white* light instead of monochromatic light, then the center of the diffraction pattern appears white because all colors have a maximum there. The outer rim of this center disc is colored red because at the distance

$$s_1 = 0.61 \cdot \frac{\lambda_{\text{blue}}}{a} \tag{7 a}$$

the blue component of the light is missing. At approximately twice this distance we expect a bluish tint because there the red component is extinguished. As we proceed outward from the center, the coloration and the intensity become progressively weaker. If the diffracting particles are *not circular* in shape, then of course the intensity J_0 depends on both α and β. However, the intensity J resulting from all N particles retains its circular symmetry so long as not only the positions but also the orientations of the particles are random, for then a summation over all possible orientations must be added to the summation over all positions in the expression (3) for S.

If the particles are *not of uniform size* but are, for example, water droplets of various radii, then according to (7) the rings of zero amplitude become diffuse under monochromatic illumination. Under white light illumination the color pattern becomes less distinct, but the white coloration of the center portion of the pattern remains. The size of this white disc can be estimated from eq. (7 a) if a is replaced by an average radius $\bar{a}$.

Our statement about the statistical average value of S is only approximately valid. Under monochromatic illumination with a carefully limited source the diffraction patterns exhibit a certain "granulation", namely a *radial fiber structure*. This is due to fluctuations about the statistical average which are stronger in the radial direction than in the azimuthal direction perpendicular to it. These fluctuations have been investigated in detail by M. von Laue[1] both experimentally and theoretically.

We now turn to the *meteorological* applications of this theory. Because of the sizes of the light sources involved (sun or moon), and because of the white nature of the light, the above-mentioned fluctuations obviously do not enter into consideration. The real *coronae* about the sun and the moon are due to

[1] Preußische Akademie 1914, p. 1144.

diffraction at cloud particles, hence primarily at water droplets. Because of the different sizes of the droplets, the color phenomena are rather weak. The sun and the moon are surrounded by a white or bluish-white field. As was remarked in the introduction, a frequently observed reddish rim indicates the diffractive nature of this phenomenon. From the angular radius of this rim, which differs for different observations, the average droplet diameter $2\bar{a}$ is found by means of (7 a) to vary from 0.01 to 0.03 mm. After the eruption of Krakatoa, a much larger red-brown sun ring was observed. This was due to the volcano's dust particles which had drifted as far as Europe. The angular radius of this ring amounted to 20° to 25°, which corresponds to the much smaller particle diameter of 0.002 mm.

The situation is different with the ice crystals of cirrus clouds. It can be assumed that they, too, contribute to the diffraction phenomenon of coronae, but the characteristic phenomena resulting from these crystals are the *halos* which are not due to diffraction but are caused by *refraction*. This fact is proved by the color arrangement in halos: violet outside, red inside. Moreover, the halos have definite radii which do not depend on the (varying) sizes of the particles but rather on the crystalline structure of these particles. The most frequently occuring angular radius is 22°, which corresponds to refraction of the light in the hexagonal cylinders of ice crystals (edge angle 60°). If because of gravity the ice crystals are oriented mostly vertically, then the light of the halo concentrates at the two points on its circumference which are at the same elevation as the sun. This is the origin of the two *parhelia*. In addition, a halo with about 45° angular radius occurs.

34. Huygens' Principle

Huygens' principle may be visualized as follows: the future shape of any given wave surface can be determined by assuming that each point of this surface emits a *spherical wave* and by constructing the *envelope* of all these spherical waves. In a homogeneous medium this construction yields a surface which is parallel to the original wave surface (possible boundaries of the original surface form an exception). We have already seen in fig. 37 that this procedure leads to the usual refraction at plane interfaces. The usual reflection is also obtained in this way.

Kirchhoff proved that Huygens' principle is an *exact* consequence of the differential equations of optics. This principle constitutes the foundation of the classical theory of diffraction, which has proved its fruitfulness in countless problems. Nevertheless, this theory is only an *approximation which is valid only for sufficiently small wavelengths*. This is so because the *boundary condi-*

tions which must be used in conjunction with Huygens' principle are not known precisely. The classical theory, moreover, does not take the vectorial character of the optical field into account. This deficiency will not be discussed until Sec. 38 et seq.

A. The spherical wave

We are familiar with the *scalar* spherical wave of acoustics from Vol. II, Sec. 13. Like the plane wave representation, that of the spherical wave is a solution of the wave equation $\Delta u + k^2 u = 0$. If it is assumed that u depends only on the coordinate x, one obtains (disregarding a constant complex factor)

$$u = e^{ikx}. \tag{1}$$

If, on the other hand, it is assumed that u is a function only of the distance r from the origin of the coordinate system, then one finds

$$\Delta u = \frac{1}{r}\frac{d^2(r\,u)}{dr^2}, \qquad \frac{d^2(r\,u)}{dr^2} + k^2 r\,u = 0, \qquad r\,u = e^{\pm ikr}$$

and hence

$$u = \frac{1}{r}\,e^{ikr}, \tag{2}$$

where the time dependence is assumed to be of the form $\exp(-i\,\omega\,t)$. (With a time factor of $\exp(+i\,\omega\,t)$ one would obtain an incoming rather than an outgoing wave.)

The *vectorial* spherical wave of electrodynamics is not so simple. The expression for this wave assumes its most convenient form if the *Hertz vector* is introduced as the characteristic function u. In particular, the special case of the Hertz vector which represents the radiation emitted by a linearly oscillating dipole, see Vol. III, Sec. 19 B, is useful here. Though the analytic expression of this vector is again given by (2), that is, by a spherically symmetric expression, this is not true of the field which is derived from the Hertz vector. The magnetic field lines are circles about the direction of oscillation, while the electric field lines are in the meridian planes of that direction. Only the phase of the field is spherically symmetric. Its amplitude depends on direction. The electric amplitude, for instance, vanishes in the direction of oscillation of the dipole at distances large compared to the wavelength.

A physical light source (point-like bulb or candle) contains all possible directions of oscillation. Such a source emits an average field in which no directions are preferred. The field intensity is therefore spherically symmetric. If we represent such a field by (2), we must realize that we thereby relinquish the possibility of representing the finer details of the light, such as its polarization.

B. GREEN'S THEOREM AND KIRCHHOFF'S FORMULATION OF HUYGENS' PRINCIPLE

Green's theorem suffices for the integration of the scalar wave equation; the reader is referred to the first introduction of this theorem in Vol. II, eq. (3.15) and to its repeated use in Vols. III and VI:

$$\int (u \,\Delta\, v - v \,\Delta\, u)\, d\tau = \int \left(u \frac{\partial v}{\partial n} - v \frac{\partial u}{\partial n} \right) d\sigma. \tag{3}$$

Let u be the function (2) of the spherical wave and let v be the desired solution of the equation $\Delta\, v + k^2\, v = 0$. The surface σ separates space into two regions. One of these regions will be called the interior of σ and the other region its exterior. If, as will usually be the case, σ extends to infinity, then the point at infinity belongs both to the interior and to the exterior. For the region of integration on the left-hand side of (3) we choose the exterior of σ. The source of the wave u is assumed to lie at the point P in the exterior and is to be excluded from the region of integration; this may be accomplished by means of a sphere K of arbitrarily small radius; see fig. 56. Then, because of the differential equation which u and v satisfy, the left-hand side of (3) vanishes. The integral on the right-hand side must be evaluated over the two boundary surfaces σ and K[1]; dn is the normal to these surfaces directed into the interior. The same considerations as in Vol. II, Sec. 20, 1 a yield for the integral over K the value $-4\pi\, v_P$, where v_P is the value of v at the center of K. From eq. (3) it follows, therefore, that

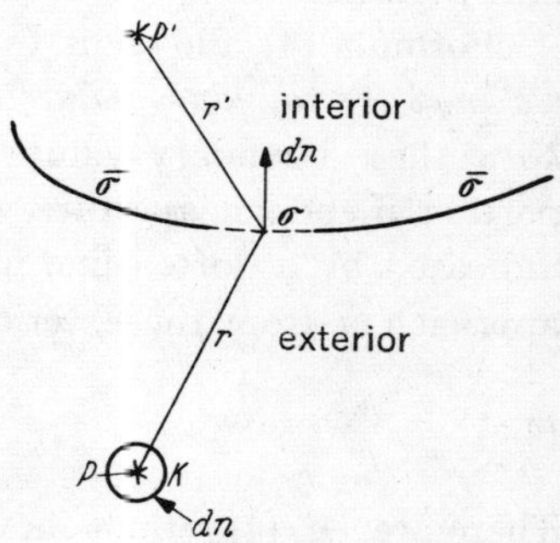

Fig. 56.

Regions of integration for Green's theorem. The surfaces σ and $\bar{\sigma}$ together form a closed surface.

$$4\pi\, v_P = \int_\sigma \left(\frac{\partial v}{\partial n} \frac{e^{ikr}}{r} - v \frac{\partial}{\partial n} \frac{e^{ikr}}{r} \right) d\sigma \tag{4}$$

---

[1] Actually a third boundary surface should be added, namely a sphere of very large radius with its center at P which excludes the point at infinity. Denoting the surface element by $d\sigma = r^2\, d\omega$ and combining the r^2 with the integrand, the integral over this surface becomes

$$\int \left\{ r \left(\frac{\partial v}{\partial n} - i\, k\, v \right) + v \right\} e^{ikr}\, d\omega.$$

Because of the radiation condition, see Sec. 38, eq. (1 d), and because v vanishes as $r \to \infty$, the bracket { } vanishes. Hence, the integral also vanishes.

It should be emphasized that in this calculation, as in the previous applications of Green's theorem in potential theory, the spherical wave u plays the role of a *mathematical auxiliary function.* It is, so to speak, a "probe" which we use for investigating the optical field v. This "virtual" spherical wave has nothing to do with the real spherical wave which we shall introduce in (4 b) as the source function of the optical field. In eq. (4) we have in a sense removed the probe from the field v under investigation by substituting for u and $\partial u/\partial n$ their values as given by (2). From now on we shall forget the origin of these quantities and shall rather regard them as describing spherical waves which are radiated by the surface elements $d\sigma$ and which arrive at the point P at the distance r from $d\sigma$. Only with this interpretation of eq. (4) have we gained the basis of Huygens' principle.

Formula (4) allows us to calculate v at every point P of the exterior *if we know the boundary values of v and $\partial v/\partial n$ on σ* (or more appropriately, if we *knew* these boundary values!). Let us assume that σ consists of an opaque portion $\overline{\sigma}$ and an aperture which we shall henceforth call σ. The latter is indicated by a dotted line in fig. 56. It is reasonable to assume that as we approach $\overline{\sigma}$ from the exterior, we shall find there the values

$$v = 0, \qquad \frac{\partial v}{\partial n} = 0. \tag{4 a}$$

Therefore, eq. (4) retains its validity even with our new definition of σ. Again, it is reasonable to assume that v has the same values in the aperture as those it would have if $\overline{\sigma}$ were absent. If, for instance, v were due to the radiation from a luminous point P' of strength A, then the boundary values in the aperture would be

$$v = A\,\frac{e^{ikr'}}{r'}, \qquad \frac{\partial v}{\partial n} = A\,\frac{\partial}{\partial n}\frac{e^{ikr'}}{r'} \tag{4 b}$$

where r' is defined in the figure.

But strictly speaking the assumptions (4 a, b) are mathematically inadmissible. A well-known theorem in Riemann's theory of functions says that if a two-dimensional potential v vanishes together with its normal derivative along a finite curve segment s, then v vanishes identically in the whole plane. This theorem can be extended to cover solutions of the two-dimensional wave equation[1]. It is also true that any solution of the three-dimensional potential or wave equation vanishes in the whole space if the condition (4 a) is satisfied on any finite surface element σ. Therefore, (4 a) would seem to imply that $v = 0$ everywhere.

[1] Heinrich Weber, Mathem. Ann. Vol. 1, 1869, p. 1.

Applying this theorem, on the other hand, to the difference $w = v - v'$ between any two analytic solutions v and v' of the three-dimensional wave equation, it follows that v and v' must be identical in the whole space if the conditions $v = v'$ and $\partial v/\partial n = \partial v'/\partial n$ are satisfied on any finite surface element σ. Therefore, the assumptions (4 a) and (4 b) not only contradict the known physical situation but also contradict one another.

As a matter of fact, we would not even obtain the boundary values (4 a) or (4 b) if we calculated them from (4) by placing P on $\bar{\sigma}$ or σ. Thus eq. (4) gives the *correct* values of v_P only if we know the *correct* boundary conditions v and $\partial v/\partial n$.

C. Green's function, simplified formulation of Huygens' principle.

This mathematical contradiction is avoided by substituting for the spherical wave u in the original eq. (3) the *Green's function belonging to our surface*. This function is defined by the following conditions[1]:

$$\Delta G + k^2 G = 0 \quad \text{in} \quad \tau, \tag{5 a}$$

$$G = 0 \quad \text{on} \quad \sigma, \tag{5 b}$$

$$G \to u \quad \text{as} \quad r \to 0, \tag{5 c}$$

$$r\left(\frac{\partial G}{\partial n} - i\,k\,G\right) \to 0 \quad \text{as} \quad r \to \infty. \tag{5 d}$$

As before, r is the distance from the point P and (5 d) is the so-called radiation condition of Vol. VI, Sec. 28. Equation (5 c) states that, like u, G shall have a singularity only at the point P and shall be continuous everywhere else in the exterior. G differs from u because of the additional condition (5 b). As a result of this condition the term containing $\partial v/\partial n$ in eq. (4) disappears and that equation becomes[2]

$$4\pi\, v_P = -\int_\sigma v\,\frac{\partial G}{\partial n}\,d\sigma. \tag{6}$$

[1] See Vol. VI, Secs. 10 E and F. In the nomenclature used there the spherical wave u is not a Green's function but is the "principal solution" of the differential equation $\Delta u + k^2 u = 0$.

[2] The spherical surface which excludes infinity (see footnote 1, p. 197) contributes now

$$\int \left(G\,\frac{\partial v}{\partial n} - v\,\frac{\partial G}{\partial n}\right) r^2\,d\omega = \int r\left(\frac{\partial v}{\partial n} - i\,k\,v\right) r\,G\,d\omega$$

which again vanishes.

Now we need to prescribe only the boundary values of v itself. As in (4 a, b) it is reasonable to assume that

$$v = 0 \quad \text{on} \quad \bar{\sigma}, \tag{6 a}$$

$$v = A \exp \frac{i k r'}{r'} \quad \text{on } \sigma. \tag{6 b}$$

These assumptions are mathematically consistent. Furthermore, according to the theory of the Green's function, the boundary values (6 a, b) are actually assumed by the function v_P as computed by (6) when the point P is placed on the screen or in the aperture.

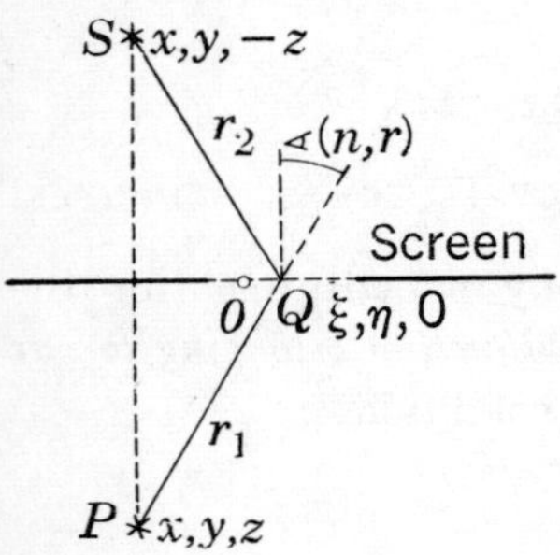

Fig. 57.
Construction of the Green's function for a plane screen.

The question remains, are these assumptions also physically justifiable? The answer again is that *they are only approximations for sufficiently small wavelengths*[1]. The field does not vanish completely behind the screen, nor is the field in the aperture entirely unaffected by the presence of the screen, at least not within distances of the order of magnitude of a wavelength from the edge of the screen.

The introduction of the Green's function therefore involves no final justification of the method but it has the practical advantage of leading to the simpler form of the integral (6) as compared to (4). However, the applicability of the Green's function method is restricted to the special case of the *plane screen.* This is the only case for which the Green's function can be conveniently expressed, namely by means of the elementary *method of images.*

In fig. 57 we construct the mirror image S of the point P with respect to the plane of the screen $z = 0$. For an arbitrary point $Q = \xi, \eta, \zeta$ where $\zeta > 0$, we form

$$G = \frac{e^{ikr_1}}{r_1} - \frac{e^{ikr_2}}{r_2}, \qquad \begin{aligned} r_1^2 &= (\xi - x)^2 + (\eta - y)^2 + (\zeta - z)^2 \\ r_2^2 &= (\xi - x)^2 + (\eta - y)^2 + (\zeta + z)^2. \end{aligned} \tag{7}$$

x, y, z and ξ, η, ζ are measured from the same origin O which lies in the plane of the screen. This function of ξ, η, ζ satisfies all the conditions (5 a) to (5 d). It should be noted that the singularity of G at the image point S does not violate these conditions because S lies on the other side of the screen $\zeta = 0$.

[1] We use here and in the following the word "wavelength" and the notation λ even though these are really only defined for plane waves and lose their simple meaning for the more complicated wave types encountered in diffraction. However, we can always interpret λ as the length $2\pi c/\omega$ which is defined for all monochromatic radiation processes and which for plane waves is identical with the actual wavelength.

From (7) we calculate

(7 a)
$$\frac{\partial G}{\partial \zeta} = \frac{d}{dr_1}\left(\frac{e^{ikr_1}}{r_1}\right)\frac{\partial r_1}{\partial \zeta} - \frac{d}{dr_2}\left(\frac{e^{ikr_2}}{r_2}\right)\frac{\partial r_2}{\partial \zeta}.$$

If we now place Q on the screen, see fig. 57, we have

$$r_1 = r_2 = r, \qquad \frac{\partial r_1}{\partial \zeta} = -\frac{\partial r_2}{\partial \zeta} = \cos(n, r)$$

and hence

(8)
$$\frac{\partial G}{\partial n} = -\frac{\partial G}{\partial \zeta} = 2\frac{\partial}{\partial r}\left(\frac{e^{ikr}}{r}\right)\cos(n, r).$$

This can be further simplified for all positions of P which are not too close to the screen. For then we have $kr = \frac{2\pi r}{\lambda} \gg 1$ and therefore

(8 a)
$$\frac{\partial}{\partial r}\left(\frac{e^{ikr}}{r}\right) = ik\frac{e^{ikr}}{r}\left(1 - \frac{1}{ikr}\right) \sim \frac{2\pi i}{\lambda}\frac{e^{ikr}}{r}.$$

Substituting (8 a) in (8) and (8) in (6) we obtain

(9)
$$i\lambda v_P = \int_\sigma \frac{e^{ikr}}{r}\cos(n, r)\, v\, d\sigma.$$

Thus we have gained an expression which is equivalent to Huygens' principle and which formulates it exactly. *A light wave falling on the aperture σ propagates as if every element $d\sigma$ emitted a spherical wave the amplitude and phase of which are given by that of the incident wave v.* The factor $\cos(n, r)$ which multiplies $d\sigma$ is of interest. It corresponds to Lambert's law of surface brightness and was used earlier by Fresnel in his qualitative considerations. The factor λ on the left-hand side of (9) is understandable because of the dimensions of the right-hand side ($d\sigma/r$ = length).

If one substitutes for v the value given in (4 b) which corresponds to illumination by a point source, then (9) becomes

(10)
$$i\lambda v_P = A\int e^{ik(r+r')}\frac{\cos(n, r)}{r r'}\, d\sigma.$$

D. Fraunhofer and Fresnel diffraction

Let the dimensions of the diffraction aperture in the screen be small compared to the distances r and r' of observer and light source. Then the factor $\frac{\cos(n, r)}{r r'}$ *varies but little* inside the opening. Hence we may place this factor in front of the integral sign, setting it equal to the value it assumes at the origin O of our integration variables ξ, η. Calling R and R', respectively, the values of r and r' at O, we thus obtain instead of (10)

$$i\lambda v_P = \frac{A}{R R'} \cos(n, R) \int e^{ik(r+r')}\, d\xi\, d\eta. \tag{11}$$

To simplify the remaining portion of the integrand, which because of the magnitude of k is a *rapidly varying function*, we first develop r in powers of ξ and η:

$$r = \sqrt{(x-\xi)^2 + (y-\eta)^2 + z^2} = \sqrt{R^2 - 2(x\xi + y\eta) + (\xi^2 + \eta^2)}$$

$$= R - \frac{x}{R}\xi - \frac{y}{R}\eta + \frac{\xi^2 + \eta^2}{2R} - \frac{(x\xi + y\eta)^2}{2R^3} = R - \alpha\xi - \beta\eta + \frac{\xi^2 + \eta^2 - (\alpha\xi + \beta\eta)^2}{2R},$$

where α and β are the direction cosines of the diffracted ray $O \to P$ with respect to the ξ- and η-axes. If we call the direction cosines of the incident ray $P' \to O$ α_0 and β_0 (hence the direction cosines of $O \to P'$ are $-\alpha_0$ and $-\beta_0$), then we find correspondingly

$$r' = R' + \alpha_0\xi + \beta_0\eta + \frac{\xi^2 + \eta^2 - (\alpha_0\xi + \beta_0\eta)^2}{2R'}.$$

From this follows

$$e^{ik(r+r')} = e^{ik(R+R')}\, e^{-ik\Phi} \tag{12}$$

with the abbreviation

$$\Phi = (\alpha - \alpha_0)\xi + (\beta - \beta_0)\eta - \left(\frac{1}{R} + \frac{1}{R'}\right)\frac{\xi^2 + \eta^2}{2} + \frac{(\alpha\xi + \beta\eta)^2}{2R} + \frac{(\alpha_0\xi + \beta_0\eta)^2}{2R'}, \tag{13}$$

and formula (11) becomes

$$i\lambda v_P = \frac{A}{R R'} \cos(n, R)\, e^{ik(R+R')} \int e^{-ik\Phi} d\xi\, d\eta. \tag{14}$$

The expansion (13) clearly presupposes that the linear dimensions of the diffraction opening are small compared to R and R'.

The evaluation of the remaining integral in (14) is simplest to perform for the case of *Fraunhofer diffraction*

$$R \to \infty, \qquad R' \to \infty \tag{14 a}$$

which obtains for the meteorological phenomena and which can also be best realized experimentally. In this case only the *linear* terms in Φ remain, and we have only to deal with a superposition of *plane* waves.

If one or both of the conditions (14 a) are not satisfied, one speaks of *Fresnel diffraction*. By proper choice of the origin O (for details see Sec. 37) one can then make $\alpha = \alpha_0$ and $\beta = \beta_0$, so that the linear terms of Φ vanish. The integration of the quadratic terms (Fresnel's integrals) provides us with a complete picture of the entire diffraction field behind the screen, while in the Fraunhofer case we confine ourselves to the limit of the diffraction field at large distances from the screen.

Fresnel and Fraunhofer diffraction are also called *microscopic* and *telescopic diffraction*, respectively. For, in the Fresnel case a magnifying glass may be used to project the field at a given point P onto an observation screen where the intensity can then be measured. In the Fraunhofer case insufficient intensity would reach the naked eye if it were placed at a very large distance from the screen (especially in the case of a single small diffraction opening). Therefore, all parallel ray bundles emerging from the diffraction opening are focused by means of a lens[1] L on a point P in the focal plane E of L. For similar reasons a point source P' at a finite distance from the screen is used, and its rays are sent through the opening as parallel ray bundles by means of a lens L' (collimator lens). P' must, of course, lie in the focal plane E' of L'; see fig. 58.

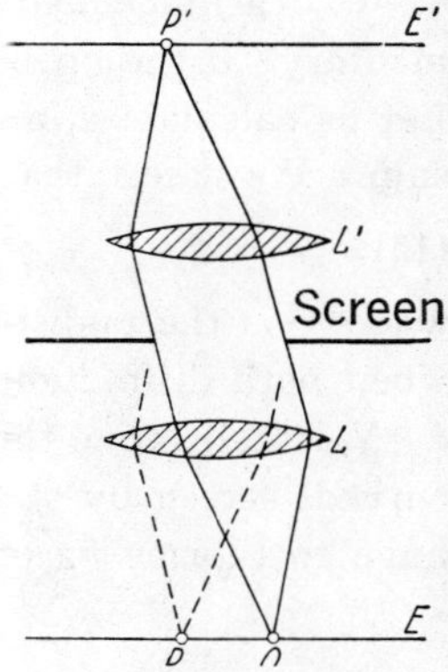

Fig. 58.
The Fraunhofer arrangement for diffraction observations.

If O is the image of P' according to geometrical optics, then the coordinates of P in the plane E (with O as the origin of the coordinate system) are proportional to the quantities $\alpha - \alpha_0$ and $\beta - \beta_0$. According to (14) and (13) the intensity at P depends only on these two quantities. (The factor in front of the integral in (14) is a constant since $|\exp\{i\,k\,(R + R')\}| = 1$ and since A must be thought of as tending to infinity like $R\,R'$.) L is the objective lens used in this "telescopic" observation; the eyepiece through which the diffraction pattern produced in E is viewed is not shown in fig. 58. The diffracted rays which are focused at P are represented in the figure by dotted lines which, in accordance with the notion of Huygens' principle, are drawn as if they originated in the diffraction opening[2].

[1] We may disregard the diffraction phenomena caused by the rims of the lenses L and L'.

[2] Fresnel diffraction can also be observed with a telescope if the eyepiece is not focused on the focal plane E but on an arbitrary extrafocal plane. Instead of observing the patterns on these planes with the eye, one can of course record them by means of a photographic plate.

When there is a very large number N of diffraction openings as in Secs. 32 and 33, the telescopic arrangement becomes unnecessary. The amplification factor N^2 in Sec. 32 and N in Sec. 33 make it possible to observe the diffraction patterns with the naked eye even at large distances.

E. Babinet's principle

Two arrangements of diffraction openings 1 and 2 are called "complementary" if the opening of 1 is congruent to the screen of 2 and vice versa. Let us calculate v_1 and v_2 for the same primary illumination and form their sum. We assert that "within the framework of Huygens' principle"

$$v_1 + v_2 = v, \tag{15}$$

where v is the undisturbed primary illumination at the point of observation when both diffraction screens are absent.

We shall prove this theorem for the general case of an arbitrary (possibly curved) screen by starting from eq. (4). When forming the desired sum, we have to replace v_P on the left-hand side of that equation by

$$(v_1 + v_2)_P.$$

On the right-hand side of (4) v has the same meaning in both summands, namely the undisturbed primary illumination. In contrast to (4) the integration must now be carried out over the *entire* surface σ because every point on σ belongs to the diffraction opening of either 1 or 2. But we obtain precisely the same integral if we apply (4) to the primary illumination v with no screens present. In that case we have just this v on the left-hand side and the integration over the whole surface σ on the right-hand side with the meanings of the symbols v and $\partial v/\partial n$ in the integrand the same as before. Equation (15) has thus been proved: $v_1 + v_2$ and v are equal because both are equal to the same $\int \ldots d\sigma$.

Equation (15) is valid for all points P of the exterior and therefore encompasses both Fresnel and Fraunhofer diffraction. It is called "*Babinet's principle*". The above proof was based on Huygens' principle. In Sec. 38 F we shall discuss how Babinet's principle must be modified when it is treated from the more precise viewpoint of the boundary value problem.

In the older literature[1] Babinet's principle occurs only in a much narrower form, which restricts its applicability to Fraunhofer diffraction. The reason is that the complete functional dependences of v_1, v_2 (including their phases)

[1] See, for instance, Kirchhoff, Vorlesungen über Optik, p. 96.

are not accessible to observation, but only the amplitudes $|v_1|$, $|v_2|$ or, what is equivalent, only the intensities are observable. The latter, of course, do *not* satisfy the equation

(15 a) $$J_1 + J_2 = J.$$

Rather, when forming the absolute value of (15), the left-hand side of (15 a) is found to contain the additional terms

(15 b) $$v_1 v_2^* + v_2 v_1^*.$$

Only in the case of Fraunhofer diffraction can a simple statement be made about the intensities: *the two complementary screens produce diffraction patterns of equal intensity*:

(16) $$J_1 = J_2.$$

To prove this we consider the focal plane E in fig. 58. For ideal lenses the primary light v is concentrated at the point O and is zero everywhere else. Excluding the point O, at which the singularity of the diffraction pattern renders observation impossible in any case, it follows from (15) that

$$v_1 = - v_2, \qquad |v_1| = |v_2|, \qquad \text{hence indeed} \qquad J_1 = J_2.$$

In Secs. 35 C and D we shall discuss a very elementary problem of Fresnel diffraction for which there exists no simple relationship at all between J_1 and J_2, but we shall convince ourselves that our formulation (15) of Babinet's principle is valid.

F. Black or reflecting screen

In the theory of diffraction it is customary to speak of a *black* screen. However, in actual diffraction experiments one finds that the physical nature of the screen in general does not affect the results noticeably. Thus a piece of tin foil into which a narrow slit has been scratched yields the same diffraction pattern regardless of whether the foil has been left reflecting or whether it has been blackened. Therefore we need only describe the screen as *opaque* in order to specify that in spite of arbitrary thinness it shall transmit no light. In the Maxwell theory such a screen would have to be defined as a material possessing an *infinite conductivity*. Such a screen would not be black but would be *perfectly reflecting*; its reflecting power would be $r = 1$. On the other hand, black, that is completely *non-reflecting* material, cannot even be defined in the Maxwell theory; blackening is not a property of the material but is a property of the surface. We shall take this into account in Sec. 38 where we shall try to describe the property "black" mathematically. Our presentation of Huygens' principle shows that this property is not essential to the theory of diffraction. Only very refined experiments can reveal the nature of the material of which the diffracting screen is composed.

The material composition of the screen, of course, affects the light field only in the immediate vicinity of the edge of the opening, that is, only within a distance of a few wavelengths from the edge. If the opening is fairly large, this edge zone is negligible compared to the rest of the aperture. This explains why the crude assumptions (4 a, b) or (6 a, b), which can of course be valid only outside the edge zone, have been so eminently successful. Deviations from Huygens' principle are to be expected with the usual methods of observation only for extremely small openings which are of the order of magnitude of a wavelength in size (or for experimental arrangements which correspond to such small openings in accordance with the similarity law of Sec. 35 E).

G. Two generalizations

So far we have restricted ourselves to those consequences of Huygens' principle which are directly applicable to the problems under discussion in this chapter. We shall now present two closely allied results which will be useful later.

1. Instead of the Green's function (7) which satisfies the boundary condition $G = 0$ at $z = 0$, we now form

$$G = \frac{e^{ikr_1}}{r_1} + \frac{e^{ikr_2}}{r_2}. \tag{17}$$

This is a function which satisfies the boundary condition $\partial G/\partial z = 0$ at $z = 0$. Substituting it for u in eq. (3) we obtain in place of (6)

$$4\pi v_P = + \int \frac{\partial v}{\partial n} G \, d\sigma. \tag{18}$$

However, this v_P is identical with the v_P of eq. (6) only if the integral in (18) is taken not merely over σ but also over the entire screen which contains the aperture σ [in eq. (6) this was not necessary because we assumed $v = 0$ on the opaque screen]. With this understanding of the integration in (18), and given the exact boundary values of v and $\partial v/\partial n$ in the opening and of $\partial v/\partial n$ on the screen, we have

$$\int \frac{\partial v}{\partial n} G_+ \, d\sigma = - \int v \frac{\partial G_-}{\partial n} \, d\sigma, \tag{19}$$

where G_+ is the Green's function defined in eq. (17), G_- is that defined in eq. (7). If we substitute for these Green's functions their values on the plane $z = 0$, we obtain

$$\int \frac{\partial v}{\partial n} \frac{e^{ikr}}{r} \, d\sigma = - \int v \frac{\partial}{\partial n} \frac{e^{ikr}}{r} \, d\sigma. \tag{20}$$

2. If the screen is not a plane but a curved (e. g. spherical) surface and has an aperture σ, then doubtless there are again two functions G_- and G_+ which satisfy the conditions $G_- = 0$ and $\partial G_+/\partial n = 0$ on σ. Every continuous solution v of the wave equation can be represented in two ways by means of these functions. Therefore, the remarkable identity (19) is valid in the sense described above also for curved surfaces σ.

However, even for the simplest case of the sphere, the analytic representations of $G_\pm$ lead to infinite series containing the eigenfunctions of the sphere. Therefore, the simplification which resulted from the introduction of the Green's function in the case of the plane screen becomes illusory for curved screens, not to mention the fact that the requirement concerning the knowledge of the exact boundary values is not fulfilled in either case.

35. The Problem of the Shadow in Geometrical and in Wave Optics

Geometrical optics constitutes our day-to-day guide to the outside world; it is the basis for the construction of the image-forming devices (spectacles, telescopes, photographic lenses). We shall here treat geometrical optics as the limiting case of wave optics as $\lambda \to 0$; see also the introduction to Sec. 34.

A. The eikonal

As in Sec. 34 we start with the scalar wave equation

$$\text{(1)} \qquad \Delta u + k^2 u = 0, \qquad k = \sqrt{\varepsilon\mu}\,\omega = \frac{2\pi}{\lambda},$$

but now we do not assume ε to be constant but rather to be a (continuously or discontinuously varying) function of position. As $\lambda \to 0$, and hence $k \to \infty$, this differential equation degenerates. In order nevertheless to be able to draw quantitative conclusions from the equation, we make the following assumption[1] as to the form of the solution:

$$\text{(2)} \qquad u = A\, e^{ik_0 S}, \qquad k_0 = \sqrt{\varepsilon_0 \mu_0}\,\omega = \frac{2\pi}{\lambda_0}.$$

A is an amplitude factor. We call S the *eikonal*, an expression introduced by H. Bruns. While u is a rapidly varying function of position (because $k_0 \to \infty$), we consider A and S as slowly varying functions of the coordinates x, y, z which do not go to infinity with k_0. By differentiating (2) we get

[1]After P. Debye, in a paper Ann. d. Phys. **35** (1911) by Sommerfeld and Iris Runge.

$$\frac{\partial u}{\partial x} = i\,k_0\,u\,\frac{\partial S}{\partial x} + u\,\frac{\partial \log A}{\partial x},$$

$$\frac{\partial^2 u}{\partial x^2} = -k_0^2\,u\left(\frac{\partial S}{\partial x}\right)^2 + 2\,i\,k_0\,u\left(\frac{1}{2}\frac{\partial^2 S}{\partial x^2} + \frac{\partial \log A}{\partial x}\frac{\partial S}{\partial x}\right) + \dots$$

$$\Delta\,u + k^2\,u = -k_0^2\,u\left[\left(\frac{\partial S}{\partial x}\right)^2 + \left(\frac{\partial S}{\partial y}\right)^2 + \left(\frac{\partial S}{\partial z}\right)^2 - \frac{k^2}{k_0^2}\right]$$
$$+ 2\,i\,k_0\,u\left[\frac{1}{2}\,\Delta\,S + \mathbf{grad}\log A\cdot\mathbf{grad}\,S\right] + \dots$$

where the terms indicated by ...... do not become infinite as $k_0 \to \infty$.

Hence eq. (1) is satisfied approximately if S and A satisfy the differential equations

$$D\,(S) = n^2, \qquad n = \frac{k}{k_0}, \tag{3}$$

$$\mathbf{grad}\log A\cdot\mathbf{grad}\,S = -\frac{1}{2}\,\Delta\,S. \tag{4}$$

D is the notation for the "first differential parameter"

$$D = \left(\frac{\partial}{\partial x}\right)^2 + \left(\frac{\partial}{\partial y}\right)^2 + \left(\frac{\partial}{\partial z}\right)^2$$

which has already been used in Vol. II, eq. (3.9 c); n is the usual index of refraction; Eq. (3) which is the *differential equation of the eikonal, is an inhomogeneous equation of first order and second degree.* Once (3) has been integrated, (4) yields the component of the gradient of log A in the direction of the gradient of S. Equation (4) makes no statement about the gradient of A in a direction perpendicular to the gradient of S. Therefore (4) permits discontinuities of A in these directions.

According to the definition (2) the surfaces S = constant are surfaces of constant phase of u. Hence they represent *wave surfaces.* The normals to these surfaces are given by the gradient of S and represent the *ray directions.* In general, if n varies in space, the rays are *curved.* In an optically inhomogeneous medium the integration of (3) is the simplest method for determining the wave surfaces and the ray directions.

In an optically homogeneous medium with n = constant, one obtains as the simplest solution of (3) the *linear function*

$$S = n\,(\alpha\,x + \beta\,y + \gamma\,z) \qquad \text{where} \qquad \alpha^2 + \beta^2 + \gamma^2 = 1. \tag{5}$$

This function contains two arbitrary constants, e. g. α and β. The *wave surfaces* determined by this solution are *planes,* the *rays* are *parallel straight lines* in the direction $\alpha:\beta:\gamma$, since, indicating the three components,

$$\mathbf{grad}\,S = n\,(\alpha, \beta, \gamma). \tag{5 a}$$

For constant n the simplest solution with one singular point is the *spherical wave*

$$S = n\, r, \qquad r = \sqrt{x^2 + y^2 + z^2}, \qquad \text{grad}\, S = \frac{n}{r}(x, y, z). \tag{6}$$

The simplest solution with one singular straight line corresponds to the cylindrical wave

$$S = n\, \rho, \qquad \rho = \sqrt{x^2 + y^2}, \qquad \text{grad}\, S = \frac{n}{\rho}(x, y). \tag{7}$$

In both these cases, and quite generally in homogeneous media, the rays are *straight lines*.

The general solution is obtained by starting with any arbitrary surface and constructing a family of parallel surfaces to it (surfaces of constant infinitesimal spacing).

We have thus obtained the simplest mathematical scheme for the understanding of the formation of shadows. We consider the light source as given. From it there emerge rectilinear rays. A screen shall be called opaque if it absorbs all rays falling on it and does not itself emit any rays. Then the shadow behind the screen is bounded by straight ray directions which emerge from the light source. In the direction perpendicular to the limit of the shadow A decreases discontinuously to zero which, as we have seen, is compatible with eq. (4). *In the limiting case as* $\lambda \to 0$ *there is no diffraction.* The rays which do not meet the screen continue unobstructed along straight lines. If several light sources are present, then there are, of course, half-shadow regions.

Geometrical optics has become second nature with us to such an extent that we suppose it to be valid even in cases where we know the rays to be curved. Thus we *see* the sun over the horizon for about 5 minutes after it has actually set. For we project the sun's rays, which because of the inhomogeneity of the earth's atmosphere are curved, along straight lines tangential to the directions of the rays as they meet the eye. The situation is similar in the case of certain diffraction phenomena; see Sec. 38 D. The edge of a screen appears to us as a luminous line because we extrapolate rectilinearly backwards the rays of the cylindrical wave which meet our eye; but in reality the field in the vicinity of the edge of the screen is continuous.

With eqs. (1) to (7) we have made the transition from wave optics to geometrical optics. Schrödinger proceeded in the opposite direction when, guided by the comprehensive ideas of Hamilton, he accomplished the transition from classical mechanics to wave mechanics. As described in Vol. I, Sec. 44, Hamilton started with the theory of optical instruments and several years later applied it to general dynamics. The differential equation (3) of the eikonal is a very simple case of Hamilton's partial differential equations of dynamics. In the same sense our eq. (5 a) is a very simple specialization of Hamilton's momentum equation $p_k = \partial S/\partial q_k$. Of course the way was cleared for Schrödinger's theory only after the discovery of the quantum of action by Planck. It should also be noted that the useful W.K.B. method (Wentzel-Kramers-Brillouin approximation) in which the same hypothesis as in (2) is made, also corresponds to the transition from wave optics to geometrical optics.

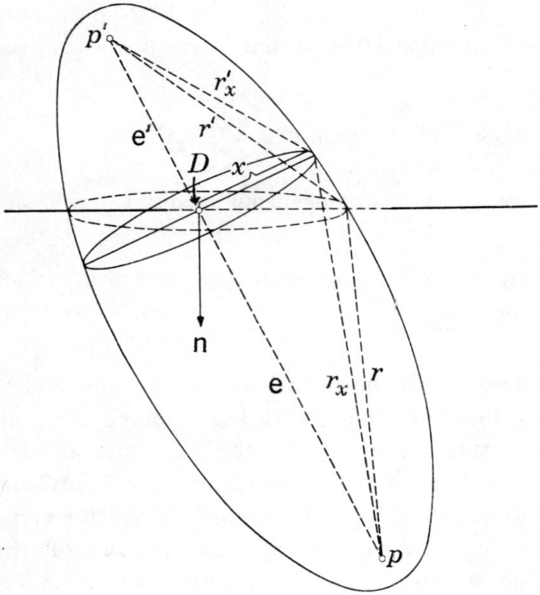

Fig. 59.
Construction of surfaces of constant phase.

B. THE ORIGIN OF THE SHADOW ACCORDING TO WAVE OPTICS

We must now seek a solution of the problem of the shadow by means of *wave optics* instead of, as before, by its asymptotic form, namely *ray optics*. For this purpose we turn again to Huygens' principle.

Let us consider the expression under the integral sign in eq. (34.10). We construct the surfaces of constant phase

$$r + r' = \text{const.} \tag{8}$$

These are ellipsoids of rotation with the common focal points P (point of observation) and P' (light source). We shall call the focal distance $\rho + \rho'$ where, see fig. 59,

$$\rho = P\,D, \qquad \rho' = P'\,D,$$

and where D is the point where the focal line intersects the plane of the screen. Let x be the radius of that circular section of the ellipsoid under consideration which passes through D; the distances of a point on this circle from P and P' shall be r_x and r_x', respectively. Provided P and P' are sufficiently far away, it is seen from the figure that

$$r_x^2 = \rho^2 + x^2, \qquad r_x'^2 = \rho'^2 + x^2, \tag{9}$$

$$r_x = \rho\left(1 + \frac{1}{2}\frac{x^2}{\rho^2} + \ldots\right), \qquad r_x' = \rho'\left(1 + \frac{1}{2}\frac{x^2}{\rho'^2} + \ldots\right). \tag{9 a}$$

From this and (8) it follows that

$$r + r' = r_x + r_x' = \rho + \rho' + p \tag{10}$$

where

$$p = \left(\frac{1}{\rho} + \frac{1}{\rho'}\right)\frac{x^2}{2} + \ldots, \qquad dp = \left(\frac{1}{\rho} + \frac{1}{\rho'}\right) x\,dx, \tag{10 a}$$

and where p is the parameter of the system of ellipses formed by the intersection of the system of ellipsoids of rotation and the plane of the screen. To the circular rings

$$d\sigma_k = 2\pi\, x\, dx \tag{10 b}$$

there correspond in the plane of the screen the elliptical rings $d\sigma_e$, whose areas are proportional to dp. We write

$$d\sigma_e = f\, dp. \tag{11}$$

These $d\sigma_e$ are the proper area elements to be used in the integration with respect to $d\sigma$ in eq. (34.10).

We call $\varphi\,(p)$ the fraction of the elliptical ring $d\sigma_e$ which falls inside the aperture and distinguish the two cases illustrated in fig. 60 a and b:

a) D lies in the screen,

b) D lies in the aperture.

In case a) the integration over p extends from p_1 to p_2. For p_1 and p_2, $\varphi\,(p) = 0$. In case b) the integration starts at $p = 0$. Between $p = 0$ and $p = p_1$, $\varphi\,(p) = 1$. From p_1 to p_2, $\varphi\,(p)$ decreases from 1 to 0.

We combine the factor φ and the factor f in (11) with the factor $\dfrac{\cos\,(n, r)}{r\,r'}$ in (34.10) and call their product $F\,(p)$. Then, using (10) and (11), (34.10) becomes for the case a)

$$i\,\lambda\, v_P = A\, e^{ik(\rho+\rho')} \int_{p_1}^{p_2} F\,(p)\, e^{ikp}\, dp. \tag{12}$$

Integration by parts[1] gives

$$\int_{p_1}^{p_2} F(p)\, e^{ikp}\, dp = \frac{1}{ik} F(p)\, e^{ikp}\Big|_{p_1}^{p_2} - \frac{1}{ik}\int_{p_1}^{p_2} F'(p)\, e^{ikp}\, dp. \tag{12 a}$$

For the case illustrated in fig. 60 a the first term on the right-hand side of (12 a) vanishes because F contains the factor φ and $\varphi(p_1) = \varphi(p_2) = 0$. As k goes to infinity, the integral in the second term likewise goes to zero. Therefore the right-hand side of (12) vanishes even after division by $i\lambda = 2\pi i/k$ and from (12),

$$v_P \to 0, \text{ shadow.} \tag{13}$$

The shadow is brought about by interference of the waves originating at the surface elements $d\sigma_e$.

In case b) we obtain in place of (12)

$$i\lambda v_P = A\, e^{ik(\rho+\rho')} \int_0^{p_2} F(p)\, e^{ikp}\, dp. \tag{14}$$

Integration by parts yields instead of (12 a)

$$\int_0^{p_2} F(p)\, e^{ikp}\, dp = \frac{1}{ik} F(p)\, e^{ikp}\Big|_0^{p_2} - \frac{1}{ik}\int_0^{p_2} F'(p)\, e^{ikp}\, dp. \tag{14 a}$$

The second term on the right-hand side again vanishes for the case illustrated in fig. 60 b as $k \to \infty$. The first term vanishes at its upper limit because $\varphi(p_2) = 0$. At the lower limit we have, according to fig. 59

$$r = \rho, \quad r' = \rho', \quad \frac{\cos(n, r)}{r r'} = \frac{(\cos n, \rho)}{\rho\rho'}, \quad \varphi(0) = 1,$$

$$d\sigma_e = \frac{d\sigma_k}{\cos(n, \rho)} = f\, dp,$$

and according to eqs. (10 a, b)

$$dp = \left(\frac{1}{\rho} + \frac{1}{\rho'}\right) x\, dx = \frac{1}{2\pi} \frac{\rho+\rho'}{\rho\rho'}\, d\sigma_k,$$

$$f = \frac{1}{\cos(n, \rho)} \frac{d\sigma_k}{dp} = \frac{2\pi}{\cos(n, \rho)} \frac{\rho\rho'}{\rho+\rho'}. \tag{14 b}$$

Hence

$$F(0) = \varphi(0)\, f\, \frac{\cos(n, \rho)}{\rho\rho'} = \frac{2\pi}{\rho+\rho'}.$$

---

[1] Although the derivative $F'(p)$ which occurs in (12 a) can under certain circumstances become infinite at the limits of integration, a closer investigation in Sec. 36 D will show that the convergence of the integral is nevertheless preserved.

This lower limit therefore yields for the value of the right-hand side of (14 a)

$$-\frac{2\pi}{ik}\frac{1}{\rho+\rho'}=\frac{i\lambda}{\rho+\rho'}.$$

Dividing (14) by $i\lambda$ we obtain therefore

$$v_P = A\,\frac{e^{ik(\rho+\rho')}}{\rho+\rho'}. \tag{15}$$

This is the incident spherical wave at the distance $\rho+\rho'$ *from the light source.*

Equations (15) and (13) contain the Fresnel theory of the phenomenon of "light and shadow"; they make it understandable from the optical point of view that light "in general" propagates along *straight lines.*

"In general" means that there are exceptions, as we shall see below under C and D, and in particular in Sec. 36 D, where we shall investigate Fraunhofer diffraction caused by screens with straight edges.

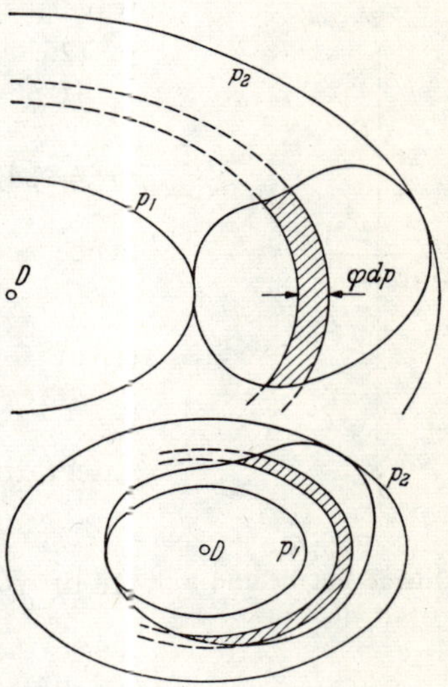

Fig. 60 a, b.

Intersection of the surfaces of constant phase with the plane of the screen.

a) The point D defined in fig. 59 lies in the screen; upper figure.

b) D lies in the opening, lower figure.

In each case the irregular curve represents the edge of the diffraction opening; P_1 and P_2 are, respectively, the parameters of the smallest and largest ellipses which touch the edge of the opening.

C. Diffraction behind a circular disc

The result derived in (13) suffers an exception if a finite portion of the edge of the screen coincides with one of the two bounding ellipses p_1 = constant or p_2 = constant. For then $\varphi(p_1)$ or $\varphi(p_2)$ is not zero and the first term on the right-hand side of eq. (12 a) does not vanish. Therefore (13) is no longer valid. *There is no shadow*; we may speak of *diffraction at the elliptically curved portion of the edge of the screen.*

A particular example of this situation is a screen which consists of a circular disc with the points P and P' lying on the perpendicular through the disc's center. Then the point D is at the center of the disc. The ellipses p = constant become the circles x = constant (notation the same as in fig. 59). The diffraction opening consists of the whole exterior of the circular disc $a < x < \infty$ (a = radius of the disc). Equation (34.10) becomes then

$$(16) \qquad i\lambda v_P = A \int_a^\infty e^{ik(r+r')} \frac{\cos(n,r)}{r r'} 2\pi x\,dx.$$

For purposes of a later application we let $\rho' = \rho$ and $r' = r$, which will also simplify the evaluation of (16). We must emphasize, however, that this specialization does not affect the result. It would be just as convenient to choose $\rho' = \infty$, that is, to have an incident plane wave instead of a spherical wave; cf. footnote [1] on p. 125.

If $\rho' = \rho$, we find (see fig. 61)

$$r^2 = \rho^2 + x^2 = r'^2, \qquad x\,dx = r\,dr, \qquad \cos(n,r) = \frac{\rho}{r};$$

hence according to (16)

$$(16\text{ a}) \qquad i\lambda v_P = 2\pi A\,\rho \int_{\sqrt{\rho^2+a^2}}^\infty e^{2ikr}\frac{dr}{r^2}.$$

Integrating by parts one obtains

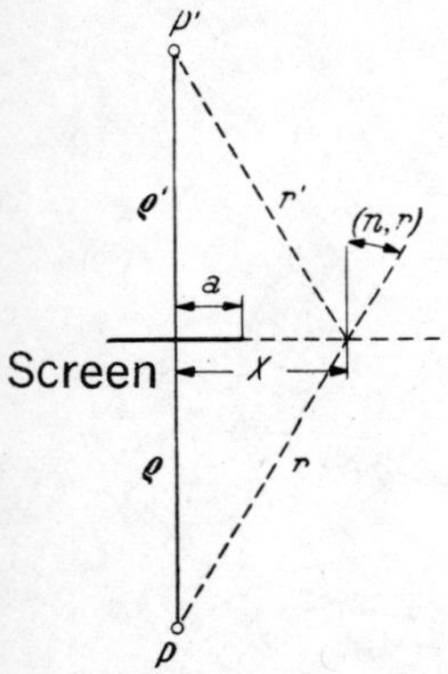

Fig. 61.
Diffraction behind a circular disc.

$$(16\text{ b}) \qquad i\lambda v_P = \frac{2\pi A\,\rho}{2ik}\left\{\frac{e^{2ikr}}{r^2}\Bigg|_{\sqrt{\rho^2+a^2}}^{\infty} + 2\int_{\sqrt{\rho^2+a^2}}^\infty e^{2ikr}\frac{dr}{r^3}\right\}.$$

If the second term in the bracket { } is again integrated by parts, it will contain the factor $\dfrac{1}{2ikr}$, which shows that this term is almost completely eliminated by interference. Disregarding, therefore, the second term, we obtain from (16 a)

$$(16\text{ c}) \qquad i\lambda v_P = -\frac{2\pi A\,\rho}{2ik}\frac{e^{2ik\sqrt{\rho^2+a^2}}}{\rho^2+a^2}.$$

If we introduce the following notation for the primary excitation at the edge of the disc

$$v_{P'} = A\frac{e^{ikr'}}{r'} = A\frac{e^{ik\sqrt{\rho^2+a^2}}}{\sqrt{\rho^2+a^2}},$$

then we can simplify (16 c), after cancelling the factor $i\lambda$, as follows

$$v_P = \frac{1}{2}\frac{\rho}{\sqrt{\rho^2+a^2}}e^{ik\sqrt{\rho^2+a^2}}\,v_{P'}.$$

Writing this in terms of the intensities $J = |v_P|^2$ and $J_0 = |v_{P'}|^2$, we obtain

$$J = \frac{1}{4} \frac{\rho^2}{\rho^2 + a^2} J_0. \tag{17}$$

This paradoxical result is represented graphically in fig. 62. *There is no darkness anywhere along the central perpendicular behind an opaque circular disc* (except immediately behind the disc). The relative intensity increases with increasing distance between the light source and the point of observation. For very large distances the intensity at the point of observation approaches one-fourth of the intensity at the edge of the disc[1]. The primary light waves pass around the edge of the disc along its whole circumference, and because of the symmetry of the arrangement, they meet along the central perpendicular with equal phases. The result is in striking contradiction with the rectilinear ray paths postulated by geometrical optics and with the shadow boundary to be expected according to the latter. We must note, however, that the intensity given by (17) is to be expected only in the immediate vicinity of the central perpendicular, because only there do the lines p = constant coincide with the edge of the disc. At a small distance from this central line the complete shadow predicted by (13) will be observed.

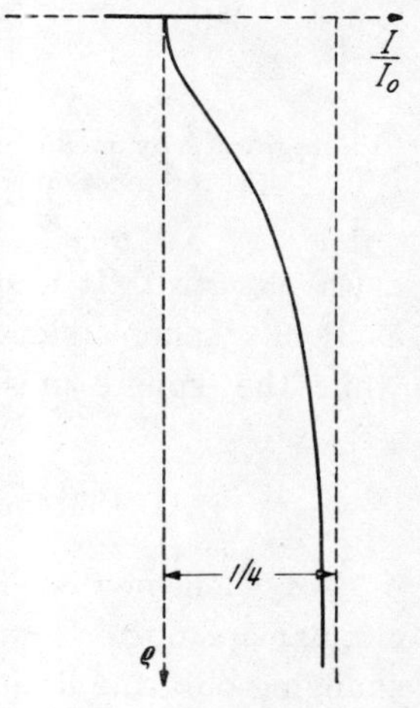

Fig. 62.
Relative light intensity J/J_0 along the axis behind a circular disc.

Poisson predicted the brightness along the axis as a consequence of Fresnel's theory of the shadow and cited it as an *objection* to that theory[2]. Therefore, or perhaps nevertheless, this phenomenon is called *Poisson diffraction.* It takes place behind an opaque sphere as well as behind a circular disc. In the case of radio waves it has been possible to detect an increase in the strength of the signal at the point of the earth which is antipodal to the primary antenna.

---

[1] If we illuminate the disc with a plane wave [$\rho' \to \infty$ in eq (16)] instead of with a spherical wave, then the factor 1/4 in (17) disappears. Therefore $J \to J_0$ as $\rho \to \infty$. At a sufficiently large distance the disc cannot be seen; the primary light wave appears undisturbed.

[2] The crucial experiment was performed by Arago and Fresnel. One therefore often speaks of an *Arago spot* instead of a *Poisson spot.* The reader is also referred to the experiment of W. Kossel who obtained stronger intensities and whose experimental arrangement had a deeper significance, Z. f. Naturforschung, Vol. 3a, p. 496 (1948).

"A photographic objective can be replaced by a steel sphere." This conclusion was drawn by R. W. Pohl in his *Einführung in die Optik* and proved by means of his illustration 185. We are fortunate in being able to show in fig. 63 a photograph taken by our colleague E. von Angerer. A sheet metal disc 50 mm. in diameter served as a "lens". The distance of the object and the plate from the disc was 35 m. each. The object used is considerably more complex and rich in detail than the simple monogram which had been used by Pohl. Though contrasts are considerably weakened in the picture, it is surprisingly true to the original; v. Angerer found that in order to obtain a sharp picture it was essential that the edge of the disc should be a precise circle (theoretically even to within the order of magnitude of a wavelength!). It is surprising that the circular disc should perform as well as Pohl's sphere, since the disc can depict only the central ray precisely while the sphere presents a circular cross section to all rays.

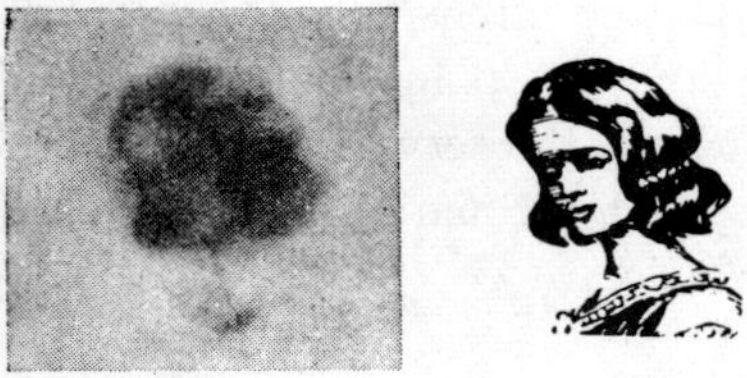

Fig. 63.
"Photograph" by means of a circular sheet metal disc.

D. The circular opening and Fresnel zones

We shall now consider the complementary arrangement, namely the circular opening. Retaining all of the above assumptions and notations and changing only the limits of integration in eq. (16) from $x = a$ and $x = \infty$ to $x = 0$ and $x = a$, we obtain in place of eq. (16 a)

$$i\lambda v_P = 2\pi A\rho \int_{\rho}^{\sqrt{\rho^2+a^2}} e^{2ikr}\frac{dr}{r^2}. \tag{18}$$

If we integrate this by parts and retain only the first order term, we get

$$i\lambda v_P = \frac{2\pi A\rho}{2ik}\frac{e^{2ikr}}{r^2}\Bigg|_{\rho}^{\sqrt{\rho^2+a^2}} \tag{18 a}$$

$$= -\frac{2\pi A}{2ik\rho}e^{2ik\rho}\left\{1 - \frac{\rho^2}{\rho^2+a^2}e^{2ik(\sqrt{\rho^2+a^2}-\rho)}\right\}.$$

In order to put this expression into a more convenient form, we shall neglect a^2 in comparison with ρ^2 in the factor $\frac{\rho^2}{\rho^2+a^2}$. The exponent, however, which contains the factor k must of course be evaluated more precisely. Accordingly, we put in the exponent

$$\sqrt{\rho^2 + a^2} - \rho = \rho \left\{ \sqrt{1 + \frac{a^2}{\rho^2}} - 1 \right\} = \frac{1}{2} \frac{a^2}{\rho}.$$

After dividing, as before, by the factor $i\lambda$, (18 a) becomes

$$v_P = \frac{A}{2\rho} e^{2ik\rho} \left\{ 1 - e^{\frac{ika^2}{\rho}} \right\} \tag{18 b}$$

$$= \frac{A}{2\rho} e^{2ik\rho} e^{ik\frac{a^2}{2\rho}} \left\{ -2i \sin \frac{ka^2}{2\rho} \right\}.$$

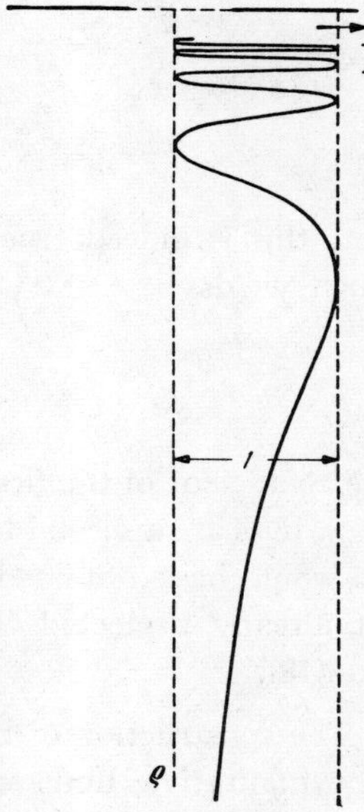

Fig. 64.
Relative intensity behind a circular opening.

Introducing again the primary intensity calculated at the edge of the screen $J_0 = \dfrac{A^2}{\rho^2 + a^2} \sim \dfrac{A^2}{\rho^2}$, we find the intensity $J = |v_P|^2$ to be

$$J = J_0 \sin^2 \frac{ka^2}{2\rho}. \tag{19}$$

The relative intensity J/J_0 is plotted in fig. 64. This quantity has an infinite number of maxima and minima which have their limit in the vicinity of the screen. All maxima have the magnitude unity, and all minima are zero. Thus the paradox represented by fig. 62 has been aggravated. *While the central axis behind a circular screen is nowhere dark, the central axis behind a circular opening has an infinite number of dark places.*

This last statement is of course only valid when the illumination is monochromatic. If white light is used, the central axis appears colored, the color alternating along its length.

The fundamental difference between the formulae (17) and (19) immediately shows that there exists no simple relation between the *intensities* of the two complementary cases of the disc, J_1 eq. (17), the opening, J_2 eq. (19), and the primary intensity J_0. However, the general relation (34.15) between the *amplitudes* v_1, v_2, and v_0 which we called the "Babinet principle" is valid even on the very singular central axis of our diffraction problem. For, when forming the sum of v_1 as given by eq. (16 a) and v_1 as given by (18), we obtain

$$i\lambda (v_1 + v_2) = 2\pi A \rho \int_\rho^\infty e^{2ikr} \frac{dr}{r^2}.$$

On the other hand, for our special arrangement ($\rho' = \rho$) the primary amplitude at the point P is given by

$$v_0 = A \frac{e^{2ik\rho}}{2\rho}.$$

Hence, according to eq. (34.15) the following equality should hold:

$$\int_{\rho}^{\infty} e^{2ikr} \frac{dr}{r^2} = \frac{i\lambda}{4\pi} \frac{e^{2ik\rho}}{\rho^2}.$$

That this is indeed true can be shown by differentiating with respect to ρ, which yields

$$-\frac{e^{2ik\rho}}{\rho^2} = -\frac{2k\lambda}{4\pi} \frac{e^{2ik\rho}}{\rho^2} + \dots$$

The coefficient of the first term on the right-hand side is equal to 1. The dots ..... indicate a second term which vanishes as $1/k$ compared to the first term. We would have obtained the precise equality demanded by (34.15) if we had not already neglected the corresponding higher order terms in eqs. (16 a) and (18).

The construction of the *Fresnel zones* provides us with a pictorial, though only qualitative, understanding of these results. About the light source P' as a center we construct a set of spheres which intersect the plane of the screen in a set of circles $K_1, K_2, \dots, K_n, \dots$ We choose the radii of the spheres in such a way that the light paths from P' via K_n to the point of observation P and from P' via K_{n+1} to P differ by $\lambda/2$. The distances of P' and P from the plane of the screen (which we called ρ' and ρ before) shall be a and b, respectively; r_n' and r_n shall be the light paths $P' K_n$ and $K_n P$, respectively. The straight line of length $a + b$ which connects P' with P intersects the plane of the screen in a point K_0 (circle of radius 0), which is also the common center of the family of circles K_n. According to Fresnel's procedure the following equalities characterize the circles $K_1, K_2, \dots$:

$$r_1' + r_1 = a + b + \frac{\lambda}{2}, \qquad r_2' + r_2 = r_1' + r_1 + \frac{\lambda}{2}, \dots$$

By adding the first n of these equations we have for K_n:

$$r_n' + r_n - a - b = n\frac{\lambda}{2}. \tag{20}$$

The radius x_n of this n^{th} circle is calculated as in (9), (9 a) in the following way:

$$r_n'^2 = a^2 + x_n^2, \qquad r_n^2 = b^2 + x_n^2,$$

$$r_n' = a + \frac{1}{2}\frac{x_n^2}{a} + \ldots, \qquad r_n = b + \frac{1}{2}\frac{x_n^2}{b} + \ldots$$

$$r_n' + r_n - a - b = \frac{1}{2}\left(\frac{1}{a} + \frac{1}{b}\right) x_n^2 + \ldots$$

Hence, according to (20)

$$x_n = \sqrt{n\,\lambda\, f} \quad \text{where} \quad \frac{1}{f} = \frac{1}{a} + \frac{1}{b}. \tag{21}$$

This expression f (which agrees with the definition of the f used in (14 b) except for the factor 2π) reminds us of the well-known formula for the focal length f of a lens. For the present, however, we shall consider f only as a convenient abbreviation.

Figure 65 shows a system of Fresnel zones which consist of the sequence of circular rings K_n, K_{n+1}. These rings are alternately denoted with the signs $+$ and $-$. If we consider the phase in the central zone to be positive, then because of the path difference $\lambda/2$ the phase in the second zone is negative, and so on. All waves falling on the central zone reinforce one another; they are attenuated by the waves falling on the second zone, reinforced by the third zone, and so on.

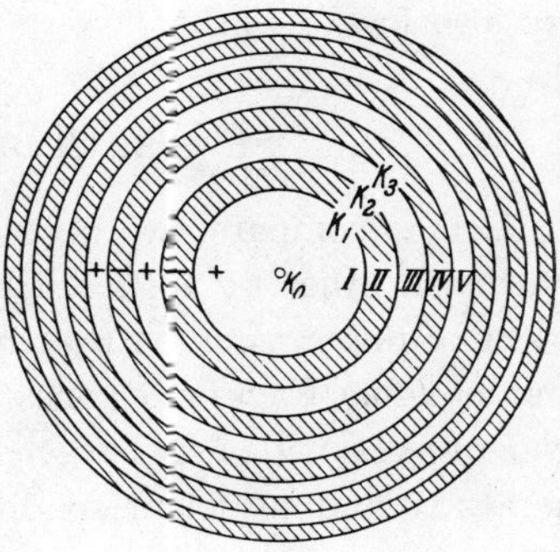

Fig. 65.
The Fresnel zones.

We now compare this process with our formula (19) where we replace a by x_n. This formula yields a *maximum* for

$$\frac{2\pi}{\lambda}\frac{x_n^2}{2\rho} = n\frac{\pi}{2} \quad \text{for } odd\ n,$$

hence for

$$x_n = \sqrt{\frac{n\,\lambda\,\rho}{2}}. \tag{21 a}$$

This result agrees with (21) because in (19) we had assumed that $\rho = \rho'$, and therefore in our present notation $a = b = \rho$ and hence $f = \rho/2$. The same result is obtained for the *minima* if n is even. The fact that all the maxima have the same value and that all the minima are zero can also be checked, though not without some arbitrariness, by summing the contributions of successive zones.

The following consequence was considered particularly surprising in Fresnel's time: a diaphragm which consists only of the central zone yields the same intensity as a very large opening; that is, it gives the full intensity of the incident light. If the opening does not coincide with any of the circles K_n, or if its shape is not circular, then the contributions from partial zones must, of course, be taken into account.

Figure 65 also shows that a "zone plate" (J. L. Suret, 1875) acts like a *lens*. In order to illustrate this the negative zones have been shaded. If these negative zones are covered up or blackened, then all the remaining positive zones act to reinforce one another and produce an intensity which is four times as large as the incident intensity. This resulting zone plate has the focal length f. Since f, like the zones themselves, depends on the wavelength, our "lens" has a strong "chromatic aberration". The submultiples f/n of f are also focal lengths.

E. The similarity law of diffraction

Let us compare two objects (openings or screens) which can be mapped into each other by a similarity transformation. We arrange the source and point of observation so that both objects contain the same number of zones and possible fractional zones. Then the diffraction patterns caused by the two objects will also be geometrically similar. According to eq. (21) the necessary and sufficient condition for this is that the dimensionless quantity

$$\text{(22)} \qquad \frac{x}{\sqrt{\lambda f}} \qquad (x = \text{an arbitrary linear dimension of the object})$$

shall have the same numerical value for both arrangements. This will be called the *similarity law of diffraction*.

It is often said that diffraction phenomena are noticeable only for very small objects. However, the similarity law says: the same diffraction phenomena observed with a small object are also observed with an object magnified by a similarity transformation, provided only that the distances of the source and the point of observation from the object are correspondingly magnified. To a magnification factor q of the linear dimensions of the object there corresponds a magnification factor q^2 of these distances. Conversely, if one wishes to observe the diffraction phenomena due to a large object at large distances in the laboratory where the distances are reduced by a factor q, then the dimensions of the object need be reduced only by a factor $\sqrt{q}$. On the basis of this law W. Arkadiew[1] performed a set of very

[1] Physikal. ZS. Vol. 14, 1913, p. 832.

interesting model experiments. As an example let us consider the following macroscopic object: a dinner plate of ordinary size held by a hand. In a laboratory at Moscow a distance $a + b = 40$ m. between the light source and the photographic plate was available. At that distance the picture of the shadow (suitably reduced to the dimensions of the photographic plate) shows, of course, no diffraction pattern but corresponds to the shadow of geometrical optics.

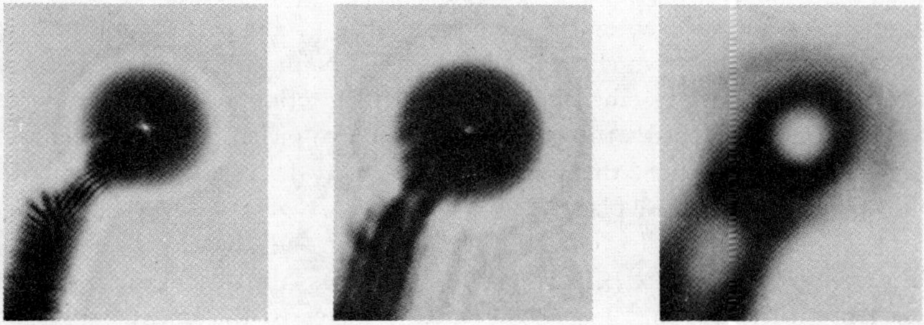

Fig. 66 a, b, c.

Illustration of the similarity law of diffraction. Photographs by Arkadiew.
a) $a + b = 7$ km, b) $a + b = 29$ km, c) $a + b = 235$ km.

We now inquire about the appearance of the shadow at a distance $a + b = 7$ km. In order to discover this pattern in the laboratory, we must use the reduction factor

$$q = \frac{40}{7000}, \qquad \sqrt{q} \sim \frac{1}{13},$$

where q applies to the distances a, b and hence also to f; $\sqrt{q}$ applies to all linear dimensions of the object. Arkadiew cut a model of the macroscopic object reduced by 1/13 out of thin sheet metal. The photographic plate showed the image pictured in fig. 66 a: the plate has received a hole (Poisson spot) and a white edge; the wrist contains bright fringes; the sleeve below the wrist is fringed.

Pictures of the shadow for $a + b = 29$ km. and 235 km. are produced by models with the reduction factors

$$\sqrt{q} = \sqrt{\frac{40}{29{,}000}} \sim \frac{1}{27}, \qquad \sqrt{q} = \sqrt{\frac{40}{235{,}000}} \sim \frac{1}{77}, \quad \text{respectively.}$$

In fig. 66 b the whole arm contains diffraction fringes. Figure 66 c shows only slight similarity to the original: the Poisson spot in the center of the plate has become enlarged and a second bright spot has appeared in the sleeve.

36. Fraunhofer Diffraction by Rectangles and Circles

To observe Fraunhofer diffraction (fig. 58) one looks through a diffracting aperture at an infinitely far removed light source with the aid of a telescope which is focused at infinity. As shown in fig. 58 such a source can be realized by placing a point source or an illuminated slit in the focal plane of a collimator lens. The position of the diffraction opening with respect to lens and telescope is in principle immaterial. However, in practice the opening is placed directly in front of the objective of the telescope so that waves which are diffracted at large angles will also enter the telescope. The eyepiece is focused on the focal plane of the objective (E in fig. 58). Every point P on this plane corresponds to a plane wave emerging from the diffraction opening. A corresponding plane wave enters the eye through the eye piece. (As was remarked in connection with fig. 58, the visual observer may be replaced by a photographic plate in the focal plane E.)

Because all ray bundles entering and leaving the opening are parallel, we must set $R = R' = \infty$ in eq. (34.13). As has already been noted in connection with (34.14 a), the phase Φ then reduces to the linear expression

$$\Phi = a\,\xi + b\,\eta, \qquad a = \alpha - \alpha_0, \qquad b = \beta - \beta_0 \tag{1}$$

and the evaluation of the integral in (34.14) becomes elementary.

Let ξ, η be the cartesian coordinates of an arbitrary point in the diffraction opening (which we shall assume to be plane as before); α, β, γ are the direction cosines of a diffracted bundle of rays; $\alpha_0, \beta_0, \gamma_0$ are the direction cosines of the incoming rays which are all parallel because of the collimator lens. If, in particular, the light source lies in the direction normal to the plane of the opening, then $\alpha_0 = \beta_0 = 0, \ \gamma_0 = 1$.

A. Diffraction by a rectangle

Let the sides of the rectangle be $2\,A$ and $2\,B$ in length. The coordinates of the center shall be $\xi = 0, \eta = 0$. Then we can write for (34.14)

$$v = C\,k \int_{-A}^{+A} e^{-ika\xi}\,d\xi \int_{-B}^{+B} e^{-ikb\eta}\,d\eta, \tag{2}$$

where C is a complex constant which is proportional to the amplitude of the incident light and is independent of the angle between the central ray and the direction of observation. The factor k outside the integral corresponds to the factor λ on the left-hand side of (34.14), which has now been transferred to the right. Performing the integrations we obtain

$$v = C\,k\,\Delta\,\frac{\sin x}{x}\,\frac{\sin y}{y}, \qquad \begin{cases} \Delta = 4\,A\,B = \text{Area of the rectangle}, \\ x = k\,a\,A, \qquad y = k\,b\,B. \end{cases} \tag{3}$$

From this we obtain for the intensity $J = |v|^2$

$$\frac{J}{J_0} = \left(\frac{\sin x}{x}\right)^2 \left(\frac{\sin y}{y}\right)^2. \tag{3 a}$$

As is evident from (3), $J_0 = (C\,k\,\Delta)^2$ is the intensity at the center $a = 0$, $b = 0$ of the diffraction pattern. At that point also $x = 0$, $y = 0$.

The behavior of the function

$$X = \left(\frac{\sin x}{x}\right)^2 \tag{4}$$

is well known: its principal maximum lies at $x = 0$ and has the value 1. The minima of magnitude $X = 0$ are located equidistantly at the points $x = \pm\,\pi$, $\pm\,2\,\pi$, $\pm\,3\,\pi, \ldots$ There are subsidiary maxima at the points where $\tan x = x$, that is, at points which approach $\pm 3/2\pi$, $\pm\,5/2\,\pi, \ldots$ more and more closely as x increases. The values of the subsidiary maxima are in the same order

$$X = 0.047,\ 0.017,\ 0.008, \ldots \tag{4 a}$$

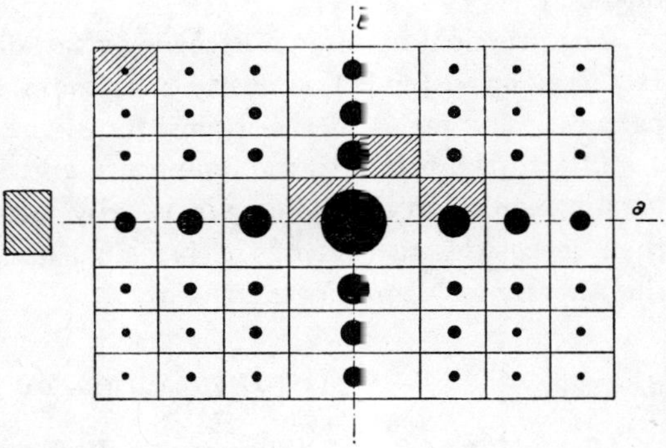

Fig. 67.

Distribution of light resulting from diffraction by a rectangle. The maxima lying between the lines of minimum intensity are symbolized by black circles. Actually they would look more like rectangles than like circles. The shape and orientation of the diffraction opening is shown at the left (not to the same scale, because the dimensions of the diffraction pattern and the opening are not commensurable).

Measured in terms of x the distance between the principal maximum and the first minimum is equal to π. We shall now express this distance in terms of the angular measure $a = \alpha - \alpha_0$. From eq. (3) we obtain

$$\pi = k\,a\,A, \qquad \text{hence} \qquad a = \frac{\lambda}{2\,A}. \tag{5}$$

The smaller the side 2 A of the rectangle is, the larger does the angular distance a become. The same is, of course, true for b and B.

At the left of fig. 67 we have drawn a diffraction opening in the shape of an upright rectangle, $2\,A < 2\,B$. The diffraction pattern on the right is subdivided into rectangles which are geometrically similar to the above but which are in an oblong position. Four of these elementary rectangles make up the rectangular field of the central principal maximum, which is bounded by lines of zero intensity; two each belong to the fields which are bisected by the axes a and b (these axes are shown dotted because they do not form part of the system of lines of zero intensity). A single such rectangle is also indicated in the upper left-hand corner bounded by the two systems of equidistant minimum lines whose spacings are π in the scales both along the x and y directions. In accordance with the equation $\tan x = x$ the subsidiary maxima lie approximately in the centers of their respective fields.

The principal maximum exceeds by far all subsidiary maxima in intensity. It forms an extended intensity maximum in the center of the diffraction pattern. The ratios of the intensities of the subsidiary maxima on the axes a and b to the principal maximum are given by the sequence (4 a). The remaining subsidiary maxima are usually barely visible because the ratios of their intensities to that of the principal maximum are given by products of the already small numbers in (4 a).

B. Diffraction by a slit

Our rectangle $2A$, $2B$ becomes a slit if we increase B until it is very much larger than A. As we increase B, the diffraction pattern parallel to the b-axis will contract more and more. We shall assume that the light source is a distant luminous line whose separate line elements emit *incoherent* light. Therefore, we shall have to add the *intensities* resulting from different line elements. Since the direction of the incoming rays is given by α_0, β_0, we shall have to perform an integration with respect to $b = \beta - \beta_0$ between certain limits $\pm b_1$ which correspond to the length of the collimator slit. Hence, according to (3), we must write

$$\int_{-b_1}^{+b_1} \left(\frac{\sin y}{y}\right)^2 db = \frac{1}{kB} \int_{-y_1}^{+y_1} \left(\frac{\sin y}{y}\right)^2 dy \qquad \text{where} \qquad y_1 = k\, b_1\, B.$$

Because of the large value of k, the limits of integration $\pm y_1$ may be treated as very large numbers, even though the values of B and b are experimentally limited. Therefore, except for terms which vanish as $1/y_1$, the above integral can be replaced by[1]

$$\int_{-\infty}^{\infty} \left(\frac{\sin y}{y}\right)^2 dy = \pi.$$

[1] This value is most easily obtained by the method of complex integration; see Vol. VI, exercise I.5, where Dirichlet's discontinuous factor $\int (\sin y/y)\, dy$ is treated by this method.

We see, therefore, that the intensity of the diffraction pattern is only a function of $x = k\,a\,A$ and is identical with the expression X in (4) except for a constant factor. Thus the slit likewise produces a principal maximum at $x = 0$ and almost equidistant subsidiary maxima, the intensities of which are scarcely noticable in comparison with the intensity of the principal maximum.

We shall now use this result to fill in a gap in Sec. 32. In eq. (5) of that paragraph we separated the intensity of the grating spectrum into two factors, the second of which was calculated from the sequence of grating lines. The first factor $f^2(\alpha)$ which resulted from the width and shape of each individual grating line was left undetermined. At least in certain very simple cases this factor is given just by our expression X in (4); by definition of x, X is clearly a function of $a = \alpha - \alpha_0$. We shall now investigate the influence of this factor on the intensity distribution of a grating (we may now set the intensity of the incident light equal to 1).

For this purpose we write down the more complete form of eq. (32.5):

$$(6) \qquad J = \frac{\sin^2 x}{x^2} \frac{\sin^2 N \dfrac{\Delta}{2}}{\sin^2 \dfrac{\Delta}{2}} \qquad \left\{ \begin{aligned} x &= \frac{2\pi\, a\, A}{\lambda} \\ \Delta &= \frac{2\pi\, a\, d}{\lambda}. \end{aligned} \right.$$

The quantity $2\,A$, which above was the width of the slit, is now the width of each individual grating line; d is again the spacing of the grating lines. In the gratings which Fraunhofer originally made, d was very large compared to $2\,A$. If this is so, then according to (6) x increases only by the small quantity A/d while Δ changes by 1, and the first factor on the right-hand side of eq. (6) varies slowly compared to the second factor. As was mentioned at the end of Sec. 32 A, this first factor has the effect of weakening the grating spectra of higher orders in comparison with the first order spectrum. The intensity pattern given by the second factor in (6) as represented in fig. 53 remains qualitatively unchanged.

Just as the results on diffraction of a slit serve to complete our previous theory of the line grating, so the results on diffraction of a rectangle as given by eq. (3) yield the function $f(\alpha, \beta)$ which was left undetermined in the theory of the cross grating, eq. (32.8).

C. The circular aperture

The circular opening is obviously of tremendous importance to the theory of the telescope, the microscope, and the photographic lens, as well as to the process of vision.

Clearly, we must introduce polar coordinates to replace the rectangular coordinates ξ, η and a, b of (1). We set

$$\xi = r\cos\varphi, \qquad a = s\cos\psi,$$
$$\eta = r\sin\varphi, \qquad b = s\sin\psi.$$

r is the distance from the center of the opening; s is the sine of the deflection angle between the diffracted ray and the perpendicularly incident ray. Denoting the radius of the aperture by the, again available, letter a, we we obtain instead of (2)

$$v = C\,k \int_0^a r\,dr \int_{-\pi}^{+\pi} e^{-ikrs\cos(\varphi-\psi)}\,d\varphi. \tag{7}$$

The φ integral cannot be evaluated by elementary methods, but it is well known to us from Vol. II, Sec. 27 and Vol. III, Sec. 22 as the Bessel function J_0. For further details see Vol. VI, Chap. IV. We recall here the formulae

$$J_0(\rho) = 1 - \frac{1}{(1!)^2}\left(\frac{\rho}{2}\right)^2 + \frac{1}{(2!)^2}\left(\frac{\rho}{2}\right)^4 - \frac{1}{(3!)^2}\left(\frac{\rho}{2}\right)^6 + \dots \tag{8}$$
$$= \frac{1}{2\pi}\int_{-\pi}^{+\pi} e^{\pm i\rho\cos\alpha}\,d\alpha,$$

$$J_1(\rho) = \frac{\rho}{2}\left(1 - \frac{1}{1!\,2!}\left(\frac{\rho}{2}\right)^2 + \frac{1}{2!\,3!}\left(\frac{\rho}{2}\right)^4 - \dots\right) = -\frac{d}{d\rho}J_0(\rho) \tag{8 a}$$

and the differential equation

$$\frac{d}{d\rho}\left(\rho\frac{dJ_0}{d\rho}\right) + \rho\,J_0 = 0 \tag{8 b}$$

from which the following relation is obtained:

$$\int_0^\rho \rho' J_0(\rho')\,d\rho' = \rho\,J_1(\rho). \tag{8 c}$$

We are also acquainted with the asymptotic representations for large ρ

$$J_0(\rho) = \sqrt{\frac{2}{\pi\rho}}\cos\left(\rho - \frac{\pi}{4}\right), \qquad J_1(\rho) = \sqrt{\frac{2}{\pi\rho}}\sin\left(\rho - \frac{\pi}{4}\right). \tag{8 d}$$

Using these results, eq. (7) can be written simply as

(9) $$v = 2\pi C k \int_0^a J_0(k r s)\, r\, dr = \frac{2\pi C k}{k^2 s^2} \int_0^{ksa} J_0(\rho')\, \rho'\, d\rho'$$
$$= \frac{2\pi C a}{s} J_1(k s a).$$

For $s = 0$ (center of the diffraction pattern, $\alpha = \alpha_0$, $\beta = \beta_0$) (8 a) yields

(10) $$v = \pi a^2 C k.$$

The zeros of v are given by the zeros of $J_1(\rho)$. The first of these is at

(11) $$\rho_1 = 3.95 = 0.61 \times 2\pi, \quad s_1 = 0.61 \frac{\lambda}{a}.$$

This zero and all the more so the following zeros $\rho_2, \rho_3, \ldots$ are given with sufficient precision by the asymptotic formula (8 d):

$$\sin\left(\rho - \frac{\pi}{4}\right) = 0, \quad \rho_n = \left(n + \frac{1}{4}\right)\pi,$$

(11 a) $$s_n = \left(n + \frac{1}{4}\right)\frac{\lambda}{2a}.$$

The corresponding graph of v is shown in fig. 68. The resulting intensity pattern $|v|^2$ has again a towering maximum at the center which is surrounded by almost equidistant dark rings. Between the dark rings are weaker maxima which rapidly decrease in intensity.

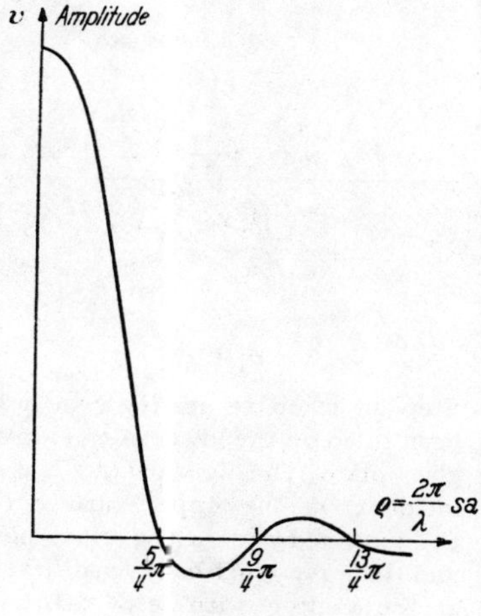

Fig. 68.
The amplitude v behind a circular opening as a function of ksa.

This central intensity maximum which is bounded by the first zero-ring determines the size of the central field produced by the droplets of Sec. 33. Indeed, we used the above expression (11) to calculate the sizes of the coronae about the sun and moon. Because of their weak intensities, the outer maxima indicated by fig. 68 generally do not affect these results.

In Chap. VI we shall discuss the fundamental importance of eq. (11) for the theory of the microscope.

In principle it is not difficult to predict the diffraction patterns of other, particularly polygonally bounded, openings. This was first done by Schwerd[1] in an exemplary fashion.

[1] F. M. Schwerd, Die Beugungserscheinungen aus den Fundamentalgesetzen der Undulationstheorie analytisch entwickelt, Mannheim 1835. Schwerd was a high school teacher in Speyer. He painstakingly colored all the figures in the whole edition of his book by hand.

D. Phase gratings

In formula (32.3) we assumed the grating lines to be linear light sources which radiated in all directions when excited by an incident wave. The nonuniform distribution of this radiation for the various directions α was taken into account by the function $f(\alpha)$ which remained undetermined there. Huygens' principle has now enabled us to calculate this function [eq. (6)] for slit openings of arbitrary widths by setting the field excitation in the opening equal to the unperturbed incident wave. In this way it has been possible to determine the diffraction field of wire gratings or of gratings which are ruled on a silver layer deposited on a glass plate. Fraunhofer produced such gratings which shall be called "*amplitude gratings*", because for perpendicular incidence the phase is constant over the plane of the grating — the plane over which the integration is to be extended, according to Huygens' principle. On this plane only the amplitude varies between zero at points which are on the metal and some constant value at points on the glass. The situation is different with the modern very closely ruled gratings. In these, groove follows upon groove in such a way that one cannot speak of a plane surface. These gratings are illuminated fairly *uniformly*, that is, with essentially constant amplitude, over their whole extent. But the *phase varies* because the points of the grating surface penetrate to varying depths into the optical field in the medium of different index of refraction. A plane wave-surface which is perpendicularly incident on the grating meets these different points at *different* times. Therefore, different points on the grating surface *radiate their elementary Huygens' waves with different phases*. Devices of this type are called "*phase gratings*". The general grating properties derived in Sec. 32 remain unchanged for phase gratings, but the directional distribution of the radiation from any one grating element, that is, the function $f(\alpha)$, can be affected in many different ways. It is, for instance, possible to divert the major part of the incident energy into a single spectrum of a given order on one side and to suppress almost completely all other spectra, in particular that of zeroth order.

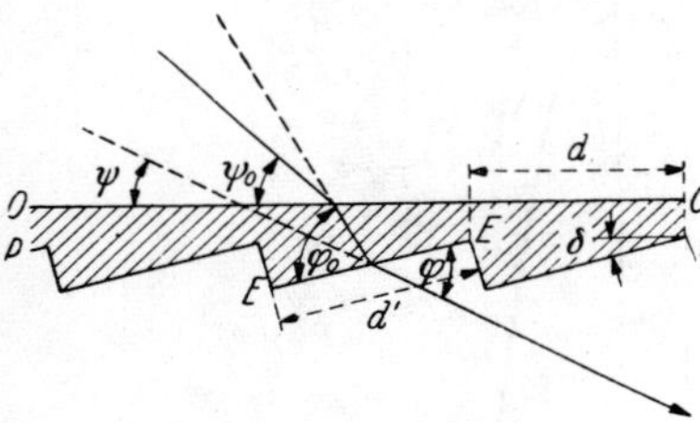

Fig. 69.

Step or echelette grating which has been ruled on the lower side of a plane glass plate. The figure shows the ray incident on the upper surface OO, the refracted ray in the glass plate, and the wave which has been diffracted at the angle φ with respect to $E\,E$.

We shall now calculate the function $f(\alpha)$ for several shapes of grating surfaces. Since the *surfaces are no longer plane*, and we therefore do not know their Green's functions, we are forced to use Huygens' principle in the old Kirchhoff formulation of Sec. 34 B.

Kirchhoff's assumption that the incident wave proceeds *unperturbed* up to the surface of the grating limits our calculations to *large grating constants* $d \gg \lambda$, to grating elements which are *not too deeply cut*, and to moderate angles of incidence and diffraction. Otherwise the radiation proceeding from one part of each grating element to another would affect the results of the calculation. In order to be able to apply previous results, we shall also limit ourselves to grating elements which are bounded by *plane* surfaces.

First we consider a step profile $P\,P$ which has been cut into the lower surface of a plane glass plate; see fig. 69. $O\,O$ shall be the upper surface of the glass plate; $E\,E$ is the grating element under consideration. We illuminate the plate from above and observe the light below the plate. Just as Kirchhoff used the unperturbed *unbounded* wave as an approximation for the wave in the bounded diffraction opening, so we must use an unperturbed plane wave emerging from an unbounded glass plane to approximate the wave originating from our closely bounded plane surface $E\,E$ (of width d'). As a result we shall use in the Kirchhoff formula (34.4) the values v and $\partial v/\partial n$ of the *refracted* wave emerging from the glass plate. The ray which is refracted upon incidence on the upper surface $O\,O$ of the glass plate determines the angle of incidence φ_0 on the step $E\,E$. The angle of diffraction with respect to the step surface shall be φ. We call the direction cosines of the incident and diffracted waves

$$a_0 = \cos\varphi_0, \qquad a = \cos\varphi,$$

and φ_1 the angle at which the refracted ray (not drawn in the figure) would emerge if the plane $E\,E$ were infinite: $\cos\varphi_1 = n\,a_0$.

Referring to our very first treatment of the problem of refraction in Sec. 3, we rewrite eq. (3.1a) in our present notation $\frac{\cos}{\sin}\varphi_1$ in place of $\frac{\sin}{\cos}\beta$; because the ray emerges into air, we write k instead of k_2; the x-axis lies now in the plane $E\,E$; hence $y = 0$ on $E\,E$:

$$\mathsf{E} = B\, e^{i k (x \cos\varphi_1 - y \sin\varphi_1)}.$$

Identifying v and $\partial v/\partial n$ with E and $\partial \mathsf{E}/\partial y$, we obtain at $y = 0$

$$v = B\, e^{i k n a_0 x} \qquad \text{and} \qquad \frac{\partial v}{\partial n} = -\sin\varphi_1\, i\, k\, v.$$

The other wave function u which occurs in Green's theorem as a "probe" can be written for the Fraunhofer mode of observation [limit as $r \to \infty$ in (34.1)] as

$$u = e^{-ikax} \quad \text{and} \quad \frac{\partial u}{\partial n} = + \sin\varphi \, i \, k \, u.$$

Substituting this in (34.4), one obtains for the relative amplitude distribution at infinity

$$f(a) = 4\pi v_P = -i\,k\,B\,(\sin\varphi_1 + \sin\varphi) \int_{-\frac{d'}{2}}^{+\frac{d'}{2}} e^{ik(na_0 - a)x}\,dx$$

$$= i\,k\,d'\,B\,(\sin\varphi_1 + \sin\varphi)\,S, \qquad S = \frac{\sin\left\{k\,(n\,a_0 - a)\,\dfrac{d'}{2}\right\}}{k\,(n\,a_0 - a)\,\dfrac{d'}{2}}.$$

The factor in front of the sine quotient S is slowly varying and causes a moderate attenuation for large angles of diffraction; we can disregard this factor. The function $f(a)$ is, therefore, essentially given by S. The curve for S is similar to that of diffraction by a slit (36.6). Its principal maximum is at $a = n\,a_0$ or, what is the same, at $\varphi = \varphi_1$, which is precisely the direction of the refracted ray as determined by geometrical optics. The zeros are arranged symmetrically around the principal maximum at

$$a = n\,a_0 \pm \nu \frac{\lambda}{d'} \qquad \nu = \text{integer}.$$

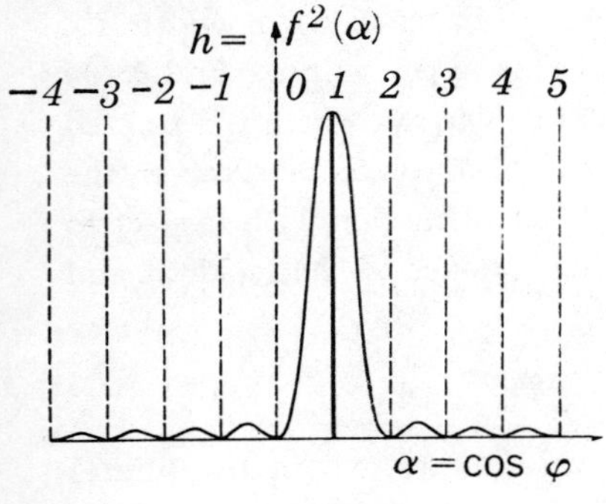

Fig. 69 a.

Intensity distribution for the step grating shown in fig. 69. The curve represents the diffraction pattern due to of a single step. The ordinate at $h = 1$ also gives the intensity of the first order *grating spectrum* which is the only spectrum emitted by the step grating.

On the other hand, let us now consider the grating spectrum which is produced according to (32.4) by the regular sequence of such grating elements at a spacing d. With the definition of d given in fig. 69 we must write in (32.1)

$$\alpha = \cos\psi = \cos(\varphi - \delta) \quad \text{and} \quad \alpha_0 = \cos\psi_0 = n\cos(\varphi_0 - \delta).$$

The grating maxima (which, because of the large number N of grating elements, are very sharp) are at the positions $\Delta/2 = 0 \pm h\pi$, thus at $\alpha = \alpha_0 \pm h\,\lambda/d$.

Their amplitudes are given by the function $f(\alpha)$, which differs from $f(a)$ only because the origin of the angles φ, φ_0 is shifted by δ with respect to the origin of ψ and ψ_0. First we see that for $d \sim d'$, that is, for small angles ψ_0, ψ, and δ, the spacing of these spectra is the same as the spacing of the zeros of $f(\alpha)$. Therefore, if by suitable choice of δ (or if δ is given by suitable choice of ψ_0), one causes the first order maximum, $h = +1$, for instance, to coincide with the principal maximum of $f(\alpha)$, then all other grating spectra coincide with zeros of $f(\alpha)$ and are completely suppressed, including the zero order spectrum. This is illustrated in fig. 69 a where the *intensity* of the spectra, that is, the *square* of $f(\alpha)$, is plotted.

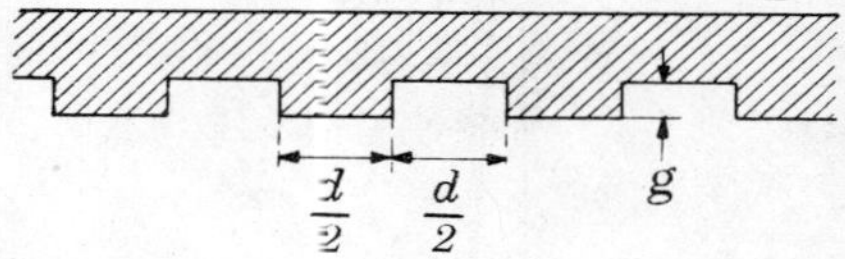

Fig. 70.
A so-called "laminary profile" with grating constant d and depth of slits g.

A reflection grating which is ruled on metal can also be treated by means of the above formulae if we formally set $n = -1$. Then the principal maximum lies, independently of wavelength, in the direction of the geometrically reflected ray. Such gratings were used for the analysis of long infrared waves for which no suitable refracting material is available. Their grating constants amount to fractions of a millimeter. With these spacings the desired step profile is quite easily attainable; these are the so-called echelette gratings. Even in gratings suitable for visible light it is possible to favor one order quite strongly over all others by using a cutting diamond of suitable shape.

As a second example we choose the *rectangular laminary profile*, fig. 70. This profile is produced by evaporating a transparent substance on a plane plate and ruling regular slits into the layer, so that the deposited substance is removed along equidistantly spaced lines. Let us call the thickness of the layer g and its index of refraction n. Then for small angles of incidence the wave falling on *half* of the grating element d is retarded by $2\Theta = (n-1)\, g\, k$. At a large distance the amplitude distribution $f(a)$ arising from a grating element extending from $-d/2$ to $+d/2$ is given by

$$f(a) = \frac{1}{d}\int_{-\frac{d}{2}}^{0} e^{i(\Phi x - \Theta)}\, dx + \frac{1}{d}\int_{0}^{+\frac{d}{2}} e^{i(\Phi x + \Theta)}\, dx = \frac{\sin\left(\Phi\frac{d}{2} + \Theta\right) - \sin\Theta}{\Phi\frac{d}{2}}$$

where $\Phi = k(n\alpha_0 - \alpha)$, and where all nonessential terms have been omitted. This is the diffraction pattern of a *single* step element. Because the step elements are parallel, see fig. 70, the distinction between a, a_0 and α, α_0 becomes meaningless. Therefore we shall henceforth write $f(\alpha)$ in place of the above $f(a)$. The asymmetry of this function with respect to the direction $\Phi = 0$ averages out in the spectrum produced by the whole grating; for, with perpendicular incidence, the directions $\Phi\, d/2 = \pm h\pi$ which go with the two spectra of equal order ($\pm h$) always produce the same contribution

$$f(\alpha_{\pm h}) = \frac{1}{\pi h} [(-1)^h - 1] \sin \Theta.$$

(For $h = 0$ the limit as $\Phi \to 0$ must be taken which yields $f(\alpha) = \cos \Theta$.) The curve $f^2(\alpha)$ is shown in fig. 70 a, and from it the intensities of the grating spectra of different orders can be determined.

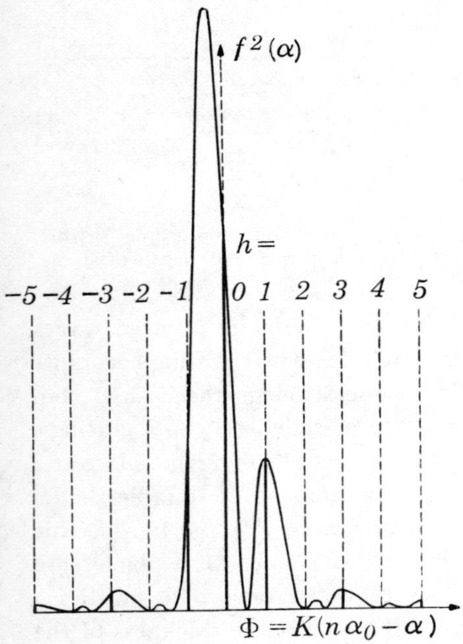

Fig. 70 a.

Intensity distribution for the laminary grating shown in fig. 70. The curve represents the diffraction pattern due to a single grating element. The heavily drawn ordinates indicate the intensities of the grating spectra of orders h which are emitted by the entire array of grating elements.

Because of its application to the microscope, a similarly designed *amplitude grating* is also of interest. Such a grating is obtained if, instead of a transparent material, an absorbing metallic layer is applied to the plane base. Mathematically this means that we must set $\Theta = i\,\Theta'$ (Θ' is real). The ratio of the transmissivities of the two halves of a grating element is then $e^{4\Theta'}$, and Θ' is directly proportional to the thickness of the absorbing layer. Thus we obtain for the zeroth order spectrum $f(\alpha) = \cosh \Theta'$, and for the higher order spectra $f(\alpha_{\pm h}) = [(-1)^h - 1] \cdot \cdot \frac{i}{\pi h} \sinh \Theta'$. The factor i in this equation means that the light in the higher order spectra differs in phase by $\pi/2$ from the zeroth order spectrum. *Therefore, the diffraction pattern of a phase grating can be changed into that of an amplitude grating simply by increasing or decreasing the phase difference between the zeroth order spectrum and all higher order spectra by* $\pi/2$.

This method can be used to replace the formerly very important *staining process* which used to be necessary in the microscopic observation of transparent tissues whose constituents differ by very little in index of refraction but absorb, owing to their chemical difference, different amounts of dye. It is seen that actually the staining method also amounts to *changing a phase grating into an amplitude grating*.

E. Supplement to section 35 B. Light fans arising from polygonally bounded apertures

What is the relationship between the special results derived in A and B of this section and the general theory of the *shadow* which was discussed in Sec. 35 B? In order to answer this question we must first of all specialize eq. (35.12) to the present case of Fraunhofer observation with perpendicular incidence. In that case the point D in fig. 60a lies infinitely far away (if we disregard the center point $\alpha = \beta = 0$). The former surfaces of constant phase are now planes, and the former intersection ellipses have become a family of parallel straight lines. The former parameter p of the system of ellipses is now proportional to the spacing of these straight lines as measured from one of them, for instance, the one passing through the center of the diffraction opening.

Let us first consider a *rectangular* opening and determine the shadow boundary in the direction perpendicular to one of the sides of the rectangle. The straight lines $p =$ constant are parallel to this side, and therefore the segments cut out of these lines by the rectangle are all equal. Hence the fraction $\varphi(p)$ which was introduced in fig. 60 a becomes independent of p. The same is true of the function $F(p)$ occurring in (35.12). Thus $F'(p) = 0$ and the second term on the right-hand side of (35.12 a) vanishes. If, for convenience, we normalize p so that $p_2 = -p_1 = \bar{p}$, then the first term yields

$$\text{(12)} \qquad \frac{F}{i\,k}\left(\epsilon^{ik\bar{p}} - \epsilon^{-ik\bar{p}}\right) = \frac{\lambda F}{\pi} \sin k\,\bar{p}\,.$$

Eq. (35.12) (the factor λ on both sides cancels) leads to the result that in the direction defined above a *light fan of finite intensity* is radiated and *no* shadow appears. It should be emphasized that the sinusoidal variations in the amplitude which are indicated by (12) are blurred out if the light source is extended and not monochromatic. Our name "light fan" is descriptive of this fact. The same is obviously true for the light fans in the directions perpendicular to the other sides of the rectangle; it is *not* true, however, in any other direction which is inclined with respect to these sides. For such directions the values of p_1 and p_2 in (32.12 a) correspond to the corners of the rectangle at which the fraction $\varphi(p)$ continuously decreases to zero. Now the first term on the right-hand side of (32.12 a) vanishes; the second term has a finite but small value. This is illustrated in our (very schematic) fig. 71 a by the quadrants which are left unhatched; these quadrants indicate *shadows*. In the singly hatched strips a finite part of the boundary of the diffraction opening coincides with an effective wave zone, and therefore a *noticeable intensity of diffracted light* is present. This situation is similar to that of the circular opening in fig. 68. Figure 71a should be compared with fig. 67 where the same results are expressed more precisely for monochromatic light.

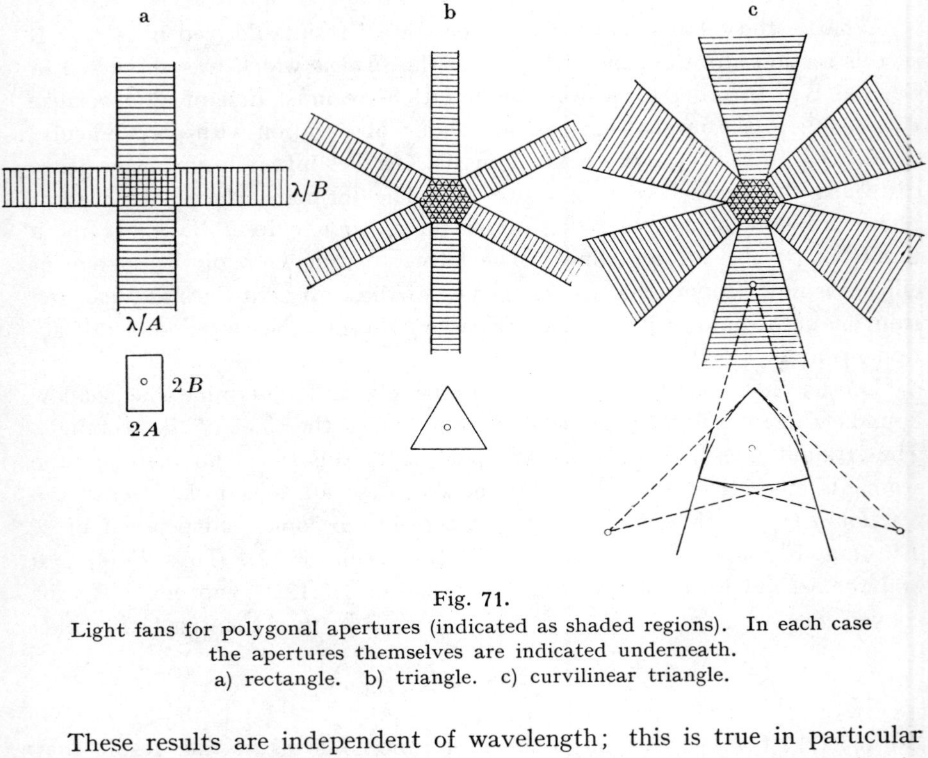

Fig. 71.
Light fans for polygonal apertures (indicated as shaded regions). In each case the apertures themselves are indicated underneath.
a) rectangle. b) triangle. c) curvilinear triangle.

These results are independent of wavelength; this is true in particular for the intensities along the axes a, b in our figure. This seems to be in contradiction to geometrical optics, i. e. the limiting case as $\lambda \to 0$, according to which only the central field should be illuminated. The apparent contradiction is resolved by the fact that in this limiting case our light fans become infinitely narrow. This is indicated by the widths λ/A and λ/B which are taken from eq. (5) and are shown in the figure. Therefore in the limit also the energy taken from the central field and radiated in these directions becomes infinitely small. The situation is here the same as in the case of the Poisson spot which also remains present in the limiting case of geometrical optics but is reduced in size to a geometrical point.

If the opening is a parallelogram, then the two bright strips are not perpendicular to each other but are perpendicular to the sides of the parallelogram. In the case of a triangular opening there are three strips which are perpendicular to the three sides of the triangle, thus there are six light fans altogether, as shown in fig. 71b. A rectilinear polygonal opening with n sides has in general $2n$ such ray directions.

If one looks at a light source through a small hole shaped like a parallelogram, and if the eye is not focused on the source (or the source is not sufficiently small), then the source appears as a star with four rays. Under the same conditions, a triangular opening yields a six-cornered star. An irregular opening usually yields a many-cornered star. It is to be noted that the diffraction pattern is very sensitive to small irregularities in the shape of the opening. It is because of small irregularities in the iris of the eye that a star seen in the night sky does not appear to be surrounded by circular rings, which would be the case with an ideal circular aperture; rather, the stars appear to us stellate and thus they have traditionally been represented in the art of all ages. The five-cornered star which is preferred in heraldry and in depicting the Christmas star is, incidentally, wave-optically impossible because the light fans must, necessarily, occur in pairs.

So far we have limited ourselves to diffraction openings which are bounded by straight lines. What happens with curved boundaries? To treat these we must again return to fig. 60a, but for Fraunhofer diffraction we must replace the elliptic arcs in that figure by a system of parallel straight lines. For every position of the (infinitely distant) point D, there are two straight lines p_1, p_2 of the system which are tangent to the edge of the diffraction opening. Thus there are now no *finite* line segments along which the edge and a wave front coincide, as there were in the case of rectilinear edges. With curved openings there are only infinitesimal points of coincidence, i. e. points of *tangency*. For this reason the intensities of the light fans are of a smaller order of magnitude than those produced by piecewise rectilinear edges. Let us estimate this order of magnitude.

As was noted following eq. (35.12) $\varphi(p_1) = \varphi(p_2) = 0$ at the points of tangency; therefore the first term on the right-hand side of (35.12 a) vanishes. In order to estimate the second term we replace the boundary curve at the point of tangency by its circle of curvature (radius ρ). We denote by 2ψ the central angles of the arcs cut out of this circle by the system of p-lines. We can choose the parameter p so that it measures the distance from the center of the circle; then $p = \rho\cos\psi$. The length of each chord is then

$$\varphi(p) = 2\rho\sin\psi = 2\sqrt{\rho^2 - p^2} \tag{13}$$

and therefore

$$\varphi'(p) = -2\,p\,(\rho^2 - p^2)^{-\frac{1}{2}}. \tag{13 a}$$

At the point of tangency ($p = \rho$) $\varphi'(p)$ and therefore also $F'(p)$ become infinite, but slowly enough so that the integral in question remains finite. An approximate evaluation shows that the intensity is proportional to $\lambda\,\rho$ and thus vanishes in the limiting case of geometrical optics. On the other hand, we saw in (12) that for rectilinear edges the diffracted intensity is of the same order of magnitude as the intensity of the incident light. Though an opening with a curved edge also radiates diffracted light in a direction perpendicular to the tangent of the edge, its intensity is of a smaller order of magnitude than that of the diffracted light from a rectilinear opening. The diffracted intensity decreases as the curvature $1/\rho$ of the edge at the point in question increases. If we consider a curvilinear polygon[1], see fig. 71 c, then each corner E of the polygon has an adjoining shadow region; the intensities radiated in directions which are within these shadows are of the same order of magnitude as the intensities in the shadow regions of rectilinear polygons, and they decrease to zero with decreasing wavelength.

Summarizing and completing the quantitative relationships we can say: every diffracting aperture produces a diffraction pattern which fans out from the central image; this pattern consists of light fans separated by shadows. If the angular separation from the central image is denoted by the (dimensionless) number a, and if A is used to denote the length of one side of the diffraction opening in the case of a rectilinearly bounded opening, and the same letter A denotes the radius of curvature (formerly ρ) in the case of a curvilinear edge, then the diffracted intensity is

in the light fans of rectilinear edges	$\frac{A^2}{a^2}$	$\frac{\lambda^2}{a^2 A^2}$
in the light fans of curvilinear edges	$\frac{A\,\lambda}{a^3}$	$\frac{\lambda^3}{a^3 A^3}$
in the shadow regions of rectilinear or curvilinear edges	$\frac{\lambda^2}{a^4}$	$\frac{\lambda^4}{a^4 A^4}$

[1] The diffraction opening in fig. 71c is a curvilinear triangle formed by three circular arcs. The three centers (centers of curvature) are indicated in the figure. The tangents which are drawn at the three corners are used to construct the boundaries of the light fans which appear in the upper drawing.

The expressions in the first column are the intensities expressed as fractions of the light energy incident per unit area of the opening; they have, therefore, the dimensions of a length squared. The expressions in the second column are the intensities relative to the intensity in the middle of the central image; they are, therefore, dimensionless.

We have neglected the interference fringes which traverse the diffraction image. As we have said before, these fringes are blurred if the light source is not a point or is not monochromatic.

37. Fresnel Diffraction by a Slit

The course to be followed in deriving the theory of pure Fresnel diffraction has already been indicated in Sec. 34 D. The procedure is somewhat cumbersome and does not always lead to its intended goal. The points in the opening are described by coordinates ξ, η, the origin of which lies at the point D at which the straight line connecting the light source P' to the point of observation P intersects the plane of the screen S; see fig. 72. Then $\alpha = \alpha_0$, $\beta = \beta_0$, $\gamma = \gamma_0$ (α_0, β_0, γ_0 = direction $P'D$; α, β, γ = direction DP), and the linear terms in the expression (34.13) for the phase Φ vanish. This results, however, in the following difficulty: or very eccentric positions of the point P [indicated by (P) in fig. 72], D will lie outside the opening, which must be assumed to be small, and then the coordinates ξ, η are no longer small as required in the series expansion of Φ. We must therefore restrict the position of P to a region (indicated by the curly bracket in fig. 72) which does not extend too far into the geometrical shadow. Outside that region it is impossible to represent Φ by the quadratic terms alone. The pattern produced by the quadratic terms must then be supplemented by a diffraction pattern to be calculated in the Fraunhofer manner. A further inconvenience is that, even with this restriction on P, the position of D varies with that of P so that every separate position of P requires its separate coordinate system ξ, η.

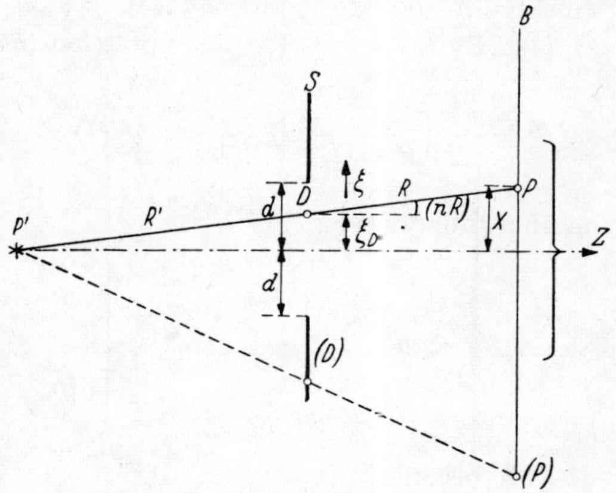

Fig. 72.
Fresnel diffraction of a slit.

The particular problem to be treated now is the diffraction pattern of a narrow rectangle (called a slit in the heading of this section). The plane of fig. 72 passes through the center of the rectangle and is parallel to its short side $2\,d$; the long side $2\,h$ is perpendicular to the plane of the drawing. The coordinates ξ and η are measured parallel to the sides of the rectangle. The observation screen which receives the diffraction pattern is parallel to the diffraction screen S; both are perpendicular to the plane of the drawing. We shall limit ourselves to points of observation P which are in the plane of the drawing. The light source P' will be assumed to lie directly in front of the center of the rectangle. Then the line $P'\,P$ lies in the plane of the drawing, and since the y-axis and also the η-axis are perpendicular to that plane, we have

$$\beta - \beta_0 = 0, \qquad \alpha^2 = \alpha_0{}^2 = 1 - \gamma^2; \tag{1}$$

as indicated in the figure, γ is equal to $\cos\,(n, R)$, a fact which will be used in eq. (4). By means of (1), the expression (34.13) reduces to

$$\Phi = -\frac{1}{2}\left(\frac{1}{R} + \frac{1}{R'}\right)\{\xi^2 + \eta^2 - \alpha^2\,\xi^2\} = -\frac{1}{2}\left(\frac{1}{R} + \frac{1}{R'}\right)(\gamma^2\,\xi^2 + \eta^2). \tag{2}$$

Using the abbreviations

$$k\,\Phi = -\Phi_\xi\,\xi^2 - \Phi_\eta\,\eta^2, \qquad \left\{\begin{aligned} \Phi_\xi &= \frac{1}{2}\gamma^2\,k\left(\frac{1}{R} + \frac{1}{R'}\right) \\ \Phi_\eta &= \frac{1}{2}\,k\left(\frac{1}{R} + \frac{1}{R'}\right) \end{aligned}\right. \tag{3}$$

eq. (34.14) becomes

$$i\,\lambda\,v_P = \frac{A\,\gamma}{R\,R'}\,e^{i\,k\,(R+R')}\int\limits_{-d-\xi_D}^{+d-\xi_D} \exp\,(i\,\Phi_\xi\,\xi^2)\,d\xi \int\limits_{-h}^{+h} \exp\,(i\,\Phi_\eta\,\eta^2)\,d\eta. \tag{4}$$

Let us remember the meanings of R and R': $R =$ distance $D\,P$, $R' =$ distance $D\,P'$, where D, being the point where $P'\,P$ pierces the plane S, itself depends on the position of P; according to (3) the abbreviations Φ_ξ, Φ_η therefore also depend on the position of P. This dependence also affects the limits of integration given in (4). Since ξ is to be measured not from the center of the slit but from the point D which has the coordinate ξ_D, these limits are not $\pm\,d$ but $\pm\,d - \xi_D$. Because of symmetry this dependence on P does not affect the η-integral in (4).

A. Fresnel's integrals

In order to conform to the historically established notation we consider the integral

$$F(w) = \int_0^w e^{i\frac{\pi}{2}\tau^2}\, d\tau. \tag{5}$$

We call it *Fresnel's integral*. Ordinarily this name is reserved for the two real integrals

$$C(w) = \int_0^w \cos\left(\frac{\pi}{2}\tau^2\right) d\tau, \qquad S(w) = \int_0^w \sin\left(\frac{\pi}{2}\tau^2\right) d\tau \tag{5 a}$$

which evidently form the real and imaginary parts of F:

$$F = C + i\,S. \tag{5 b}$$

We wish to emphasize, however, that the separation of F into real and imaginary parts is absolutely of no advantage [we did not separate the plane wave $\exp(i\,k\,x)$ into a cosine and a sine either!]. The two integrals in (4) can be reduced to F by simple substitutions. One obtains

$$\int_{-d-\xi_D}^{+d-\xi_D} \exp(i\,\Phi_\xi\,\xi^2)\,d\xi = \sqrt{\frac{\pi}{2\,\Phi_\xi}}\,[F(w_2) - F(w_1)], \qquad \left.\begin{matrix} w_2 \\ w_1 \end{matrix}\right\} = \sqrt{\frac{2\,\Phi_\xi}{\pi}}\,(\pm d - \xi_D)$$

$$\int_{-h}^{+h} \exp(i\,\Phi_\eta\,\eta^2)\,d\eta = \sqrt{\frac{2\,\pi}{\Phi_\eta}}\,F(W), \qquad W = \sqrt{\frac{2\,\Phi_\eta}{\pi}}\,h.$$

For a slit $h \gg d$ and $W \gg w_{2,1}$. We shall now convince ourselves that W may be set equal to infinity [1]. To show this it is convenient to introduce the quasi-focal length f which was introduced in connection with the similarity law, Sec. 35 E. Thus, in our case we set

$$\frac{1}{f} = \frac{1}{R} + \frac{1}{R'}, \qquad \Phi_\eta = \frac{\pi}{\lambda\,f}, \qquad W = \sqrt{\frac{2}{\lambda\,f}}\,h \gg \sqrt{\frac{2}{\lambda\,f}}\,d. \tag{6}$$

In accordance with the similarity law, if the magnitude of d is just barely suitable for diffraction experiments, then W is so large that it does not produce

---

[1] In the beginning of Subsection C we shall return to the question of the admissibility of such limit processes.

an appreciable diffraction effect of its own, and we can therefore go to the limit $\lambda \to 0$, $W \to \infty$ as in geometrical optics. Introducing at the same time the value of Φ_η from (6), we thus set

$$\int_{-h}^{+h} e^{i\Phi_\eta \eta^2} d\eta = \sqrt{2\lambda f}\, F(\infty). \tag{6 a}$$

Correspondingly, we find

$$\int_{-d-\xi_D}^{+d-\xi_D} \exp(i\Phi_\xi \xi^2)\, d\xi = \frac{1}{\gamma}\sqrt{\frac{\lambda f}{2}}\{F(w_2) - F(w_1)\}, \qquad \left.\begin{matrix} w_2 \\ w_1 \end{matrix}\right\} = \gamma \frac{\pm d - \xi_D}{\sqrt{\frac{\lambda f}{2}}}. \tag{6 b}$$

Hence, according to (4)

$$i v_P = f A \frac{e^{ik(R+R')}}{R R'} \{F(w_2) - F(w_1)\} F(\infty).$$

This expression simplifies if we immediately introduce the value $F(\infty) = \frac{1+i}{2}$, a result which will be derived later, and return to the original definition (6) of f. For then

$$v_P = \frac{1-i}{2} A \frac{e^{ik(R+R')}}{R+R'} \{F(w_2) - F(w_1)\}, \tag{7}$$

where

$$A \frac{e^{ik(R+R')}}{R+R'} = v_0$$

is the optical field amplitude which would be observed at the point P if the intervening screen were removed entirely. Thus we can write for (7)

$$v = \frac{1-i}{2} v_0 \{F(w_2) - F(w_1)\}. \tag{7 a}$$

If we disregard the first factor which is of no interest for the present, we can say:

The pattern produced on the screen B differs from the primary undiffracted field by a factor which is equal to the difference between the Fresnel integrals $F(w_2)$ *and* $F(w_1)$.

We can be brief in our description of the analytic properties of the function $F(w)$. They correspond entirely to those of the Gaussian error integral

$$F(x) = \int_0^x e^{-\tau^2} d\tau.$$

a) $F(w)$ is an entire transcendental function of w; by its definition (5), $F(w)$ can therefore be expanded in the following series which converges everywhere in the finite plane

$$F(w) = w\left(1 + \frac{i}{1!\,3}\frac{\pi}{2}w^2 - \frac{1}{2!\,5}\left(\frac{\pi}{2}w^2\right)^2 - \frac{i}{3!\,7}\left(\frac{\pi}{2}w^2\right)^3 + \dots\right). \tag{8}$$

This expansion follows directly from the exponential series. From it one obtains the respective series for $C(w)$ and $S(w)$.

b) Of greater importance is the divergent (so-called asymptotic) series development of $F(w)$ which yields a sufficiently exact approximation of the function for large values of w, provided that only a limited number of the terms of this series are summed. We obtain this development by setting

$$F(w) = F(\infty) - \int_w^\infty e^{\frac{i\pi\tau^2}{2}}\,d\tau = F(\infty) - \int_w^\infty \frac{d}{d\tau}\left(e^{\frac{i\pi\tau^2}{2}}\right)\frac{d\tau}{i\pi\tau}.$$

Upon integrating by parts this becomes

$$F(w) = F(\infty) + \frac{e^{\frac{i\pi}{2}w^2}}{i\pi w} - \int_w^\infty e^{\frac{i\pi}{2}\tau^2}\frac{d\tau}{i\pi\tau^2}$$

$$= F(\infty) + \frac{e^{\frac{i\pi}{2}w^2}}{i\pi w} - \int_w^\infty \frac{d}{d\tau}\left(e^{\frac{i\pi}{2}\tau^2}\right)\frac{d\tau}{(i\pi)^2\tau^3},$$

and continuing the process of integrating by parts we get

$$F(w) = F(\infty) + \frac{e^{\frac{i\pi}{2}w^2}}{i\pi w}\left(1 + \frac{1}{i\pi w^2} + \frac{1\cdot 3}{(i\pi w^2)^2} + \frac{1\cdot 3\cdot 5}{(i\pi w^2)^3} + \dots\right). \tag{8 a}$$

From this follow the respective asymptotic series for $C(w)$ and $S(w)$.

c) In order to calculate $F(\infty)$ we recall the well-known Laplace integral

$$\int_0^\infty e^{-\alpha\tau^2}\,d\tau = \frac{1}{2}\sqrt{\frac{\pi}{\alpha}}.$$

We need only set $\alpha = -\frac{i\pi}{2}$ to find

$$F(\infty) = \frac{1}{2}\sqrt{\frac{2}{-i}} = \frac{1}{\sqrt{-2i}} = \frac{1}{1-i} = \frac{1+i}{2}. \tag{8 b}$$

This result can be checked by considering the integral in the complex plane of the variable τ which we shall, however, omit here.

B. Discussion of the Diffraction Pattern

We now investigate the intensity extrema (maxima and minima) in the diffraction pattern. That is, we seek those points on the observation screen which under monochromatic illumination will correspond to bright and dark fringes. These points are defined by the condition $\frac{d\,|v|^2}{dx} = 0$, where x is the distance of a point on the observation screen from the center of the screen. According to fig. 72, x is related to the coordinate ξ_D which measures distance from the center on the diffraction screen S. Therefore we may discuss $\frac{d\,|v|^2}{d\xi_D} = 0$ instead of $\frac{d\,|v|^2}{dx} = 0$. Since ξ_D occurs only in the limits of integration w_2 and w_1 in eq. (6 b), and since $\frac{dw_2}{d\xi_D} = \frac{dw_1}{d\xi_D} = -\gamma\left(\frac{\lambda f}{2}\right)^{-1/2}$, the following condition for the extrema[1] results:

$$\frac{d}{d\xi_D}\{F(w_2) - F(w_1)\} = -\frac{\gamma}{\sqrt{\lambda f/2}}\{F'(w_2) - F'(w_1)\} = 0, \tag{9}$$

so that

$$\exp\left(\frac{i\pi}{2}w_2^2\right) = \exp\left(\frac{i\pi}{2}w_1^2\right),$$

$$\frac{\pi}{2}(w_2^2 - w_1^2) = -2\pi g, \qquad (w_2 - w_1)(w_2 + w_1) = -4g \tag{10}$$

where g is a (positive or negative) integer. Now, according to (6 b) we have

$$w_2 - w_1 = \frac{2\gamma d}{\sqrt{\lambda f/2}}, \qquad w_2 + w_1 = \frac{-2\gamma\xi_D}{\sqrt{\lambda f/2}}. \tag{10 a}$$

From (10) and (10 a) follows

$$\frac{2\gamma^2\xi_D d}{\lambda f} = g, \qquad \xi_D = \frac{\lambda f g}{2\gamma^2 d},$$

and the distance between two successive extrema is

$$\Delta\xi_D = \frac{\lambda f}{2\gamma^2 d}. \tag{10 b}$$

[1]Using temporarily the abbreviation $f(x) = F(w_2) - F(w_1)$, then

$$|v|^2 = C f f^*, \qquad C = \frac{1}{2}|v_0|^2, \qquad \frac{d}{dx}|v|^2 = C\left(f\frac{df^*}{dx} + f^*\frac{df}{dx}\right)$$

where f^* is the complex conjugate of f. In eq. (9) we have satisfied the condition $df/dx = 0$, but at the same time also the condition $df^*/dx = 0$ is fulfilled (interchange of $+i$ with $-i$ and $-g$ with $+g$). Therefore the condition $d|v|^2/dx = 0$ is fulfilled as well. Hence eq. (9) is not only the extremal condition for the amplitude v but also for the intensity $|v|^2$.

The separation between extrema decreases with increasing d and increases with increasing λ and f. The same is true of the separation Δx of the fringes on the observation screen.

The discussion of the diffraction pattern can be well illustrated by means of *Cornu's spiral* which is constructed by the following mapping process:

We interpret $F = C + i\,S$ as a point on the complex F-plane, that is, as the point with the cartesian coordinates C and S. In addition we consider a complex w-plane in which, however, only the real axis is of interest. The equation $F = F\,(w)$ represents a conformal (angle-preserving) mapping of the w-plane onto the F-plane. The real axis of the w-plane, which is the only part of that plane which will enter into consideration, is mapped onto a certain curve in the F-plane. We claim that this mapping is length-preserving. For we have

$$\frac{dF}{dw} = e^{\frac{i\pi}{2} w^2}, \qquad \text{hence} \qquad \left|\frac{dF}{dw}\right| = 1, \qquad |dF| = |dw|. \tag{11}$$

Hence the w-axis and the F-curve are mapped on each other without stretching. We already know three points of this map, see eqs. (8) and (8 b):

$$w = 0, \qquad w = \infty, \qquad w = -\infty,$$

$$F\,(0) = 0, \qquad F\,(\infty) = \frac{1+i}{2}, \qquad F\,(-\infty) = -\frac{1+i}{2}.$$

The length of the F-curve between the two end points $F\,(\pm\,\infty)$ is infinite as is the length of the w-axis. The curve is symmetrical with respect to the origin of the F-plane; for by eq. (8)

$$F\,(-w) = -F\,(w).$$

The tangent at the origin is horizontal; the curve has an inflection point there; for according to (8), we have at $w = 0$

$$\frac{dF}{dw} = 1, \qquad \frac{d^2F}{dw^2} = 0.$$

For $w = \pm\,\infty$ the direction of the tangent is indeterminate, according to (8 a). Asymptotically the curve approaches these points as a spiral. The entire curve is plotted in fig. 73.

Not only does this curve illustrate the whole range of values which F assumes (for real w), but at the same time it also represents all of the amplitude ratios $|v|/|v_0|$ in the diffraction pattern. For from (7 a)

$$\sqrt{2}\; |v|/|v_0| = |F\,(w_2) - F\,(w_1)|, \tag{12}$$

that is, the amplitude ratio (times $\sqrt{2}$) equals the *length of the chord* which connects the two points representing w_2 and w_1 on Cornu's spiral. By (10 a) the difference between the two w-values is

$$w_2 - w_1 = \frac{2\gamma d}{\sqrt{\lambda f/2}} = \text{const.}$$

hence it is independent of both ξ_D and of the coordinate x of the point of observation. $w_2 - w_1$ is a certain segment on the real w-axis. The *arc of Cornu's spiral* between the endpoints of our chord has this same constant length.

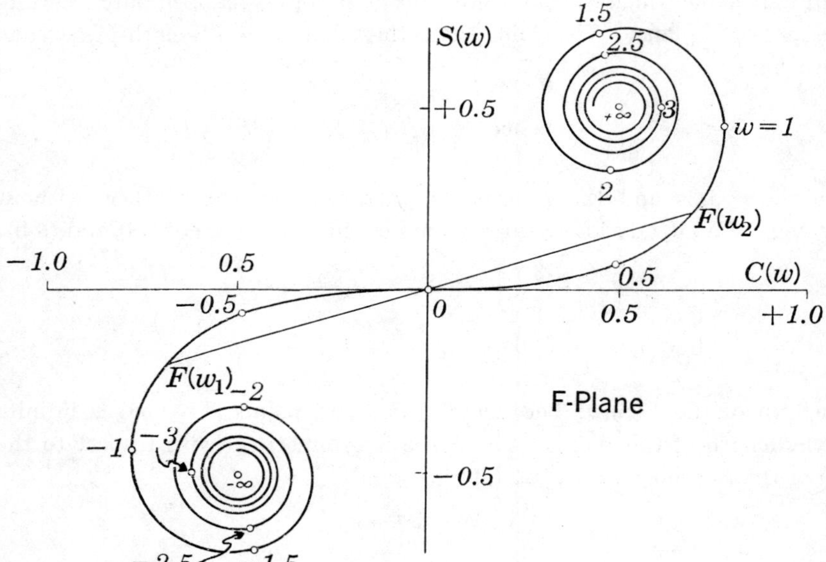

Fig. 73.
Cornu's Spiral.

In fig. 73 we have drawn the chord which corresponds to the point $x = 0$ on the diffraction pattern. This chord passes through the origin of the F-plane and ends at two diametrically opposite points on the spiral which belong to the arguments $w_2 = \dfrac{d}{\sqrt{\lambda f/2}}$, $w_1 = -\dfrac{d}{\sqrt{\lambda f/2}}$. If we shift the starting point of the chord by a certain distance, then we must shift the end point by so much that the arc of the spiral has the same length as before. In this way the length of the chord is changed. This change implies a changed amplitude $|v|$ at the new point of observation x which corresponds to the new position of the chord. If we approach the upper limit point of the spiral with the starting point of the chord, then the end point of the chord will also approach this limit; the chord becomes progressively smaller, and so does the amplitude $|v|$ which, in the process, goes through an infinite number of extrema of continually decreasing magnitudes.

C. Diffraction by a straight edge

If we make the slit infinitely wide ($d \to \infty$) by keeping one of the edges, e. g. the right-hand one, fixed and moving the left-hand edge off to infinity, then we have the simpler problem of the straight edge. To begin with it is to be noted that the various limiting processes pile up and seem to exclude one another; in the series development (34.13) we had assumed the opening to be "small". In treating the slit we assumed $h \gg d$ and put $h = \infty$. Now we also let $d \to \infty$. In order to be mathematically precise, we would have to conduct a careful appraisal of these limit processes. However we shall omit this here, because the problem of the half-plane will be treated again with all desirable accuracy in Sec. 38.

We further simplify the problem by letting the incident wave be plane instead of spherical; that is, we move P' to infinity. We are, however, still dealing with Fresnel diffraction (see p. 206) if we observe the pattern on an observation screen B which is placed at a finite distance a from the diffraction screen. In this case $f = a$ (because $b = \infty$ and $1/f = 1/a + 1/b$). If the light is perpendicularly incident the coordinates ξ on S equal the coordinates x on B and $\gamma = 1$. If we place the origin of x on the boundary of the geometrical shadow, then

$$d - \xi_D = \xi = x, \qquad w_2 = \frac{x}{\sqrt{\lambda a/2}}, \qquad w_1 = -\infty. \tag{13}$$

Setting $w_2 = w$, we obtain instead of (7 a)

$$\left|\frac{v}{v_0}\right| = \frac{1}{\sqrt{2}}|F(w) - F(-\infty)|. \tag{13 a}$$

In the Cornu spiral construction the starting point of the chord is now fixed at the lower limit point of the spiral. Only the end point of the chord changes with x. In the region of the geometrical shadow $(-\infty < x < 0)$ the length of the chord increases steadily, as indicated by the sequence of chords ending at the points a, b, c, d, e in fig. 74. The point c corresponds to the boundary of the geometric shadow. At that point $w = 0$ and $F(w) = 0$ and

$$\left|\frac{v}{v_0}\right| = \frac{1}{\sqrt{2}}|F(-\infty)| = \frac{1}{\sqrt{2}}\left|-\frac{1+i}{2}\right| = \frac{1}{2}, \text{ according to eq. (8 b).} \tag{13 b}$$

From there on the length of the chord keeps increasing up to the first maximum which is attained at the point f in the figure. Then the chord decreases to the first minimum at the point g, and after that the chord oscillates between alternating extrema of decreasing heights. The asymptotic value of $|v/v_0|$ for $w = \infty$ is twice its value (13 b) on the boundary of the geometric shadow; it is given by

$$\frac{1}{\sqrt{2}}|F(+\infty) - F(-\infty)| = 1, \tag{13 c}$$

which corresponds to the full intensity of the incident light. The intensity at the shadow boundary is one-fourth of the incident intensity. The variations in the amplitude are shown in fig. 75.

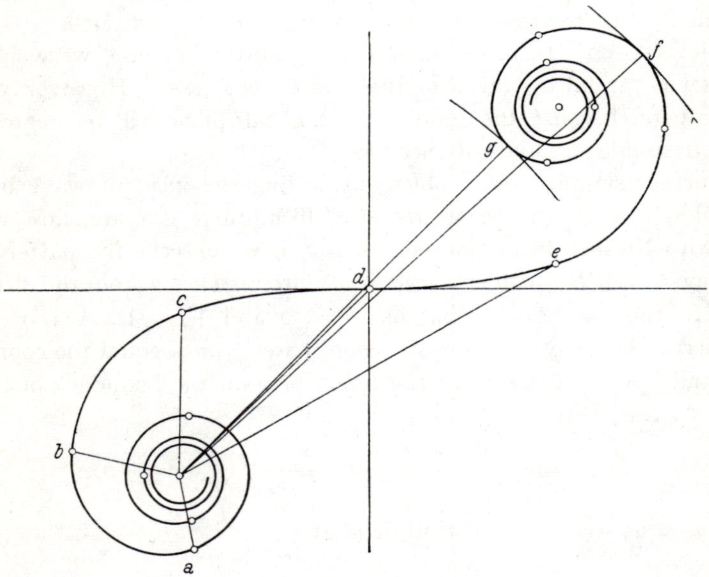

Fig. 74.

Determination of the diffraction pattern of the straight edge by means of Cornu's spiral.

We have assumed the diffraction screen to be infinitely thin and at the same time opaque. Therefore these results cannot be realized experimentally. Under a microscope even the edge of a razor looks more like a parabolic cylinder than like a sharp half-plane. However, it is very remarkable that the patterns on precise diffraction photographs (see for instance, Arkadiew loc. cit. p. 225) exhibit almost no dependence on the material and shape of the diffraction edge. Even a bent glass plate whose radius of curvature is several meters and which may or may not be blackened yields essentially the same diffraction fringes as the edge of a razor. In each case the pattern is that shown in fig. 75.

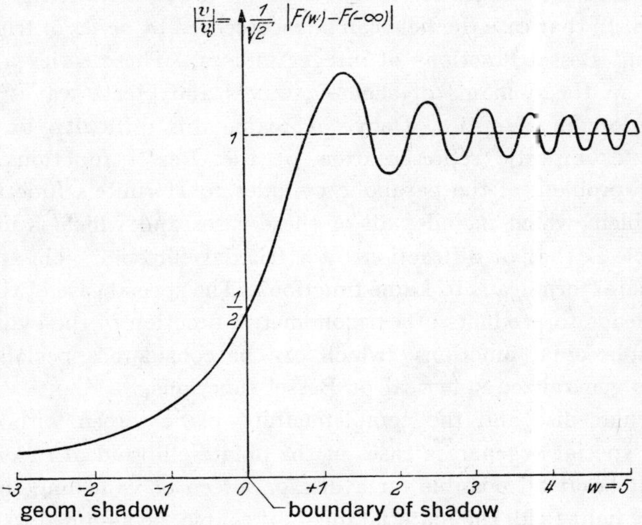

Fig. 75.
Amplitude $|v|$ behind a straight edge.

38. Rigorous Solutions of Certain Diffraction Problems

We shall call a solution of a diffraction problem exact only if it satisfies Maxwell's equations both outside and inside the diffracting object and if it satisfies the proper boundary conditions on the surface of that object. The solution must, furthermore, correspond to a given type of excitation (plane wave or point source). Such a solution can be found only for special shapes of diffracting objects, and certainly only if the wave equation can be "separated" in a coordinate system which is suited to the shape of the object.

The simplest example of such an object is a sphere. The field outside a sphere can be represented by series of spherical harmonics and Bessel functions of half-integer indices. These series have been discussed by G. Mie[1] for colloidal particles of arbitrary compositions. But even there a mathematical difficulty develops which quite generally is a drawback of this "method of series

[1] Ann. d. Phys. **25**, p. 377. 1908.

development": for fairly large particles ($k a > 1$, a = radius, $k = 2\pi/\lambda$) the series converge so slowly that they become practically useless. Except for this difficulty we could in this way obtain a complete solution of the problem of the rainbow[1], the difficulty of which was pointed out on p. 179.

What is true for the sphere is also true for a cylindrical wire of circular cross section. In that case the field can be represented by series of trigonometric functions and Bessel functions of integer indices. These series are entirely satisfactory in the domains of acoustic waves and Hertz waves[2] but they fail in the domain of optics. Debye overcame this difficulty by means of his famous asymptotic representation of the Bessel functions. Epstein[3] reduced the problem of the parabolic cylinder to Hermite's functions.

The problem which includes all of these cases and which is in principle still separable is that of diffraction by a triaxial ellipsoid. This problem in its most general form leads to Lamé functions. The special case of the ellipsoid of rotation leads to products of a trigonometric function of the cylinder angle and two "spheroidal functions" which can be considered specialized Lamé functions or generalized spherical or Bessel functions.

The circular disc and the complementary plane screen with a circular opening are special degenerate cases of the oblate ellipsoid of revolution. In order that it be at all possible for a disc or screen of vanishing thickness to influence the light field, the material must, of course, be assumed to be opaque (perfectly conducting). The general Maxwell boundary conditions reduce then to the requirement that $E_{\text{tangential}} = 0$ and consequently that $B_{\text{perpendicular}} = 0$. With these boundary conditions the treatment of the problem can still be made mathematically rigorous, but it is no longer rigorous in the physical sense as defined above, for the diffracting material is no longer physically realizable. The solution to such a problem can be considered physically rigorous only in the case of acoustic waves[4] or Hertz waves[5] (wavelengths large compared to the thickness of the diffracting object).

The series of spheroidal functions which appear in these solutions again converge sufficiently well only if the radius a of the disc or opening is not too large compared to the wavelength. Even the case $k a \sim 1$ can be computed

[1] The two papers which come closest to solving this problem are those by B. van der Pol and H. Bremmer, Phil. Mag. **24**, p. 191 and 825, 1937 and by H. Bucerius, Optik, Vol. **I**, p. 181, 1946. Debye had previously treated the two-dimensional rainbow (diffraction by a glass rod), Phys. Zeitschr. **9**, p. 775, 1908.

[2] Schaefer-Grossmann, Ann. d. Phys. (Leipzig) **31**, p. 454, 1910. Experimental verification with undamped waves: Schaefer-Merzkirch, Z. f. Phys. **13**, p. 166, 1922 and Schaefer-Wilmsen, ibid. **24**, p. 345, 1924.

[3] P. S. Epstein, Dissertation, Munich, 1914.

[4] O. J. Bouwkamp, Proefschrift, Groningen, 1941.

[5] J. Meixner, ZS. f. Naturf. Vol. **3a**, p. 506, 1948.

numerically only with the aid of tables; here again asymptotic formulae of the type of Debye's formula for the Bessel functions are needed for an approximative evaluation of the result.

The problem of the slit and the complementary problem of the strip leads to Mathieu functions and has been solved numerically by Morse and Rubenstein[1] with the aid of tables of Mathieu functions

It is impossible to discuss these function-theoretical details here; they belong to the chapter "Eigenvalues and Eigenfunctions" of Vol. VI.

A. The problem of the straight edge

This problem also is physically not rigorous because we shall assume the screening half-plane to be infinitely thin but nevertheless opaque. We shall obtain a mathematically rigorous solution of the problem which will even be in closed form and easily applicable to all wavelength domains. With this problem it was first demonstrated[2] that Fresnel diffraction constitutes a well-defined mathematical boundary value problem. (Fraunhofer diffraction cannot be treated directly by this method but only as a limiting case of Fresnel diffraction.)

We let the edge of the screen be the z-axis of a cylindrical coordinate system r, φ, z; the front and rear surfaces of the screen shall be the surfaces $\varphi = 0$ and $\varphi = 2\pi$, respectively. We assume that in the r, φ-plane a monochromatic plane wave is incident on the front surface of the screen at an angle α (the angle of incidence measured against the normal to the screen is then $\pi/2 - \alpha$). The wave shall be linearly polarized in such a way that the electric field is directed parallel to the z-axis. Then the diffracted electric field will also be parallel to the z-axis and the problem becomes two-dimensional; only processes in the r, φ-plane are involved. Therefore we can use a scalar function u; the part of this function which describes the incident wave will be

$$u_0 = A\, e^{-ikr\cos(\varphi-\alpha)}. \tag{1}$$

The negative sign in the exponent is due to the fact that we think of the time dependence as given by $\exp(-i\omega t)$ as usual, and that the wave propagates in the direction of the half-ray $\varphi = \pi + \alpha$; see fig. 76 (the arrows originating from O pertain to the discussion in the later section D). The field u as modified by the presence of the screen must satisfy the following conditions:

(1 a) the wave equation $\Delta u + k^2 u = 0, \qquad \Delta = \dfrac{\partial^2}{\partial r^2} + \dfrac{1}{r}\dfrac{\partial}{\partial r} + \dfrac{1}{r^2}\dfrac{\partial^2}{\partial \varphi^2},$

---

[1] Phys. Rev. **54**, p. 895, 1938.

[2] A. Sommerfeld, Mathem. Ann., Vol. **47**, p. 317, 1896. A simplified presentation is to be found in chapter 20 of Vol. II of "Differentialgleichungen der Physik", edited by Frank and von Mises, second edition 1934, first edition 1927 (Vieweg, Braunschweig).

(1 b) the boundary conditions $u = 0$ for $\varphi = \begin{cases} 0 \\ 2\pi \end{cases}$, (corresponds to $E_{\tan}=0$),

(1 c) the condition u is finite and continuous everywhere, including the edge of the screen.

To these must be added the radiation condition[1] at infinity. Specialized to our case this condition must be formulated differently for the "illuminated" region $I + II$ and for the "shaded" region III of the r, φ plane (the words "illuminated" and "shaded" refer to the geometrical optics point of view). These conditions are

$$\text{(1 d)} \qquad \lim_{r\to\infty} r\left(\frac{\partial v}{\partial r} - i\,k\,v\right) = 0, \qquad v = \begin{cases} u - u_0 & \text{for} \quad 0 < \varphi < \pi + \alpha \\ u & \text{for} \quad \pi + \alpha < \varphi < 2\pi; \end{cases}$$

or expressed in words: in the illuminated region the incident portion of the field is given precisely by u_0 and the difference $u - u_0$ (reflected + diffracted wave) has the radiative character as required by (1 d); in the shaded region u itself is the radiative field.

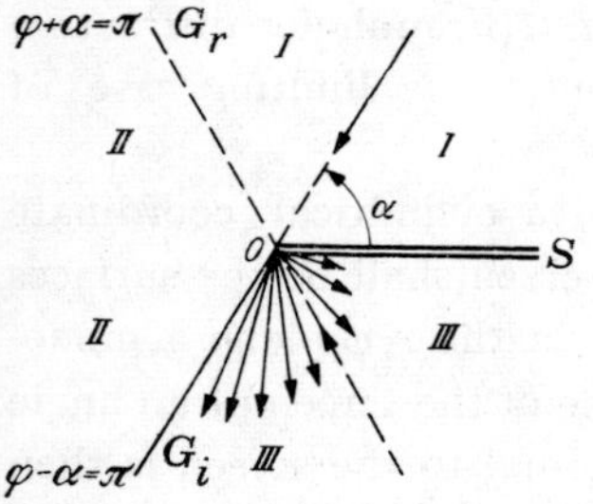

Fig. 76.
The diffraction screen S with the shadow boundary G_i of the incident ray and G_r of the reflected ray.

Finally, we must complete our requirement (1 c) with a statement about the behavior of r grad u at the edge of the screen, namely that

$$\text{(1 e)} \qquad r\,\text{grad}\,u \to 0 \qquad \text{as} \qquad r \to 0.$$

Accordingly, grad u can become infinite at $r = 0$ but only "weakly" so. In the limit, r grad u must vanish. We shall see in section C below that when this condition is satisfied, the edge of the screen neither radiates nor absorbs energy. Therefore we can characterize the requirements (1 d) and (1 e) as additional energy conditions which suffice to make the problem physically unique[2].

---

[1] This condition is fully discussed in Vol. VI, Sec. 28. The requirement is equivalent to demanding that if all light sources are situated in the finite regions of space, then the field at infinity must behave like an *outgoing spherical* wave, $\exp(ikr)/r$. This expression fulfills (1 d) everywhere when $v = u_0$. Separate formulations have to be given in (1 d) for the two regions because the incident wave is a plane wave originating at infinity.

[2] J. Meixner, ZS. f. Naturforsch., Vol. 3, p. 506 established a more general condition (that the energy density at the edge of the screen shall be integrable with respect to space). In our case this condition becomes equivalent to (1 e). In three-dimensional cases where our condition (1 c) on the finiteness of u cannot be imposed, Meixner's "edge condition" is not only necessary but also *sufficient*.

The problem obviously cannot be solved by means of the usual method of images. For, if we were to add to the incident wave (1) a reflected wave

$$u_0' = -A\, e^{-ikr\cos(\alpha+\varphi)}$$

(direction of incidence $\varphi = 2\pi - \alpha$), our condition (1 d) would be violated. Furthermore, the resulting solution would vanish not only on the half-ray $\varphi = \begin{cases} 0 \\ 2\pi \end{cases}$ but on the whole ray $\varphi = \begin{cases} 0 \\ \pi \end{cases}$, which would certainly be wrong.

However, the method of images can be retained if one uses, instead of the ordinary plane wave $u_0(r, \psi)$ of period 2π (where ψ stands for $\varphi - \alpha$), a function $U(r, \psi)$ which has the *period* 4π *in the variable* ψ and which satisfies the conditions (1 a) and (1 c) for all $-2\pi < \psi < 2\pi$ and the condition (1 d) with $v = U - u_0$ for $|\psi| < \pi$ and with $v = U$ for $|\psi| > \pi$. In the language introduced by Riemann for algebraic functions this means the following: U is a solution of our wave equation on a two-sheeted *Riemann surface* which has simple branch points at $r = 0$ and $r = \infty$. U is uniquely determined by its behavior at infinity (incident wave only in the sheet $|\psi| < \pi$, no incident wave in the sheet $|\psi| > \pi$) and by the requirement that it shall be everywhere continuous.

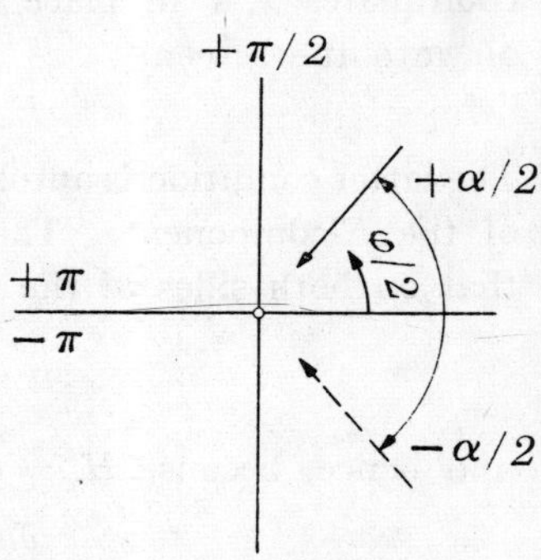

Fig. 77.
Symbolic representation of the method of images applied to the half-plane.

The usual model of this Riemann surface is familiar. It consists of two flat sheets which lie one on top of the other and are joined along the half-rays $\psi = \pm\pi$, for instance. Instead, we shall represent the surface by a single plane defined by the angle $\varphi/2$; see fig. 77. Every quadrant of this plane is, of course, a half-plane in the variable φ. The arrow[1] drawn as coming from the direction $+\alpha/2$ corresponds to the incident plane wave $U(r, \varphi - \alpha)$. The image-wave $U(r, \varphi + \alpha)$ is represented by the half-ray $-\alpha/2$ in the fourth quadrant. Since these two waves are symmetric with respect to α, they cancel one another on the symmetry lines $\varphi/2 = 0$ and $\varphi/2 = \pm\pi$ which

[1] In connection with fig. 77 it should be noted that the two straight arrows drawn there represent only the rays passing through the origin correctly rays parallel to these would in the figure have to be drawn as parabolic curves (coordinates r, $\varphi/2$ instead of r, φ). Further, it should be noted that the arrows refer only to the incident waves at infinity; our schematic figure does not represent the diffraction produced by the screen at all.

represent the two surfaces of our diffraction screen $\varphi = 0$ and $\varphi = 2\pi$. *Therefore the solution of our diffraction problem is given by the formula*

$$u = U(r, \varphi - \alpha) - U(r, \varphi + \alpha). \tag{2}$$

We now turn to the diffraction problem for the other polarization, that is, for light whose electric field is not parallel but perpendicular to the edge of the diffraction screen. All other geometrical conditions will be kept the same. In this case the magnetic vector H is parallel to the edge of the screen not only for the incident component but also for the reflected and diffracted components of the field. We now denote the magnetic vector by u and ask for the correct boundary condition which again follows from the requirement that $E_{\text{tangential}} = 0$. If we introduce, temporarily, the cartesian coordinates x, y in place of r, φ, then we must require that in going from air into the screen

$$E_x = 0 \quad \text{and} \quad E_z = 0.$$

The latter condition is automatically satisfied because of the given polarization of the E-component. The first condition requires, according to Maxwell, that on both sides of the screen

$$\text{curl}_x \mathsf{H} = \frac{\partial H_z}{\partial y} - \frac{\partial H_y}{\partial z} = 0$$

and hence, because $H_y = 0$, $H_z = u$:

$$\frac{\partial u}{\partial y} = 0 \quad \text{for} \quad \varphi = \begin{cases} 0 \\ 2\pi . \end{cases}$$

For this we can write also

$$\frac{\partial u}{\partial n} = 0 \tag{3}$$

where n stands for the normals on both sides of the screen. We can satisfy this condition immediately by means of the sum

$$u = U(r, \varphi - \alpha) + U(r, \varphi + \alpha) \tag{4}$$

which is analogous to (2). We wish to recall here that the two Green's functions G_- and G_+ of Sec. 34 G were formed by a method of images quite analogous to (2) and (4).

Fundamentally, our method has an even wider applicability. It can be extended without difficulty to the problem of the slit. In that case a Riemann surface with two branch points at the traces of the two slit edges would have to be used in place of the surface with a single branch point in the finite plane and another at infinity; besides, the ordinary polar coordinates would have to be replaced by bipolar coordinates. The method could even be extended, at least in the scalar (acoustic) case, to cover the problem of an arbitrarily bounded plane screen or a complementary opening. Here the two-sheeted Riemann surface would be replaced by a "Riemann double-space", the two "sheets" of which would have a common "branch line" on the bounding curve of the screen or opening. The difficulty with these generalizations lies in the mathematical construction of the branched solutions. It has been possible to construct these solutions only for the simplest case of the half-plane. As we shall see presently, even here very special mathematical devices are required.

B. Construction of branched solutions

The factor A in eq. (1) can be considered as an arbitrary function of the angle of incidence α. Replacing this α by a variable of integration β and integrating the expression with respect to β, we obtain the "wave bundle"

$$u = \int A\,(\beta)\, e^{-ikr\cos(\varphi-\beta)}\, d\beta. \tag{5}$$

This expression is a solution of the differential equation (1 a) for any arbitrary, possibly complex, path of integration. If the integration path is complex, then u represents an "inhomogeneous" wave of the type, for instance, which we have encountered in total reflection. Let us first choose a closed path in the complex β-plane which encloses the point $\beta = \alpha$. If we see to it that $A\,(\beta)$ has a pole of the first order with the residue $1/2\,\pi\, i$ at α, then by Cauchy's residue theorem (5) becomes the solution u_0 given by (1), normalized so that $A = 1$. (Let u_0 henceforth denote this normalized solution.) In particular, we shall chose $A\,(\beta)$ to be a periodic function of β with the period $2\,\pi$, namely

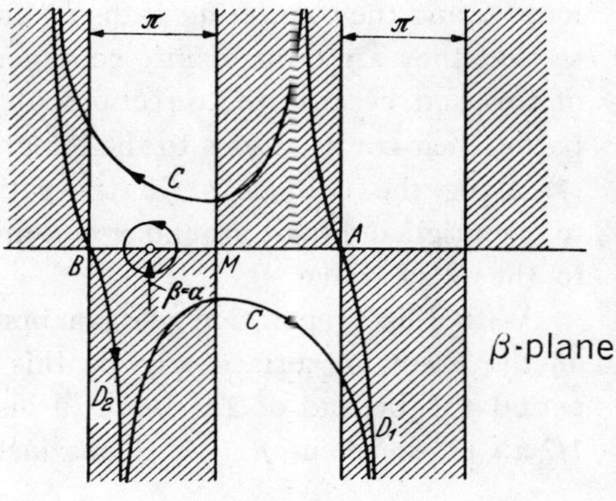

Fig. 78.
Path of integration in the β-plane used in the representation of u_0.

$$A\,(\beta) = \frac{1}{2\,\pi}\,\frac{e^{i\beta}}{e^{i\beta} - e^{i\alpha}}. \tag{6}$$

Thus we obtain

$$u_0 = \frac{1}{2\,\pi}\oint \frac{e^{i\beta}}{e^{i\beta} - e^{i\alpha}}\, e^{-ikr\cos(\varphi-\beta)}\, d\beta \tag{7}$$

where $\oint$ indicates contour integration around a closed path.

We can deform the path around the pole $\beta = \alpha$ in an arbitrary manner as long as it does not cross any other singularities of the integrand (that is, none of the points $\beta = \alpha \pm 2\pi, \alpha \pm 4\pi, \ldots$). If we wish to deform the path so that it goes to infinity, then we must make sure that the integrand vanishes in the limit along the path. In fig. 78 the regions in which $\cos(\varphi - \beta)$ has a negative imaginary part have been shaded. These regions are bounded by straight lines. For positive values of kr the real part of $-ikr\cos(\varphi - \beta)$ goes to $-\infty$ as β goes to infinity inside the shaded regions; therefore the integrand in (7) becomes vanishingly small. At the corners A, B and M of the resulting pattern we have

$$\beta = \begin{cases} \varphi - \pi \ldots . A \\ \varphi + \pi \ldots . B. \\ \varphi \quad \ldots . M \end{cases} \tag{7 a}$$

The path of integration which has been drawn in fig. 78 consists of the two loops C and the connecting paths D_1 and D_2. The latter two have been chosen so that they are brought into congruence by a displacement of 2π. Because of this and because the directions of integration are opposite for these two paths, their contributions to the integral cancel. Thus we need only integrate (7) along the two loops C; the path of integration will still be equivalent to the original circuit around $\varphi = \alpha$ and the integral (7) will still be identical to the plane wave u_0.

With this preparation we can immediately find the desired function U on the Riemann surface. To do this we give the arbitrary function A the period 4π (instead of 2π) but still insist that it possess a pole with residue $1/2\pi i$ at the point $\beta = \alpha$. Thus instead of (6) we set

$$A(\beta) = \frac{1}{4\pi} \frac{e^{i\beta/2}}{e^{i\beta/2} - e^{i\alpha/2}}. \tag{8}$$

Then we obtain in place of (7)

$$U = \frac{1}{4\pi} \int\limits_C \frac{e^{i\beta/2}}{e^{i\beta/2} - e^{i\alpha/2}} e^{-ikr\cos(\beta - \varphi)} \, d\beta, \tag{9}$$

where the path of integration is to be taken along the loops C (without the connecting paths D). This function is obviously also a solution of the wave equation because, like (7), it consists of a superposition of ordinary plane waves.

The shaded pattern in fig. 78 depends on the value of the angle φ, as is evident from (7 a). The whole pattern together with the integration paths shifts when φ is varied. This is inconvenient in the calculations to follow and can be avoided by replacing the integration variable β with

$$\gamma = \beta - \varphi, \tag{10}$$

which is advisable because the angles φ and α can then be combined as before:

(10 a) $$\psi = \varphi - \alpha.$$

Writing (9) in terms of γ and ψ, we obtain

(11) $$U = \frac{1}{4\pi} \int_C \frac{e^{i\gamma/2}}{e^{i\gamma/2} - e^{-i\psi/2}} e^{-ikr\cos\gamma} d\gamma.$$

This representation of U immediately shows us that *U has the period 4π in ψ and is therefore double-valued on the simple r, ψ plane; but U is single-valued on our Riemann surface.* U also satisfies the wave equation because it is, after all, only the function (9) written in a different form.

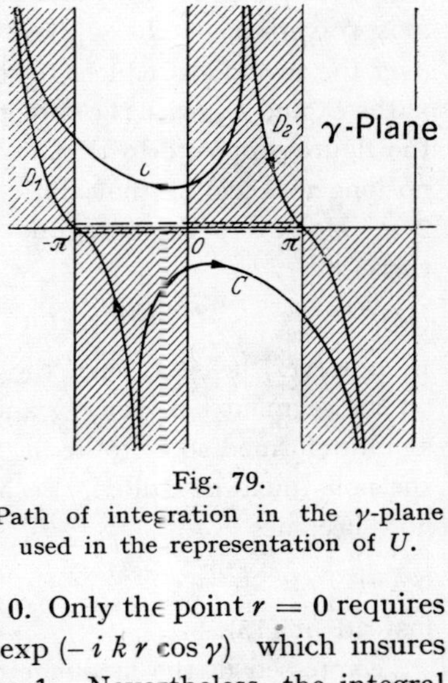

Fig. 79.
Path of integration in the γ-plane used in the representation of U.

We shall now explain fig. 79. The points marked

$$\gamma = +\pi, -\pi, \text{ and } 0$$

correspond, according to the eqs. (10) and (7 a), to the points A, B, and M in fig. 78. The branches D_1, D_2 of the path of integration drawn in fig. 79 should be disregarded for the present. The pole $\beta = \alpha$ in fig. 78 lies now at $\gamma = \alpha - \varphi = -\psi$. It has not been drawn in fig. 79 because we shall begin by considering the case $|\psi| > \pi$ for which the pole lies outside the segment $-\pi < \gamma < +\pi$. Since the loops C go to infinity inside the shaded regions, U is certainly finite and continuous for all $r > 0$. Only the point $r = 0$ requires special consideration. For here the factor $\exp(-ikr\cos\gamma)$ which insures convergence of the integral becomes equal to 1. Nevertheless, the integral converges because for $r = 0$ it becomes, except for a finite factor $-2i$,

(12) $$\int \frac{dz}{z - \zeta} = \log(z - \zeta) \Big|_{C_1}^{C_2} = \log \frac{z_2 - \zeta}{z_1 - \zeta}$$

where $z = e^{i\gamma/2}$, $\zeta = e^{-i\psi/2}$. C_1 and C_2 indicate the end points at infinity of the (upper and lower) loops C; z_1 and z_2 are the values of z at those points. But since $z = e^{i\gamma/2}$, their values are $z_1 = z_2 = 0$ for the upper loop and $|z_1| = \infty$, $|z_2| = \infty$ for the lower loop. Therefore on the lower loop we can neglect ζ in comparison to z. Thus we obtain

(12 a) $$\log \frac{z_2 - \zeta}{z_1 - \zeta} = \begin{cases} \log 1 = 0 & \text{on the upper loop,} \\ \log z_2/z_1 = i\pi & \text{on the lower loop.} \end{cases}$$

The last value follows from the fact that according to fig. 79 the γ-values of C_1 and C_2 differ by 2π. Thus (12) and (12 a) constitute the proof for the convergence of the integral (11) at $r = 0$.

Next, we must investigate the behavior of U as $r \to \infty$. As $r \to \infty$, the integrand of (11) vanishes *everywhere* in the shaded regions and not merely in their infinitely distant portions. In the case $|\psi| > \pi$, which is represented by fig. 79, the two loops C can be deformed so that they lie entirely in the shaded regions. For instance, the upper loop can be made to coincide with the two previously used by paths D_1 and D_2 plus the segment of the real axis between $-\pi$ and $+\pi$; the lower loop can be made to coincide with the lower portions of D_1 and D_2 and the same segment $+\pi$ to $-\pi$ of the real axis traversed in the opposite direction. Then the sum of the integrals over the two loops reduces to the sum of the integrals along the connecting paths D_1 and D_2, with the direction of integration indicated by the arrows in the figure (opposite to that in fig. 78). The integrals along these two paths no longer cancel as they did before because the period of γ is 4π and not 2π. However, the integrals along both paths are individually zero and therefore

$$U = 0 \quad \text{for} \quad r \to \infty \quad \text{and} \quad |\psi| > \pi. \tag{13}$$

If, however, $|\psi| < \pi$, then the pole of the integrand of (11) at $\gamma = -\psi$, lies on the segment between $-\pi$ and $+\pi$ in fig. 79. Therefore, if the loops C are again deformed so as to lie in the shaded regions, a positive circulation of the pole must be added. Because the residue is 1 at this pole the integral now becomes

$$U = e^{-ikr\cos\psi} \quad \text{for} \quad r \to \infty \quad \text{and} \quad |\psi| < \pi \tag{13 a}$$

instead of (13).

Expressed in the language of the two-sheeted Riemann surface, we have an "upper sheet" $|\psi| < \pi$ which is illuminated by the plane wave u_0 and a "lower sheet" $|\psi| > \pi$ which lies in the shadow. These sheets are connected along the "shadow boundary" $\psi = \pm\pi$. The contradistinction of "light and shadow" here finds its simplest mathematical formulation. For finite values of r the transition of U from one sheet to the other is continuous, and this transition constitutes the diffraction phenomenon. Actually the transition is also continuous for $r \to \infty$ in spite of the apparent discontinuity expressed by (13) and (13 a); (the transition region shrinks to zero only when measured in terms of the angular scale ψ).

However, we are not finished with our investigation of the infinitely distant point. We must carry our analysis a step further and show that not only the conditions (13), (13 a), but also the more stringent condition (1 d) is fulfilled. This condition must be satisfied by $v = U - u_0$ in the illuminated sheet and by $v = U$ in the shadow sheet. Only then can we be sure that the function U which we have constructed is actually the unique solution of the diffraction problem demanded by nature.

Hence, we must find an asymptotic approximation of the integral (11) for large r. In the shadow sheet we obtain from the difference between D_2 and D_1

$$4\pi U = \int_{D_2} e^{-ikr\cos\gamma}\,\Phi(\gamma)\,d\gamma, \tag{14}$$

$$\Phi(\gamma) = \frac{e^{i\gamma/2}}{e^{i\gamma/2} - e^{-i\psi/2}} - \frac{e^{i\gamma/2}}{e^{i\gamma/2} + e^{-i\psi/2}}. \tag{14 a}$$

In the second fraction which is due to D_1, the sign of $\exp(i\gamma/2)$ is reversed because γ is shifted by 2π with respect to its value in the first numerator. This has the effect of changing the sign of $\exp(-i\psi/2)$. The negative sign in front of the second fraction corresponds to the opposite directions of integration on D_1 and D_2 in fig. 79. The general method of evaluating (14) is the method of saddle-points, see Vol. VI, Sec. 19 E. We do not need to carry this out here, because we shall immediately develop a more convenient and even more precise method of evaluating this integral. We shall only make the following remark on the saddle-point method: from fig. 79 we see that in our case the critical saddle-point lies at $\gamma = \pi$, because at that point the path D_2 passes closely by two unshaded regions. For $\gamma = \pi$ the first factor of the integrand of (14) becomes $\exp(ikr)$; this factor can be taken out of the integral. The remaining integral which is to be computed only in the vicinity of the saddle-point yields the factor $\frac{1}{\sqrt{kr}}$, aside from a constant which does not interest us at present. Therefore, one finds that

$$U = \frac{C}{\sqrt{kr}}\,e^{ikr}. \tag{15}$$

This expression does indeed satisfy our radiation condition (1 d). For,

$$\frac{\partial U}{\partial r} - ikU = \frac{ikC}{\sqrt{kr}}\left(e^{ikr} - \frac{1}{2ikr}\,e^{ikr} - e^{ikr}\right),$$

and this vanishes, even when multiplied by r, in the limit as $r \to \infty$. The difference $U - u_0$ in the illuminated sheet behaves in the same manner.

We shall postpone the verification of the condition (1 e) as well as the complete proof of (15) until the end of the next section

C. Representation of U by a Fresnel integral

Our purpose in this section is to bring the above formulae into a form which is comparable to the expressions derived in Sec. 37 C. Unfortunately this transformation is of a somewhat lengthy and largely formal character.

The expression (14 a) can be rewritten

$$\Phi(\gamma) = \frac{2\, e^{1/2\, i(\gamma-\psi)}}{e^{i\gamma} - e^{-i\psi}}.$$

We introduce the new variable of integration η by setting

$$\gamma = \pi + \eta \qquad \text{and} \qquad \gamma = \pi - \eta$$

on the upper and lower half of D_2, respectively. The sum of the values of Φ at points with equal $|\eta|$ is found to be

$$\Phi(\pi+\eta) + \Phi(\pi-\eta) = \frac{-4\, i \cos\dfrac{\psi}{2}\cos\dfrac{\eta}{2}}{\cos\psi + \cos\eta}.$$

If we substitute this in (14), the latter becomes

$$\pi U = -i \cos\frac{\psi}{2}\int e^{ikr\cos\eta}\,\frac{\cos\dfrac{\eta}{2}}{\cos\psi+\cos\eta}\,d\eta, \tag{16}$$

where the range of integration is to be taken from $\eta = 0$ to a value $i\infty - \eta'$ and η' can be any arbitrary real number $< \pi$ (see fig. 79).

The expression (16) suggests that, instead of U, the quantity

$$V = \frac{U}{u_0} \tag{16 a}$$

be considered and that the factor $1/u_0 = \exp(+\,i\,k\,r\cos\psi)$ be placed under the integral, for then the denominator of (16) disappears after differentiation with respect to r. Performing this differentiation, one obtains

$$\pi\frac{\partial V}{\partial r} = k\cos\frac{\psi}{2}\int e^{i\,k\,r(\cos\psi+\cos\eta)}\cos\frac{\eta}{2}\,d\eta. \tag{17}$$

Now it is possible to perform the integration with respect to η. Since $\cos\eta = 1 - 2\sin^2\eta/2$, the integral to be evaluated is

$$\int e^{-2ikr\sin^2\frac{\eta}{2}}\cos\frac{\eta}{2}\,d\eta, \tag{17 a}$$

which, because $\cos\psi = 2\cos^2\psi/2 - 1$, is still to be multiplied by the factor

$$\exp\left(2\,i\,k\,r\cos^2\frac{\psi}{2}\right), \tag{17 b}$$

which is independent of η. Upon making the substitution

$$\sin\frac{\eta}{2} = \sqrt{\frac{\pi}{4kr}}\,\tau$$

(17 a) becomes an integral of the Fresnel type. For, introducing the above limits of integration, (17 a) is equal to

$$\sqrt{\frac{\pi}{kr}}\,F^*(\infty), \qquad F^*(\infty) = \int_0^\infty e^{-\frac{i\pi}{2}\tau^2}\,d\tau = \frac{1-i}{2}, \quad \text{see (37.8 b).}$$

Multiplying now by the factor (17 b), we obtain for the value of (17)

$$\frac{\partial V}{\partial r} = \frac{1-i}{2}\sqrt{\frac{k}{\pi r}}\cos\frac{\psi}{2}\exp\left(2ikr\cos^2\frac{\psi}{2}\right). \tag{18}$$

The right-hand side can be written as the differential quotient with respect to r of an expression which we can again write in the form of a Fresnel integral[1]. That is, (18) is equivalent to

$$\frac{\partial V}{\partial r} = \frac{1-i}{2}\frac{\partial}{\partial r}\int_{-\infty}^{\rho} e^{\frac{i\pi}{2}\tau^2}\,d\tau, \qquad \varrho = 2\sqrt{\frac{kr}{\pi}}\cos\frac{\psi}{2}. \tag{18 a}$$

This becomes, by integrating with respect to r,

$$V = \frac{1-i}{2}\int_{-\infty}^{\rho} e^{\frac{i\pi}{2}\tau^2}\,d\tau, \tag{18 b}$$

and because of (16 a)

$$U = u_0\frac{1-i}{2}\int_{-\infty}^{\rho} e^{\frac{i\pi}{2}\tau^2}\,d\tau. \tag{19}$$

Because of the definition of ρ in (18 a), this representation of U is an *analytic function of ψ with the period 4π.* Therefore, it is valid not only on the shadowed sheet, for which (19) was derived, but also on the illuminated sheet of our two-sheeted Riemann surface. On the latter we obtain for $r \to \infty$

$$U = u_0\frac{1-i}{2}\int_{-\infty}^{\infty} e^{\frac{i\pi}{2}\tau^2}\,d\tau = u_0(1-i)\int_0^\infty e^{\frac{i\pi}{2}\tau^2}\,d\tau$$

$$= u_0(1-i)\,F(\infty) = u_0\,\frac{(1-i)(1+i)}{2} = u_0,$$

as required.

---

[1] If we were to choose the lower limit of the integral as zero, the ordinary Fresnel integral $F(\rho)$ would result. But then an (easily evaluated) integration constant would have to be added to the right-hand side of (18 b). With our choice of $-\infty$ as the lower limit this constant becomes zero.

We can discuss the representation (19) conveniently for both large and small values of $|\rho|$ with the help of the approximation formulae for Fresnel's integral which were derived in Sec. 37.

a) For large $|\rho|$ in the region of the geometrical shadow ($\rho < 0$) we write

$$\int_{-\infty}^{\rho} = \int_{-\infty}^{0} - \int_{-|\rho|}^{0} = F(\infty) - F(|\rho|),$$

and hence, according to (37.8 a)

$$U = -u_0 \frac{1-i}{2} \frac{e^{\frac{i\pi}{2}\rho^2}}{i\pi\rho} \left(1 + \frac{1}{i\pi\rho^2} + \dots\right).$$

Introducing the expressions for u_0 and ρ we find

$$u_0 e^{\frac{i\pi}{2}\rho^2} = \exp\left\{-ikr\cos\psi + 2ikr\cos^2\frac{\psi}{2}\right\} = \exp(ikr),$$

and obtain

$$U = \frac{1+i}{4\sqrt{\pi k r}\cos\frac{\psi}{2}} e^{ikr}\left(1 + \frac{1}{i\pi\rho^2} + \dots\right). \tag{20 a}$$

If the correction terms inside the parentheses are neglected, this is of the same form as the asymptotic behavior which we predicted in (15); the factor C, which was there undetermined, now turns out to be a function of ψ.

b) For large ρ in the illuminated region ($\rho > 0$) we substitute in (19)

$$\int_{-\infty}^{\rho} = \int_{-\infty}^{0} + \int_{0}^{\rho} = F(\infty) + F(\rho) = 2F(\infty) - [F(\infty) - F(\rho)],$$

and consider that

$$\frac{1-i}{2} 2F(\infty) = \frac{(1-i)(1+i)}{2} = 1.$$

Using the same expansion as in a), we obtain from (19)

$$U = u_0 - \frac{1+i}{4\sqrt{\pi k r}\cos\frac{\psi}{2}} e^{ikr}\left(1 + \frac{1}{i\pi\rho^2} + \dots\right). \tag{20 b}$$

This asymptotic expression is also in agreement with the remarks made following formula (15).

c) For small ρ (illuminated or shaded sheet) we set

$$\int_{-\infty}^{\rho} = \int_{-\infty}^{0} + \int_{0}^{\rho} = F(\infty) + F(\rho)$$

and by means of (37.8) we obtain from (19)

$$U = \frac{u_0}{2}\left\{1 + (1-i)\,\rho\left(1 + \frac{i\pi}{6}\rho^2 + \ldots\right)\right\}. \tag{20 c}$$

At the branch point itself $\rho = 0$, $u_0 = 1$ and therefore

$$U = \frac{1}{2} \quad \text{and} \quad \frac{\partial U}{\partial r} = \frac{1-i}{2}\frac{\partial \rho}{\partial r} = \frac{1-i}{2}\sqrt{\frac{k}{\pi r}}\cos\frac{\psi}{2}. \tag{20 d}$$

Thus $\partial U/\partial r$ goes to ∞ as $r \to 0$ but only so weakly that the product of r and the gradient remains finite as required by (1 e):

$$r\,\mathrm{grad}\,U \to 0 \quad \text{as} \quad r \to 0. \tag{20 e}$$

Thus we have finally shown that our branched solution U satisfies all the conditions as postulated in section A.

D. The diffraction field of the straight-edge

Returning now to fig. 76 and the representations (2) and (4), we shall describe the field in the region of observation $0 < \varphi < 2\pi$. The plane is divided into three sectors I, II, III by the screen S, the shadow boundary G of the reflected wave and the shadow boundary G_i of the incident wave. These three sectors have the central angles $\pi - \alpha$, 2α, $\pi + \alpha$, respectively. I is illuminated by the incident wave $U(r, \varphi - \alpha)$ and by the reflected wave $U(r, \varphi + \alpha)$, II belongs to the illuminated sheet of the incident wave and to the shaded sheet of the reflected wave, while III is in the shadow of both the incident and reflected waves. The direction of incidence of the reflected wave does not lie in the region of observation. We must think of this direction as lying on a Riemann sheet which is connected to the region of observation along S.

Let us first consider sector III. Since we are only interested in distances $r \gg \lambda$, we need to consider only large values of $k\,r$ and large values of ρ (except in the immediate vicinity of the shadow boundary G_i where $\cos\psi/2 = 0$). Therefore, we may use the approximation formula (20 a) and thus obtain by (2) and (4)

$$u = \frac{1+i}{4\sqrt{\pi k r}}\,e^{ikr}\left(\frac{1}{\cos\frac{\varphi-\alpha}{2}} \mp \frac{1}{\cos\frac{\varphi+\alpha}{2}}\right). \tag{21}$$

The upper sign corresponds to the case where **E** oscillates parallel to the edge of the screen; the lower sign corresponds to the case where **E** oscillates perpendicular and **H** parallel to the edge of the screen (in this latter case u represents not **E** but **H**). The second term in the parentheses takes on appreciable values only at the shadow boundary G_r and can therefore be neglected[1] in comparison to the first term.

Since the value of the expression in the parentheses decreases slowly with increasing φ, we conclude that *the light is diffracted far into the region of the geometrical shadow*. The infinitely large value which this expression assumes on the shadow boundary is, of course, illusory and is due to the fact that our asymptotic approximation is not valid there. In this region the exact representation (19) should be used instead of the approximation (21); see below.

Even more interesting than the φ-dependence of eq. (21) is its r-dependence $\dfrac{e^{ikr}}{\sqrt{r}}$ *which has the character of a cylindrical wave emitted by the edge of the screen.* We have indicated this by the arrows emerging from O in fig. 76. A. Kalaschnikow[2] showed that these ray directions can be photographed. He inserted pins into a photographic plate which he then placed at an angle to the ray directions. After a sufficiently long exposure, radially directed shadows of these pins appeared on the plate.

If one focuses the eye on the edge of the screen, this edge appears as a *thin luminous line*. This effect was described very early by Grimaldi, the father of all diffraction discoveries. The explanation is that the eye performs an inadmissible extrapolation. It infers from the asymptotic field, which is correctly represented by (21), that the field becomes infinite at $r = 0$, which is not true. In fact, the energy radiated into an angular region $\delta\varphi$ per unit length of the edge per unit time is, depending on whether **E** is parallel or perpendicular to the edge,

$$\delta S = S_r\, r\,\delta\varphi = r\,\delta\varphi \left\{ \begin{array}{l} -E_z H_\varphi \\ +E_\varphi H_z \end{array} \right\} = \left\{ \begin{array}{l} \dfrac{i}{\omega\mu_0} E_z r \dfrac{\partial E_z}{\partial r}\,\delta\varphi, \\[2ex] \dfrac{i}{\omega\varepsilon_0} H_z r \dfrac{\partial H_z}{\partial r}\,\delta\varphi. \end{array} \right. \tag{22}$$

In the upper line ($E = E_z$) we have used the equation $\dot{\mathbf{B}} = -\operatorname{curl}\mathbf{E}$, and in the lower line ($H = H_z$) we have used $\dot{\mathbf{D}} = \operatorname{curl}\mathbf{H}$. The factors $\partial/\partial r$ in both lines become infinite for $r = 0$ but only so slowly that $\delta S = 0$; see (20 e). Hence the "luminous edge" is not real.

[1] Retention of this term would yield a small difference between the cases $\mathbf{E}_{\|}$ and $\mathbf{E}_{\perp}$; and hence a small polarization effect.

[2] Journal of the Russ. Phys. Soc. **44**, p. 133, 1912.

The factor

(23) $$1+i=\sqrt{2}\,e^{i\frac{\pi}{4}}$$

in (21) is also of some interest. It shows that the phases of the diffracted and incident waves, the former extrapolated to $r=0$, do not agree. The phase of u_0 is

$$-ikr\cos\psi-i\omega t, \quad \text{hence equal to } -i\omega t \text{ at } r=0,$$

while that of (21) is

$$\frac{i\pi}{4}+ikr-i\omega t, \quad \text{hence equal to } \frac{i\pi}{4}-i\omega t \text{ at } r=0.$$

Such a "phase jump" always takes place when light passes through a focal point (or focal line in the case of our cylinder wave); see Sec. 45. But as in the case of the luminous edge, the phase jump is only a result of the extrapolation and is not real. Actually the phase, like the amplitude, is continuous at the origin, as far as it is at all permissible to talk of a phase of the complicated oscillations in that vicinity.

Let us now turn to sector II. At some distance from the shadow boundaries G_i and G_r we can set

$$U(r,\varphi-\alpha)\sim u_0, \qquad U(r,\varphi+\alpha)\sim 0:$$

that is, we can disregard diffraction and obtain the pure field $u=u_0$ of the incident wave.

We must proceed differently in the vicinity of G_i. Here we set

(24) $$\varphi-\alpha=\pi-\delta, \qquad \cos\frac{\varphi-\alpha}{2}=\sin\frac{\delta}{2},$$

and call δ the "diffraction angle" which shall be reckoned positive in the direction towards II and negative in the direction towards III. Then

(24 a) $$\rho=2\sqrt{\frac{kr}{\pi}}\sin\frac{\delta}{2}$$

is finite even for large kr, provided δ is correspondingly small. Therefore $U(r,\varphi-\alpha)$ also has a finite value in comparison to which we can neglect $U(r,\varphi+\alpha)$. Using the exact representation (19) for $U(r,\varphi-\alpha)$, we obtain

(25) $$U=u_0\frac{1-i}{2}\{F(\infty)+F(\rho)\}$$

for both cases (2) and (4). Calculating the ratio U/u_0 we obtain

(25 a) $$\left|\frac{U}{u_0}\right|=\frac{1}{\sqrt{2}}|F(\infty)+F(\rho)|.$$

Formally, this agrees exactly with (37.13). For, since Cornu's spiral is symmetrical about the origin, $F(\infty) = -F(-\infty)$. There is, however, a difference in meaning between our present variable ρ and the previous variable

$$w = \frac{x}{\sqrt{\lambda a/2}}, \tag{26}$$

as defined in (37.6 b). There, x was the distance of the point of observation from the shadow boundary, which we must now denote by $r \sin \delta$; a was the distance between the observation screen and the diffraction screen for perpendicular incidence; this we may now denote by r. Thus we obtain from (26)

$$w = \frac{\sqrt{r} \sin \delta}{\sqrt{\lambda/2}} = \sqrt{\frac{k r}{\pi}} \sin \delta. \tag{26 a}$$

A comparison with (24 a) shows that (26 a) contains the factor $\sin \delta$ *in place of our previous factor* $2 \sin \delta/2$. For small values of δ, which are the only ones of interest in the vicinity of the shadow boundary, this represents only a difference of the third order. Therefore we can still use fig. 75 to represent the results of our present more rigorous theory. This figure correctly exhibits the positions and amplitudes of the diffraction maxima and minima on the illuminated side of the shadow boundary, as well as the monotonically decreasing intensity in the region of the geometrical shadow. The intensity value 1/4 on the shadow boundary itself is also in agreement with the present theory. Leaving all numerical considerations aside, we nevertheless wish to point out that the occurrence of $\sin \delta/2$ in (24 a) reflects a typical feature of our theory, namely that the diffraction angle has the period 4π.

We add one critical remark regarding the use of Huygens' principle. Let us consider in greater detail the half-plane $\varphi = \pi$ which is left open by the screen and plays the role of the "diffraction opening" in Huygens' principle. According to our prescription of Sec. 34 C the "boundary values" are chosen as the values u_0 given by the unperturbed incident wave; if, for simplicity, we assume perpendicular incidence ($\alpha = \pi/2$), then $u_0 = 1$. In contrast to this assumption eqs. (19) and (18 a) yield for $\varphi = \pi$ and $\alpha = \pi/2$

$$U = \frac{1-i}{2} \int_{-\infty}^{\rho} e^{-\frac{i\pi}{2}\tau^2} d\tau, \qquad \rho = \sqrt{\frac{2 k r}{\pi}}.$$

This expression varies from the value $U = 1/2$ at $r = 0$ to $U = 1$ at $r = \infty$ and *oscillates* in between. These values are in greatest contrast to the assumed boundary value $u_0 = 1$ which was used in applying Huygens' principle. A corresponding result holds for the reflected wave U, in which case α must be replaced by $-\pi/2$ and ρ by $-\sqrt{(2\,k\,r)/\pi}$, and therefore the contradiction applies also to the superposition of the two waves. Thus we can say that the boundary values used in Huygens' principle differ from the (in our case) exact boundary values not only in the vicinity of the edge of the screen but even at *large numerical distances* $k\,r$ from that edge. It is amazing that the classical diffraction theory nevertheless yields for all practical purposes satisfactory results.

The sector I belongs, as we know, to the illuminated regions of both the incident and reflected waves. At the boundary G_r of the latter region there occur, of course, diffraction phenomena which can be calculated in the same manner as those at the boundary G_i. But these diffraction fringes are masked by the full intensity of the incident wave in that region; they have been investigated experimentally only in the case of "Fresnel's mirror" (two half-planes inclined at a very shallow angle with respect to each other) and have been introduced in the calculations as very small perturbations added to the ordinary interference of the two reflected waves.

E. Generalization

It is easy to make the transition from the two-sheeted to an n-sheeted Riemann surface. It is only necessary to generalize eq. (8) to

$$A(\beta) = \frac{1}{2\pi\,n}\,\frac{e^{i\beta/n}}{e^{i\beta/n} - e^{i\alpha/n}}. \tag{27}$$

Analogously to (9) this leads to a function U of period $2\pi\,n$ with the help of which image problems in a sectorial space of central angle $2\pi\,n/m$ (m = integer) can be solved. Among these problems is, of course, that of the exterior of a rectangular wedge for which $n = 3$, $m = 4$. The representation in terms of Fresnel's integral which we treated in C is limited to the case where $n = 2$. W. Pauli[1] has shown that for arbitrary (even non-integer) values of n, Fresnel's integral is replaced by a confluent hypergeometric function.

The limiting case $n = \infty$ is of particular interest. It leads to the infinitely many-valued function

$$U = \frac{1}{2\pi i}\int e^{-ikr\cos\gamma}\,\frac{d\gamma}{\psi - \gamma}, \tag{27 a}$$

which is here expressed in terms of the integration variable γ of eq. (11). We regard this function as the best possible representation for the case of the conventional "*black screen*": the wave which is incident at an angle $\varphi = \alpha$

[1] Phys. Rev. **54**, p. 924, 1938.

on the front of the screen $\varphi = 0$ enters the screen and loses itself among the infinitely many sheets $\varphi < 0$. None of its energy is returned to the physical space via the infinite number of sheets $\varphi > 2\pi$ through the back side of the screen $\varphi = 2\pi$. To understand this one should recall the experimental realization of a black body used in heat radiation measurements, that is, a cavity which is kept at a constant temperature and has a small hole. All radiation entering the hole from the outside is reflected back and forth inside the cavity without ever again leaving it. The hole absorbs completely and acts, therefore, like a black surface. But the property "black" cannot be defined by boundary conditions within the realm of Maxwell's theory. Therefore diffraction by a black screen cannot be formulated as a boundary value problem. Our formulation (27 a) is by no means unique or devoid of arbitrariness.

We shall indicate only briefly other possible generalizations of our method. First, there is the case of the *cylindrical wave* (luminous line in the finite region and parallel to the edge of the screen) for which our method also yields a complete solution of the diffraction problem in closed form[1]. The generalization to three dimensions is directly possible only for scalar (acoustic) problems. For these the diffraction of a spherical wave or a plane wave which is incident not perpendicularly but at an angle with respect to the edge of the screen can be treated by our method.

F. Basic remarks on branched solutions

In electrodynamics there are two general types of problems: the summation problems and the boundary value problems; see Vol. III, Sec. 7 and Sec. 9. When the distribution of charges throughout space was given, then we only needed to sum over all these charges in the proper way in order to obtain the complete electrostatic field. The same was true in the magnetostatic case when the magnetisation was everywhere known. However, when material bodies, such as conductors, dielectrics, or magnetizable matter with *unknown* charges and magnetizations were present, certain boundary conditions had to be fulfilled. We were then faced with the mathematically much more complicated boundary value problems. An obvious requirement on the correct formulation of these boundary value problems was that of *uniqueness*.

Huygens' principle attempts to solve diffraction problems by the *summation method*. Since the boundary values to be prescribed in the diffraction opening are fundamentally unknown, and a certain plausible yet arbitrary

[1] For details see Frank-Mises, loc. cit., p. 826.

choice must be made regarding them, the correctness and uniqueness of the solutions obtained in this way may properly be doubted. In the case of the rigorously solvable diffraction problems which we formulated at the beginning of this section, the boundary conditions were clear on the basis of Maxwell's theory. Supplemented by the radiation condition at infinity, these boundary conditions insured the uniqueness of our problem. In the conventional case of an *infinitely thin* screen one must go to the limit of infinite conductivity and use the corresponding limiting form of the boundary conditions. We have already seen that the *black screen* which is preferred in theory and in experiment cannot be described in terms of boundary conditions, and therefore the diffraction caused by it cannot be described by a uniquely defined boundary value problem.

For a perfectly reflecting plane screen of arbitrary shape the method of images leads to the problem of constructing *branched solutions* of the wave equation such that the edge of the screen is the branch line. In two-dimensional problems (slit, parallel strip, half-plane) the range of values of the solutions is represented on a two-sheeted Riemann surface. In three-dimensional problems the solution is defined on a Riemann double space. The mathematical construction of the branched solutions is possible only for the case of the half-plane. Nevertheless, our method of the Riemann double space leads to quantitative results also for the problem of an arbitrary plane screen. In order to see this we need further preparation.

It has been known since Euler that functions which are symmetric in the n roots of any n^{th} degree *algebraic equation* are rational functions of the coefficients of that equation. The same is true for the branches of an *algebraic function*, i. e. for the roots of an nth degree equation whose coefficients are entire functions of a complex variable z. Such an algebraic function is defined on an n-sheeted Riemann surface. If we denote an algebraic function by $w(z)$ and its n branches by $w_1, w_2, \ldots, w_n$, then all symmetric functions of the $w_1, w_2, \ldots, w_n$ are single-valued in z and are rational functions of the coefficients of the defining equation.

This theorem is used repeatedly in two-dimensional potential theory, as for instance in the mapping problems of hydrodynamics. If the velocity potential $u(x, y)$ and the stream function $v(x, y)$ are combined in the form $w = u + i v$, one obtains a function of the complex variable $z = x + i y$ whose real and imaginary parts satisfy Laplace's equation $\Delta \begin{Bmatrix} u \\ v \end{Bmatrix} = 0$. If w is multiple-valued, then the symmetric functions of its branches $w_1, \ldots, w_n$ are single-valued in z just as in the case of algebraic functions. From these single-valued functions w, and therefore also u and v, can be calculated *algebraically*.

Here we are interested in two-valued solutions U of the *wave equation* where, in contrast to the solutions of Laplace's equation, there is no function conjugate to U. Of the symmetric functions only the linear combination $U_1 + U_2$ needs to be considered. This sum is a single-valued solution of the same differential equation and can therefore be regarded as known. (The symmetric product $U_1 U_2$ is not a solution of the wave equation; otherwise the two branches U_1 and U_2 could each be calculated algebraically, and the construction of the branched solution would be simple.) Limiting ourselves to *scalar* problems, we consider in particular a plane wave in the Riemann double space, the branch line of which coincides with the edge of the screen. $U_1(P)$ shall refer to the first branch, $U_2(P)$ to the second branch of this space; the two branches are connected in the plane of the screen. We form

$$U_1(P) + U_2(P) = u_0(P). \tag{28}$$

Then $u_0(P)$ is a single-valued solution of the wave equation in the simple space and is identical with our previous function u_0 which represented a plane wave with no screen present. This result is rigorous because the solutions of the wave equation are uniquely determined by the continuity conditions which must always be imposed and by the condition imposed on the behavior of the solution at infinity. (The same eq. (28) is obviously valid for an incident spherical or cylindrical wave as well as for a plane wave.)

We shall first confirm eq. (28) by applying it to the explicit formulae for our double space with a straight branch line. It is convenient to start with eq. (19). Denoting by ρ the quantity (18 a) which is positive on the illuminated sheet, and hence denoting by $-\rho$ the corresponding quantity on the shaded sheet, we obtain

$$U_1(P) + U_2(P) = u_0 \frac{1-i}{2} \left\{ \int_{-\infty}^{\rho} + \int_{-\infty}^{-\rho} \right\} e^{\frac{i\pi}{2}\tau^2} d\tau. \tag{29}$$

We see immediately that by changing the sign of τ in the second integral both integrals can be combined into

$$\int_{-\infty}^{\infty} e^{\frac{i\pi}{2}\tau^2} d\tau = 2 \int_{0}^{\infty} e^{\frac{i\pi}{2}\tau^2} d\tau = 2F(\infty) = 1 + i,$$

and (29) becomes identical with (28). It should be emphasized that this proof depends in no way on the transformed form (19) but can be performed equally well using the original form (9). In this latter formulation the path of integration for U_2 is obtained by displacing the integration path for U_1 by 2π. The two paths, see fig. 78, combine then into two loops C which span a distance 4π instead of 2π. Because of the periodicity of the integrand, the two loops can be transformed into a circuit around the pole $\varphi = \alpha$, so that the integral again yields u_0. In the same way it can also be seen that for the generalization embodied in (27) (n arbitrary instead of $n = 2$), the statement (28) is generalized to

$$U_1(P) + U_2(P) + \ldots + U_n(P) = u_0(P).$$

Let us now compare (28) with our earlier formulation of Babinet's principle (34.15). The formal similarity between these two expressions suggests that U_1 and U_2 be associated with the diffraction patterns of two *complementary* screens I and II[1]. This is however permissible only for the case of *black* screens which, like the branch cut of our Riemann surface, can absorb light but cannot reflect it. In addition, this association is subject to the same lack of uniqueness which the definition of the black body suffers. Therefore we seek a formulation of Babinet's principle which is valid for the well-defined *reflecting* screen and which can be considered as a more *precise formulation* of the principle. By way of preparation we again consider the simple case of the half-plane.

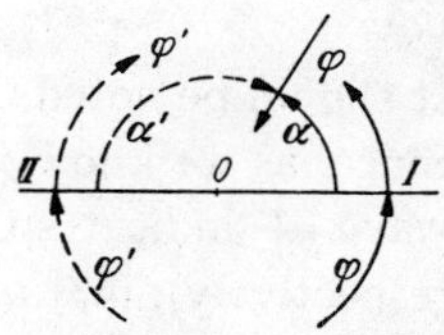

Fig. 80.

Babinet's principle: directions to be used for measuring the angles φ and φ' for the cases of the original screen $O\,I$ and the complementary screen $O\,II$.

We compare the diffraction by the original half-plane

$$(30) \quad u_I = U(r, \varphi - \alpha) \mp U(r, \varphi + \alpha), \quad 0 < \varphi < 2\pi$$

with that by the complementary half-plane which, in the corresponding notation, reads

$$(31) \qquad u_{II} = U(r, \varphi' - \alpha') \mp U(r, \varphi' + \alpha'), \qquad 0 < \varphi' < 2\pi.$$

If the direction of incidence of the plane wave is the same for both screens, then we must make

$$(31\text{ a}) \qquad \alpha' = \pi - \alpha.$$

The relationship between φ' and φ can be inferred from fig. 80. It leads to the following compilation:

Front surface of the complementary screen	$\varphi' = 0, \quad \varphi = \pi$	(31 b) $\varphi' = \pi - \varphi$,
Front surface of the original screen	$\varphi' = \pi, \quad \varphi = 0$	
Rear surface of the original screen	$\varphi' = \pi, \quad \varphi = 2\pi$	(31 c) $\varphi' = 3\pi - \varphi$.
Rear surface of the complementary screen	$\varphi' = 2\pi, \quad \varphi = \pi$	

[1] Compare with the presentation by the author in Frank-Mises, Vol. II, Chap. XX, Sec. 1, eq. (15).

Equation (31 b) is the desired relation for the front surface of both screens; eq. (31 c) is the relation which applies to the rear surface.
Substituting (31 a, b) in (31) we obtain for the *front surface*

(32) $$u_{II} = U\,(r, -\varphi + \alpha) \mp U\,(r, 2\pi - \varphi - \alpha).$$

Because of the property (28) of our branched solution

(32 a) $$U\,(r, 2\pi - \varphi - \alpha) = u_0\,(-\varphi - \alpha) - U\,(r, -\varphi - \alpha)$$

and because of the right-left symmetry of both the branched and unbranched waves

(32 b) $$U\,(r, -\varphi - \alpha) = U\,(r, \varphi + \alpha), \qquad U\,(r, -\varphi + \alpha) = U\,(r, \varphi - \alpha),$$
$$u_0\,(-\varphi - \alpha) = u_0\,(\varphi + \alpha).$$

Substituting (32 a, b) in (32) it follows that

(33) $$u_{II} = U\,(r, \varphi - \alpha) \pm U\,(r, \varphi + \alpha) \mp u_0\,(\varphi + \alpha).$$

It should be noted that the sign of $U\,(r, \varphi + \alpha)$ is now *opposite* to that in (30). Since, as we know, this sign is determined by the polarization of the incident wave ($\mp$ indicates that $\mathbf{E}$ is parallel or perpendicular to the edge of the screen, respectively), (33) tells us that we must compare the diffraction pattern of the *complimentary* screen illuminated by *parallel* polarized incident light ($\mathbf{E}_{||}$) with the diffraction pattern of the *original* screen illuminated by *perpendicular* polarized light ($\mathbf{E}_{\perp}$). Furthermore, the term $\mp u_0\,(\varphi + \alpha)$ shows that for the complementary screen we must omit the reflected light wherever it is present for the original screen and that we have to add reflected light where it is missing for the original screen. This is understandable from the viewpoint of geometrical optics.

A corresponding calculation using (31 a, c) and (31) yields for the *rear surface*

$$u_{II} = U\,(r, 2\pi - \varphi + \alpha) \mp U\,(r, 4\pi + \varphi + \alpha),$$

and applying the correspondingly modified eqs. (32 a, b) we obtain

(34) $$-u_{II} = U\,(r, \varphi - \alpha) \pm U\,(r, \varphi + \alpha) - u_0\,(\varphi - \alpha).$$

Thus we have the same interchange of polarizations as in (33); in addition, the incident wave must now be omitted for the complementary screen where it was originally present, i. e. behind the screen, and the incident wave must be added where it was missing with the original screen, i. e. in the shadow of the latter which by geometrical optics is the illuminated region of the complementary screen.

The problem of the half-plane shares with other *two-dimensional* problems (slit, grating, ...) the feature that it can be treated as a *scalar* problem. This is not the case for the *three-dimensional* problems of *optics* (e. g. circular disc or circular opening). To solve these problems *vector* calculations are necessary (or suitably defined potentials may be used). This is not so in acoustics where the scalar pressure (or the velocity potential) is treated. In the dissertation which was referred to on p. 248, Bouwkamp set up the rigorous form of Babinet's principle for the scalar three-dimensional problem of an arbitrarily bounded rigid plane screen and its coplanar complementary screen. The statement of the principle is the same as in our two-dimensional problems, and the proof again relies upon the relation (28) for branched functions. The above method can be extended to the three-dimensional scalar case by considering the straight screen edge and the cylindrical surfaces surrounding it to be deformed in the manner of topology into the given arbitrarily shaped edge curve and the corresponding toroidal surfaces surrounding it. These surfaces can be distinguished by assigning to them a parameter φ which can be chosen so as to increase by 2π as it passes from one sheet of the double space to the other. If this is done, eqs. (30) to (34) can be interpreted directly as the expression of Babinet's principle for the scalar three-dimensional case.

The completely general, rigorous formulation of Babinet's principle for the three-dimensional optical case has been given by J. Meixner[1]. Since in the incident wave $\mathsf{E}_\perp$ implies $\mathsf{H}_{||}$ and (retaining a right-handed system) $\mathsf{E}_{||}$ implies $-\mathsf{H}_\perp$, we can replace our sign inversion $\mp \to \pm$ in going from the original to the complementary screen by the following interchange rule:

$$(\mathsf{E}, \mathsf{H}) \to (\mathsf{H}, -\mathsf{E}).$$

Therefore, according to Meixner, one obtains the diffraction field E, H *of the complementary screen from the diffraction field* H, $-\mathsf{E}$ *of the original screen* (provided the incident or reflected wave is cancelled or added at the front or rear of the screens in a precisely specified manner). The proof holds for arbitrary distributions of light sources, not merely for a wave coming from infinity. It is based solely on the symmetry properties of Maxwell's equations. We wish to note in this connection that fundamentally our representation in terms of branched solutions also implies a symmetry property of Maxwell's equations.

Finally, we shall establish the connection between our energy condition (1 e) and the theory of functions of a complex variable z. We shall again limit ourselves to the scalar case. Specifically, (1 e) is related to the Puiseux expansion of a function about a branch point which replaces the otherwise valid Taylor expansion. If the number of sheets connected in the branch point $z = 0$ is n, and if this branch point is not at the same time a pole of the function $w(z)$, then

[1] Z. f. Naturforschung, Vol. 3 a, p. 508, 1948.

$$w(z) = \sum_{m=0}^{\infty} C_m z^{\frac{m}{n}} = \sum_{m=0}^{\infty} C_m r^{\frac{m}{n}} e^{i\frac{m}{n}\varphi}.$$

The corresponding expansions of solutions u of the wave equation are in terms of Bessel functions with fractional indices

$$u(r, \varphi) = \sum_{m=0}^{\infty} C_m J_{\frac{m}{n}}(k r) e^{i\frac{m}{n}\varphi}. \tag{35}$$

Since $J_{\frac{m}{n}}(\rho)$ is proportional to $\rho^{\frac{m}{n}}$ for small values of ρ, this expansion yields at the branch point $r = 0$

$$u = C_0 = \text{finite value, but } \frac{\partial u}{\partial r} = k \sum_{m=1}^{n-1} C_m J_{\frac{m}{n}}'(k r) e^{i\frac{m}{n}\varphi}. \tag{36}$$

$\partial u/\partial r$ becomes infinite at $r = 0$ but only weakly, so that

$$\lim_{r \to 0} r \operatorname{grad} u = 0. \tag{36 a}$$

This condition was verified explicitly for $n = 2$ in eq. (20 e). It is reasonable to postulate this same condition also for the case of a spatial branch line[1], in which case r would mean the shortest distance from the branch line. As a result our condition (1 e) turns out to be a *mathematical consequence* of the condition that u be everywhere continuous (also at the edge of the screen) and therefore (1 e) can be omitted as a special requirement.

[1]Compare A. Sommerfeld, Proc. London Math. Soc., Vol. 28, particularly p. 405, 1897. "Branched Potentials in Space" are treated there.

CHAPTER VI

ADDENDA, CHIEFLY TO THE THEORY OF DIFFRACTION

39. Diffraction By a Very Narrow Slit

When the dimensions of the diffracting aperture become small compared to the wavelength or even only a few times larger, Huygens' principle becomes meaningless. For in our applications of this principle (Sec. 34) we used only the unperturbed wave incident on the opening and *neglected the effect of the edge zones* entirely. Therefore our present problem in which the aperture consists, so to speak, mainly of edge zones belongs definitely to the category of *boundary value problems*; it is now necessary to determine the state of the field in the opening from the continuity conditions imposed on the total solution. Thus the distinction between the incident and diffracted wave breaks down.

Lord Rayleigh[1] was the first to tackle this problem. With masterful brevity and clarity he reduced the problem to known solutions of hydrodynamic or electrostatic problems; he did this in particular for the case of a circular opening of radius $a \ll \lambda$ or for a sufficiently narrow slit.

Bethe[2] treated the problem of the small circular aperture independently from Rayleigh from an electromagnetic point of view and obtained substantially the same result. The work of Levine and Schwinger[3] is aimed at the more difficult goal of bridging the gap between the limiting cases $a \ll \lambda$ (Lord Rayleigh) and $a \gg \lambda$ (Huygens-Kirchhoff) by means of a variational principle. For the time being this work is restricted to the scalar acoustic case.

We shall treat the experimentally important problem of the slit in which we may operate with scalar equations (see p. 277) by considering separately the two cases E and H parallel to the slit edges.

---

[1] On the Passage of Waves through Apertures in Plane Screens and Allied Problems, Phil. Mag. **43**, p. 259, 1897. Scientific Papers, Vol. IV, p. 283.
[2] H. A. Bethe, Phys. Rev. **66**, p. 163, 1944.
[3] H. Levine and J. Schwinger, Phys. Rev. **74**, p. 958, 1948, **75**, 1423, 1949.

A. The boundary value problem of the slit

Let the slit lie in the xy-plane and let the slit edges be parallel to the y-axis. The screen shall be thought of as infinitely thin and perfectly conducting. The width of the slit shall be $2a$, and the edges are to be given by $x = \pm a$, $z = 0$. A plane wave coming from the negative z direction is incident perpendicular to the plane of the slit. The problem is entirely independent of y and is therefore two-dimensional.

The incident wave will be represented by $A \exp(ikz)$. We shall assume first that **E** oscillates parallel to the edges of the slit, i. e. $\mathbf{E} \to E_y$. If there were no slit in the screen, the field would be given by

$$\text{(1)} \qquad \begin{cases} v = A\,(e^{ikz} - e^{-ikz}) & \text{for} \quad z < 0, \\ v = 0 & \text{for} \quad z > 0. \end{cases}$$

Because of the presence of the slit, (1) must be changed to

$$\text{(2)} \qquad \begin{aligned} v &= A\,(e^{ikz} - e^{-ikz}) + u_- \quad && \text{for} \quad z < 0, \\ v &= \phantom{A\,(e^{ikz} - e^{-ikz}) +{}} u_+ && \text{for} \quad z > 0. \end{aligned}$$

We denote the value of v in the slit opening by $\overline{u}$. Because of the continuity of the field v

$$\text{(3)} \qquad u_+ = u_- = \overline{u} \qquad \text{for} \qquad z = 0.$$

We call the x-coordinate of a point in the slit ξ and write therefore $\overline{u} = \overline{u}\,(\xi)$. If $\overline{u}\,(\xi)$ were known, then we could compute u_+ and u_- rigorously for all points $z \gtrless 0$ by the general method of Green's function introduced in Sec. 34 C. However, in the Green's function (34.7)

$$\text{(4)} \qquad G = \frac{e^{ikr}}{r} - \frac{e^{ikr'}}{r'}, \qquad \begin{cases} r^2 = (\xi - x)^2 + (\eta - y)^2 + (\zeta - z)^2, \\ r'^2 = (\xi - x)^2 + (\eta - y)^2 + (\zeta + z)^2, \end{cases}$$

we must now replace the spherical wave e^{ikr}/r by the cylindrical wave $H\,(kr)$, where $H = H_0^{(1)}$ is the Hankel function of the first kind of index zero. We must also interpret r in the two-dimensional sense (because of the nature of our light source, the integration over the coordinate y has, so to speak, already been performed implicitly in $H\,(kr)$). Thus we obtain in place of (4)

$$\text{(4 a)} \qquad G = \frac{i\pi}{2}\,(H\,(kr) - H\,(kr')), \qquad \begin{cases} r^2 = (\xi - x)^2 + (\zeta - z)^2, \\ r'^2 = (\xi - x)^2 + (\zeta + z)^2. \end{cases}$$

Equation (34.6) yields then

$$\text{(5)} \qquad 2\pi\, u_\pm = -\int_{-a}^{+a} \overline{u}\,(\xi)\,\frac{\partial G}{\partial n}\,d\xi\,.$$

The factor 2π replaces the 4π of (34.6) because in the two-dimensional Green's theorem the left-hand side of (5) arises from the integration over a circle of vanishingly small radius instead of over the surface of a sphere as in the three-dimensional theorem. $\partial G/\partial n$ means the derivative with respect to the outward normal, as before; in our case $\partial n = -\partial\zeta$ for u_+, $\partial n = +\partial\zeta$ for u_- and by (4 a)

$$\frac{\partial G}{\partial n} = \mp \frac{i\pi}{2}\frac{\partial}{\partial\zeta}(H(kr) - H(kr')) = \pm \frac{i\pi}{2}\frac{\partial}{\partial z}((H(kr) + H(kr')).$$

In the slit $\zeta = 0$ and hence $r = r'$ and

(5 a) $$\frac{\partial G}{\partial n} = \pm i\pi \frac{\partial}{\partial z} H(kr_0), \qquad r_0{}^2 = (\xi - x)^2 + z^2.$$

Equation (5) becomes

(6) $$u_\pm = \mp i \frac{\partial}{\partial z}\int_{-a}^{+a} \overline{u}(\xi)\, H(kr_0)\, d\xi.$$

Since $\overline{u}(\xi)$ is actually not known, eq. (6) contains no information. This equation must be supplemented by the requirement that $\partial v/\partial z$ be continuous at the slit; the continuity of v is already guaranteed by (3). This continuity condition of $\partial v/\partial z$ now takes the place of our original boundary value problem which has so far been solved only incompletely.

According to eq. (2)

$$\frac{\partial v}{\partial z} = 2ikA + \frac{\partial u_-}{\partial z} \quad \text{as} \quad z \to 0 \quad \text{from} \quad z < 0,$$

$$\frac{\partial v}{\partial z} = \frac{\partial u_+}{\partial z} \quad \text{as} \quad z \to 0 \quad \text{from} \quad z > 0.$$

Therefore we must require that

(7) $$\frac{\partial u_+}{\partial z} - \frac{\partial u_-}{\partial z} = 2ikA \qquad \text{for} \qquad z = 0.$$

From this together with eq. (6) it follows that

(8) $$-\frac{\partial^2}{\partial z^2}\int_{-a}^{+a} \overline{u}(\xi)\, H(kr_0)\, d\xi = kA.$$

This condition must be fulfilled for $z = 0$ and for all values $-a < x < +a$. It is to be noted here that according to (5 a), r_0 depends on z and therefore the limit value

(8 a) $$r_0 = |\xi - x|$$

may be substituted only after the twofold differentiation indicated in (8) has been performed.

We can simplify the mathematical situation if we take into account[1] the fact that the cylindrical wave $H(k\,r_0)$ satisfies the two-dimensional wave equation $\Delta H + k^2 H = 0$. Therefore also the integral in (8) satisfies this equation and we have

$$-\frac{\partial^2}{\partial z^2}\int \ldots = \left(\frac{\partial^2}{\partial x^2} + k^2\right)\int \ldots \tag{8 b}$$

where we are now permitted to go to the limit (8 a) on the right-hand side and may also write d^2/dx^2 in place of $\partial^2/\partial x^2$. Thus we obtain from (8)

$$\left(\frac{d^2}{dx^2} + k^2\right) X = k\,A \tag{9}$$

with the abbreviation

$$X = \int_{-a}^{+a} \overline{u}(\xi)\, H(k\,|\xi - x|)\, d\xi. \tag{9 a}$$

We integrate (9) using the rule for the integration of inhomogeneous differential equations. A particular integral of (9) is $X = A/k$; because of the symmetry of the problem, only that part of the general solution of the homogeneous equation which is even in x, that is $B \cos k\,x$, is to be used. Hence

$$X = A/k + B \cos k\,x. \tag{9 b}$$

In order to determine the constant of integration B we set $x = 0$ and find using (9 a, b), that

$$B = -A/k + \int_{-a}^{+a} \overline{u}(\xi)\, H(k\,|\xi|)\, d\xi.$$

Substituting this value on the right-hand side of (9 b) and using on the left-hand side for X its value (9 a), one obtains

$$\int_{-a}^{+a} \overline{u}(\xi)\,\{H(k\,|\xi - x|) - \cos k\,x\, H(k\,|\xi|)\}\, d\xi = \frac{A}{k}(1 - \cos k\,x). \tag{10}$$

This is a linear integral equation for the unknown function $\overline{u}(\xi)$ which must be satisfied for all values $-a < x < +a$. The "Kernel" of the integral equation is the expression inside the curly brackets in (10). A general rule is thus confirmed: *the solution of a boundary value problem can be reduced to the solution of an integral equation.* The solution of the latter can always be obtained numerically, but of course only for special values of the

[1] After the manner of Levine and Schwinger, loc. cit. eq. (A 3) for the analogous case of the circular hole.

parameters occurring in the equation (in this case the values of $k\,a$ and $k\,x$). This is of no help to us. In order to obtain a general solution one must in each case invent suitable approximation methods. In our case these have to arise from the assumption $k\,a \ll 1$. Furthermore, we note that the kernel in (10) is unsymmetric in x and ξ while mathematical theory commonly deals with symmetric kernels.

Before we proceed to solve the integral equation, we must briefly consider the other case of polarization. We now denote by v the magnetic vector H which, because $E_x = 0$, must satisfy $\partial v/\partial z = 0$ on the screen. Calling the magnetic amplitude of the incident wave A' [1], we set in place of (2)

$$v = A'(e^{ikz} + e^{-ikz}) + u \qquad \text{for} \qquad z < 0,$$
$$v = u_+ \qquad\qquad\qquad \text{for} \qquad z > 0.$$

and obtain in place of (3)

$$\frac{\partial u_+}{\partial z} = \frac{\partial u_-}{\partial z} = \omega,$$

where $\omega = \omega(\xi)$ is now the unknown function to be determined. As the Green's function we have to take

$$G = \frac{i\pi}{2}\{H(k\,r) + H(k\,r')\}. \tag{11}$$

From it one finds, in contrast to (5) (compare with Sec. 34 G):

$$2\pi\,u_\pm = \mp\int_{-a}^{+a} \omega(\xi)\,G\,d\xi = \mp\,i\pi\int_{-a}^{+a} \omega(\xi)\,H(k\,r_0)\,d\xi; \tag{11 a}$$

hence in the slit opening

$$2\pi\,u_\pm = \mp\,i\pi\int_{-a}^{+a} \omega(\xi)\,H(k\,|\xi - x|)\,d\xi. \tag{11 b}$$

Because of the continuity of v at the slit

$$\overline{u}_+ - \overline{u}_- = 2\,A';$$

from this and (11 b) it follows that

$$\int_{-a}^{+a} \omega(\xi)\,H(k\,|\xi - x|)\,d\xi = i\,A'. \tag{12}$$

The derivation and form of this integral equation is somewhat simpler than in the preceding case; also the "Kernel" $H(k\,|\xi - x|)$ of (12) is symmetric[2].

[1] The notation is the same as in Sec. 2; A and A' have different dimensions and differ by the "wave resistance".

[2] This integral equation was first obtained by G. Jaffé, Phys. Zeitschr. 22, p. 578, 1921.

B. Solution of the integral equations (10) and (12)

It will be necessary to make a hypothesis as to the form of the function $\overline{u}(\xi)$ in (10) such that it contains an infinite number of undetermined coefficients and then to attempt to determine these coefficients from (10). The choice of form is limited by the following considerations:

1. $\overline{u}(\xi)$ must vanish at $\xi = \pm a$ because the solution must approach continuously the value $v = 0$ prescribed on the screen.

2. Because of the symmetry of the problem, $\overline{u}(\xi)$ must be an even function of ξ.

3. In view of our treatment of branched wave functions and their representation at the end of Sec. 38, $u(x, z)$ must change its sign for every complete circuit around one of the two branch points $x = \pm a$. Together with the requirements 1. and 2., this leads to the form

$$\overline{u}(\xi) = \sum_{n=1}^{\infty} C_n \left(1 - \frac{\xi^2}{a^2}\right)^{n-\frac{1}{2}}. \tag{13}$$

The C_n are infinitely many available complex coefficients.

In an earlier acoustic work by the author[1], which was also used as a starting point by Levine and Schwinger (see p. 273), a result analogous to (13) was obtained by a protracted calculation. This calculation at the same time provided the values of the C_n in the form of numerically definite power series in the parameter ka, which is the only characteristic parameter. It turned out that the series for C_{n+1} starts with a power of $a\,k$ which is greater by one than the power of the first term in the series for C_n. Here we have been able to write down the form of the solution (13) directly on the basis of function theory so that it is valid for a slit of arbitrary width. For a very narrow slit $a\,k \ll 1$, the above-mentioned result shows that C_1 is larger by one order of magnitude than all the other C_n. Therefore we shall specialize (13) to

$$\overline{u}(\xi) = C_1 \left(1 - \frac{\xi^2}{a^2}\right)^{\frac{1}{2}}. \tag{13 a}$$

Then the integral equation (10) reads

$$C_1 \int_0^a \left(1 - \frac{\xi^2}{a^2}\right)^{\frac{1}{2}} K(x, \xi)\, d\xi = \frac{A}{k}(1 - \cos k\,x) \sim A\,k\,x^2/2. \tag{13 b}$$

[1] Die frei schwingende Kolbenmembran, Ann. d. Phys. (Lpz.) **42**, 389, 1943.

In contrast to (10) the interval of integration has here been restricted to $0 < \xi < a$; the interval $-a < \xi < 0$ must be taken into account by the following modification of the kernel in which x may be assumed to be positive:

(13 c) $$K(x, \xi) = H(k|\xi - x|) + H(k(\xi + x)) - 2\cos kx\, H(k\xi).$$

We decompose this kernel into two parts

(13 d) $$K_I = H(k(\xi + x)) + H(k|\xi - x|) - 2H(k\xi).$$

(13 e) $$K_{II} = 2(1 - \cos kx)\, H(k\xi) \sim k^2 x^2 H(k\xi).$$

Since $ka \ll 1$, the arguments of all the H-functions are small in the whole interval of integration, and therefore we can use everywhere the approximate formula from Vol. III, eq. (22.5):

(13 f) $$H_0(\rho) = \frac{2i}{\pi}\log\frac{\gamma\rho}{2i}; \qquad \log\gamma = 0.5772 = \text{Euler-Mascheroni constant.}$$

Then

(13 g) $$K_I = \frac{2i}{\pi}\log\frac{|\xi^2 - x^2|}{\xi^2}, \qquad K_{II} = \frac{2i}{\pi}k^2 x^2 \log\frac{\gamma k\xi}{2i}.$$

The logarithm in K_I must be expanded differently depending upon whether $\xi < x$ or $\xi > x$:

$$\log\frac{|x^2 - \xi^2|}{\xi^2} = \begin{cases} \log\dfrac{x^2}{\xi^2} - \left(\dfrac{\xi^2}{x^2} + \dfrac{1}{2}\dfrac{\xi^4}{x^4} + \ldots\right) & \text{for } \xi < x, \\[2ex] -\left(\dfrac{x^2}{\xi^2} + \dfrac{1}{2}\dfrac{x^4}{\xi^4} + \ldots\right) & \text{for } \xi > x. \end{cases}$$

Correspondingly, the integral (13 b) must be decomposed into two parts:

(14) $$\int_0^a \ldots = \int_0^x \ldots + \int_x^a \ldots = J_1 + J_2,$$

(14 a) $$J_1 = \int_0^x \left\{\log\frac{x^2}{\xi^2} - \left(\frac{\xi^2}{x^2} + \frac{1}{2}\frac{\xi^4}{x^4} + \ldots\right)\right\} d\xi,$$

(14 b) $$J_2 = -\int_x^a \left(1 - \frac{\xi^2}{a^2}\right)^{1/2}\left(\frac{x^2}{\xi^2} + \frac{1}{2}\frac{x^4}{\xi^4} + \ldots\right) d\xi.$$

Since C_1 is independent of x, it is permissible to choose x so small compared to a that the factor $(1 - \xi^2/a^2)^{1/2}$ in J_1 may be replaced by 1. This has already been done in (14 a).

By elementary integration one obtains

(15) $$J_1 = 2x - x\left(\frac{1}{1\cdot 3} + \frac{1}{2\cdot 5} + \frac{1}{3\cdot 7}\ldots\right).$$

Thus J_1 is proportional to x, while the right-hand side of our integral equation (13 b) was proportional to x^2. In Appendix 1 we shall show that J_1 is cancelled by the contribution from the lower limit of the integral J_2. It will also be shown there that the contribution from the upper limit of J_2 yields, except for terms of higher power in x/a,

$$\frac{\pi}{2}\frac{x^2}{a}. \tag{15 a}$$

According to (13 g) and (14) the entire value of K_I on the left-hand side of (13 b) becomes now

$$C_1\frac{2i}{\pi}\cdot\frac{\pi}{2}\frac{x^2}{a}=i\,C_1\frac{x^2}{a}. \tag{16}$$

The contribution of K_{II} to (13 b) follows from (13 g):

$$C_1\frac{2i}{\pi}k^2x^2\int_0^a\left(1-\frac{\xi^2}{a^2}\right)^{1/2}\log\frac{\gamma k\xi}{2i}\,d\xi. \tag{16 a}$$

If we introduce the substitution (25) from Appendix 1 and put

$$q=\frac{4}{\pi}\int_0^{\pi/2}\cos^2\varphi\log\sin\varphi\,d\varphi, \tag{16 b}$$

we obtain in place of (16 a)

$$\frac{1}{2}C_1 i k^2 a x^2\left(\log ka+\log\frac{\gamma}{2i}+q\right). \tag{17}$$

The sum of (16) and (17) is

$$i\,C_1\frac{x^2}{a}\left\{1+\frac{1}{2}\Big/(ka)^2\left(\log ka+\log\frac{\gamma}{2i}+q\right)\right\}.$$

Because $ka \ll 1$, the second term in the curly brackets can be neglected; thereupon the integral equation (10) yields

$$i\,C_1\frac{x^2}{a}=A\,k\frac{x^2}{2},\qquad C_1=-i\,A\,k\,a/2. \tag{18}$$

We now turn to the integral eq. (12). As at the end of section A, v and $u_\pm$ are magnetic vector components parallel to the y-axis, and ω is the value of $\partial u_\pm/\partial z$ at $z=0$ inside the slit. First we seek the form of ω which is analogous to (13). We assert that ω is again given by (13), except that *the lower limit in the summation must be replaced by* $n=0$. Thus in the first approximation

$$\omega(\xi)=\frac{C_0}{a}\Big/\left(1-\frac{\xi^2}{a^2}\right)^{1/2}. \tag{19}$$

To justify this statement it is only necessary to remark that near the branch points $x = \pm a$ the vector H behaves like the square root of the distance from the branch points, and therefore the gradient of H behaves like the inverse square root of that distance. We shall indeed see that (19) leads to the unique solution of our problem. The factor a in the denominator of (19) has been included in order that C_0 (as C_1 before) have the same dimension as our present u. $\omega(\xi)$ is, as $\overline{u}(\xi)$ was before, an even function of ξ.

We now rewrite the integral eq. (12) in the form

$$\int_0^a \omega(\xi)\,\{H(k\,|\xi - x|) + H(k\,|\xi + x|)\}\,d\xi = i\,A'. \tag{20}$$

The $\{\ldots\}$ is again called the *kernel* of the integral equation. In order to be able to use the above calculations we again separate this kernel into two parts

$$K_I = H(k\,|\xi + x|) + H(k\,|\xi - x|) - 2\,H(k\,\xi), \tag{20 a}$$

$$K_{II} = 2\,H(k\,\xi). \tag{20 b}$$

The first part is identical with K_I in (13 d). Therefore eqs (14 a, b) are *mutatis mutandis* again valid; these yield now only terms proportional to x, the sum of which is zero; see appendix 2. Equation (20) therefore simplifies to

$$\int_0^a \omega(\xi)\,K_{II}\,d\xi = \frac{2\,C_0}{a}\int_0^a \left(1 - \frac{\xi^2}{a^2}\right)^{-\frac{1}{2}} H(k\,\xi)\,d\xi = i\,A' \tag{21}$$

and one obtains, see appendix 2:

$$C_0 = \frac{1}{2}\,A'/p; \qquad p = \log\frac{\gamma\,k\,a}{4\,i} = \log k\,a - 0.81 - i\pi/2. \tag{22}$$

C. Discussion

In figures 81 a, b the distributions of $\overline{u}(x)$ and $\omega(x)$ are shown in comparison to the amplitudes of the incident waves A and A', respectively. From (13 a) and (18) or (19) and (22), respectively, one obtains for $\xi = 0$:

$$\frac{\overline{u}(0)}{A} = \frac{C_1}{A} = -\frac{i}{2}\,k\,a; \qquad \frac{\omega(0)}{A'} = \frac{C_0}{A'} = \frac{1}{2\,p}.$$

$|\overline{u}(x)|$ is a very flat ellipse. $|\omega(x)|$ is the corresponding reciprocal curve which has a much larger value than $|\overline{u}(x)|$ at the center of the slit and goes to infinity at the edges.

From $\overline{u}$ and ω one can compute $u_+ (x, z)$ by means of formulae (6) and (11 a), and thereby one obtains for the diffraction field v:

$$\pi v = -C_1 \frac{\partial}{\partial z} \int_{-a}^{+a} \left(1 - \frac{\xi^2}{a^2}\right)^{1/2} H(k r_0)\, d\xi,$$

and

$$\pi v = -\frac{C_0}{a} \int_{-a}^{+a} \left(1 - \frac{\xi^2}{a^2}\right)^{-1/2} H(k r_0)\, d\xi,$$

with

$$r_0^2 = (x - \xi)^2 + z^2.$$

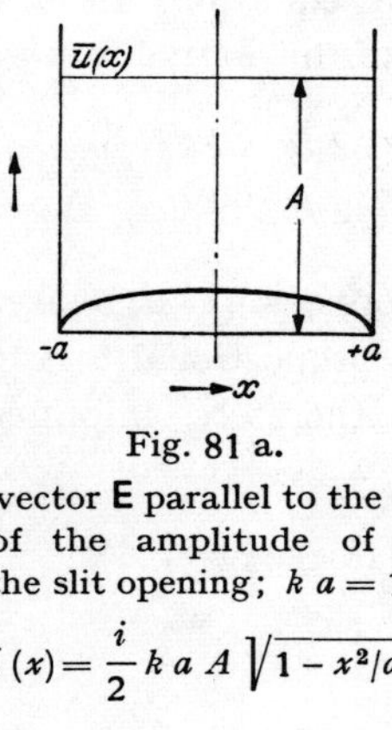

Fig. 81 a.

Electric vector **E** parallel to the slit edges. Graph of the amplitude of $\mathbf{E} = \overline{u}(x)$ in the slit opening; $k a = 1/10$,

$$\overline{u}(x) = \frac{i}{2} k a A \sqrt{1 - x^2/a^2}.$$

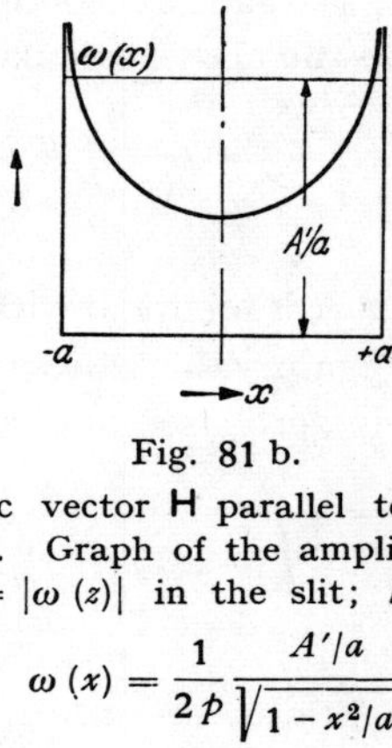

Fig. 81 b.

Magnetic vector **H** parallel to the slit edges. Graph of the amplitude of $\partial \mathrm{H}/\partial z = |\omega(z)|$ in the slit; $k a = 1/10$

$$\omega(x) = \frac{1}{2p} \frac{A'/a}{\sqrt{1 - x^2/a^2}}.$$

Since the point of observation x, z is at a distance of many wavelengths from the slit, we may use for H its asymptotic formula [see, for instance, Vol. III, eq. (22.7)], and we may take $r_0 = r = \sqrt{x^2 + z^2}$ independent of ξ in the integration. In this way we obtain

$$\pi v = -C_1 \frac{\partial}{\partial z} \sqrt{\frac{2}{\pi k r}}\, e^{i(kr - \pi/4)} \int_{-a}^{+a} \left(1 - \frac{\xi^2}{a^2}\right)^{1/2} d\xi,$$

and

$$\pi v = -\frac{C_0}{a} \sqrt{\frac{2}{\pi k r}}\, e^{i(kr - \pi/4)} \int_{-a}^{+a} \left(1 - \frac{\xi^2}{a^2}\right)^{-1/2} d\xi.$$

Making the substitution $\xi = a \sin \Phi$, as in the appendix one obtains for the values of the above integrals $a\pi/2$ and $a\pi$, respectively. Then, if the differentiation with respect to z is performed only in the exponent, one finds

$$v = -i\,a\,k\,C_1 \frac{z}{r} \frac{1}{\sqrt{2\pi k r}} e^{i(kr-\pi/4)}, \tag{23}$$

and

$$v = -C_0 \sqrt{\frac{2}{\pi k r}}\, e^{i(kr-\pi/4)} \,. \tag{23 a}$$

Thus we have obtained two cylindrical waves of different amplitudes which originate from the slit (or rather from its center line). The amplitude of the first of these waves contains the cosine factor

$$\cos \delta = \frac{z}{r}, \qquad (\delta = \text{angle of diffraction}).$$

In order to make eq. (23 a) dimensionally commensurate with (23) (the former represents H_y and not E_y as does the latter), we compute the electric components E_x, E_z belonging to H_y. We use Maxwell's equation $\dot{\mathsf{D}} = \text{curl}\, H$

$$-i\,\omega\,\varepsilon_0 E_x = -\frac{\partial H_y}{\partial z} = -i\,k\frac{z}{r} H_y, \qquad E_x = \sqrt{\frac{\mu_0}{\varepsilon_0}} \frac{z}{r} H_y,$$

$$-i\,\omega\,\varepsilon_0 E_z = +\frac{\partial H_y}{\partial x} = +i\,k\frac{x}{r} H_y, \qquad E_z = -\sqrt{\frac{\mu_0}{\varepsilon_0}} \frac{x}{r} H_y$$

from which

$$E_\perp = \sqrt{E_x^2 + E_z^2} = \sqrt{\frac{\mu_0}{\varepsilon_0}}\, H_y.$$

With this wave resistance factor $(\mu_0/\varepsilon_0)^{1/2}$ the amplitude A' occurring in C_0 has the same dimensions as the A occurring in C_1. Therefore equal intensity of illumination for both cases implies not $A' = A$, but rather $A'(\mu_0/\varepsilon_0)^{1/2} = A$.

We are now able to calculate the expected *polarization* of the diffracted light if the incident light is composed of equal intensities of the two modes of polarization. The polarization of the diffracted light is characterized by the quotient of (23) and (23 a) which, taking (18) and (22) into account, is

$$\frac{1}{2} a\,k \left|\frac{C_1}{C_0}\right| \cos\delta = \frac{1}{2}(a\,k)^2\, |p| \cos\delta \sqrt{\mu_0/\varepsilon_0}. \tag{24}$$

Therefore for small values of $k\,a$ much less light with $E_{\|}$ passes through the slit than light with $E_\perp$. The oscillations $E_{\|}$ are suppressed by the slit because of the boundary condition $E_y = 0$. The oscillations $E_\perp$ induce charges on the slit edges, and these charges travel along the screen in the manner of

Hertz waves in wires; in this way they are able to overcome the curvature of the slit edges. This polarization effect is well known from Hertz's experiments with gratings.

The intensities for the two states of polarization exhibit notably different dependences on wavelength. For, according to (23) and (18)

$$J_{\|} \text{ is proportional to } \left(\frac{a^2 k^2}{\sqrt{k r}}\right)^2 = \left(\frac{2\pi a}{\lambda}\right)^3 \frac{a}{r},$$

while according to (23 a) and (22)

$$J_{\perp} \text{ is proportional to } \left(\frac{1}{\sqrt{k r}\,|p|}\right)^2 = \frac{\lambda}{2\pi a}\frac{a}{r}\frac{1}{|\log \lambda/a + \ldots|^2}.$$

In both cases the behavior is different from Rayleigh's λ^{-4} law for the blue of the sky. The latter is based on openings which are small in *all* their dimensions (or on correspondingly small discs). Our slit which is narrow in only *one* dimension yields also a totally different behavior for different directions of polarization of the incident light.

Our expressions (23) and (23 a) for the diffraction field are in complete agreement with eqs. (53) and (47) of Lord Rayleigh, loc. cit. Rayleigh also remarked that for the complementary case of the metallic strip eqs. (53) and (47) interchange their roles. This, as we know, is the precise statement of Babinet's principle; see Sec. 38 F.

Despite its complexity our solution of the problem has the following advantages:

1. Unlike Rayleigh's method it does not require any previous knowledge of electrostatics or hydrodynamics.

2. Our method is capable of being *generalized*.

One only needs to extend our one-term expressions (13 a) and (19) by adding terms with $C_2, C_3, \ldots$ or C_1, C_2, respectively, in order to extend the results to wider and wider slits. How this may be done will be shown in appendix 3.

Appendix 1

To evaluate the integral J_2 in (14 b) the following substitution is made.

$$(25) \qquad \xi = a \sin \Phi, \qquad \left(1 - \frac{\xi^2}{a^2}\right)^{1/2} = \cos \Phi, \qquad d\xi = a \cos \Phi \, d\Phi.$$

If, at the same time, we put

$$(25\text{ a}) \qquad x = a \sin \psi; \qquad \psi \sim \frac{x}{a} \ll 1$$

in the lower limit of the integral, we are led to the following auxiliary expressions

$$(26) \qquad j_{2n} = -\int_{\psi}^{\pi/2} \frac{\cos^2 \Phi}{\sin^{2n} \Phi} d\Phi \qquad \text{for} \qquad n = 1, 2, 3 \ldots$$

Calling the corresponding indefinite integrals $\bar{j}_{2n}$, it is easily verified by differentiation that

$$(26\text{ a}) \quad \bar{j}_2 = \Phi + \cot \Phi, \qquad \bar{j}_4 = \frac{1}{3} \cot \Phi; \qquad \bar{j}_6 = \frac{1}{5} \cot^5 \Phi + \frac{1}{3} \cot^3 \Phi, \ldots$$

From this and a recursion formula for the $\bar{j}_{2n}$, one concludes that all j_{2n} vanish at the upper limit of the integral (26) with the exception of j_2 which there assumes the value $\pi/2$. Except for higher powers of the arbitrarily small quantity ψ, the j_{2n} assume the following values at the lower limit:

$$-\frac{1}{2n-1} \cot^{2n-1} \psi \sim -\frac{1}{2n-1} \psi^{-2n+1} = \frac{-1}{2n-1} \left(\frac{a}{x}\right)^{2n-1}.$$

Hence

$$j_2 = \frac{\pi}{2} - \frac{a}{x}, \qquad j_{2n} = -\frac{1}{2n-1} \left(\frac{a}{x}\right)^{2n-1}, \qquad n \geqq 2.$$

Rewriting eq. (14 b) in terms of the j_{2n} and substituting the above values, we obtain

$$J_2 = j_2 \frac{x^2}{a} + \sum_{n=2}^{\infty} \frac{1}{n} j_{2n} \frac{x^2 n}{a^{2n-1}} = \frac{\pi}{2} \frac{x^2}{a} - x \sum_{n=1}^{\infty} \frac{1}{n(2n-1)}.$$

The first term agrees with (15 a) while the second term combines with (15) to give

$$x \left\{ 2 - \sum_{n=1}^{\infty} \frac{1}{n} \left(\frac{1}{2n+1} + \frac{1}{2n-1} \right) \right\}.$$

The { } can be rewritten in the form

$$(26\text{ b}) \quad 2 - 4 \sum_{n=1}^{\infty} \frac{1}{(2n-1)(2n+1)} = 4 \left(\frac{1}{2} - \frac{1}{1 \cdot 3} - \frac{1}{3 \cdot 5} - \frac{1}{5 \cdot 7} - \ldots \right) = 0,$$

see, for instance, Vol. VI, solution to exercise I.3. Thus our statements following eq. (15) have been proved.

Appendix 2

In the second (magnetic) case eqs. (14 a) and (15) for J_1 are unchanged because the kernel K_I remains unchanged, and the difference between the assumed forms for ω and $\overline{u}$ is insignificant in the interval $0 < \xi < x$. The equations for J_2 simplify because upon making the substitutions (25) and (25 a), one finds that

$$\left(1 - \frac{\xi^2}{a^2}\right)^{-1/2} d\xi = a\, d\Phi.$$

Hence in terms of Φ eq. (14 b) becomes:

$$J_2 = -\int_{\psi}^{\pi/2} \left(\frac{1}{\sin^2\Phi}\frac{x^2}{a} + \frac{1}{2\sin^4\Phi}\frac{x^4}{a^3} + \frac{1}{3\sin^6\Phi}\frac{x^6}{a^5} + \ldots\right) d\Phi.$$

Since

$$\int \frac{d\Phi}{\sin^2\Phi} = -\cot\Phi, \qquad \int \frac{d\Phi}{\sin^4\Phi} = -\frac{1}{3}\cot\Phi\left(\frac{1}{\sin^2\Phi} + 2\right),$$

$$\int \frac{d\Phi}{\sin^6\Phi} = -\frac{1}{5}\cot\Phi\left(\frac{1}{\sin^4\Phi} + \frac{4}{3}\frac{1}{\sin^2\Phi} + \frac{2}{3}\right), \ldots$$

all terms in J_2 vanish at the upper limit $\Phi = \pi/2$. In the approximation $\psi \ll 1$, $x \ll a$ the lower limit contributes

$$-x\left(1 + \frac{1}{2\cdot 3} + \frac{1}{3\cdot 5} + \frac{1}{4\cdot 7} + \ldots\right)$$

which combines with J_1 in (15) to give zero; see (26 b).

Hence, only K_{II} contributes significantly to the integral equation which, according to (20) and using the approximation (13 f) for H, now reads

$$\frac{4}{\pi} C_0 \left\{\int_0^{\pi/2} \log\frac{\gamma k a}{2i}\, d\Phi + \int_0^{\pi/2} \log\sin\Phi\, d\Phi\right\} = A'. \tag{27}$$

The second of these integrals has the value

$$-\frac{\pi}{2}\log 2.$$

This combines with the value of the first integral to give

$$\frac{\pi}{2} p; \qquad p = \log\frac{\gamma k a}{4i} \quad \text{as in eq. (22).}$$

Hence, according to (27)

$$2 i C_0 p = -\pi A'. \tag{27 a}$$

Numerically, one obtains

$$(27\text{ b})\qquad p - \log k\,a = \log \frac{\gamma}{4} - \frac{i\,\pi}{2} = 0.577 - 1.386 - i\,\frac{\pi}{2} = -\,.81 - i\,\frac{\pi}{2}.$$

Our assertion (22) is thus verified by (27 a, b).

Appendix 3

In order to demonstrate how our method may be generalized, we shall first consider the simpler case discussed in appendix 2. We extend (19) to

$$(28)\qquad \omega\,(\xi) = \frac{C_0}{a}\left(1 - \frac{\xi^2}{a^2}\right)^{-\frac{1}{2}} + \frac{C_1}{a}\left(1 - \frac{\xi^2}{a^2}\right)^{+\frac{1}{2}}.$$

We must also use a more exact approximation of the kernel $K = K_I + K_{II}$ of (20 a, b) by retaining the terms in $k^2\,x^2$ and $k^2\,\xi^2$. The integration is again elementary and may be carried out in the manner of Appendix 2. By equating the coefficients of x^0 and the coefficients of x^2 on the left- and right-hand sides, one obtains the two conditions

$$\begin{aligned} C_0\left[p\left(2 - \frac{k^2\,a^2}{4}\right) + \frac{k^2\,a^2}{8}\right] &+ C_1\left[p\left(1 - \frac{k^2\,a^2}{16}\right) - \frac{1}{2} + \frac{3}{64}\,k^2\,a^2\right] = A' \\ -\,C_0\,\frac{k^2\,a^2}{2}\left[p + \frac{1}{2}\right] &+ C_1\left[1 - p\,\frac{k^2\,a^2}{4}\right] = 0. \end{aligned}$$

Solving these [in the second equation the approximation (27 a) for C_0 may be used], one finds

$$(29)\qquad \begin{aligned} C_1 &= \frac{i\,\pi\,A'}{4\;p}\,k^2\,a^2\left(p + \frac{1}{2}\right), \\ C_0 &= \frac{i\,\pi\,A'}{2\;p}\left\{1 + \frac{k^2\,a^2}{8}\,(1 - 2\,p)\right\}. \end{aligned}$$

which obviously agrees with the result of Appendix 2 when $k^2\,a^2$ is neglected.

In the case treated in Appendix 1, (13 a) must be extended to

$$(30)\qquad \overline{u}\,(\xi) = C_1\left(1 - \frac{\xi^2}{a^2}\right)^{\frac{1}{2}} + C_2\left(1 - \frac{\xi^2}{a^2}\right)^{3/2}.$$

In (13 b) and in the expansion of the kernel (13 c), not only $k^2\,x^2$ but also $k^4\,x^4$ must be retained. Equating the resulting coefficients of x^2 and x^4 on the right- and left-hand sides of eq. (13 b), one obtains the two conditions

$$\begin{aligned} C_1\left[1 - \frac{k^2\,a^2}{4}\,(1 - p)\right] &+ C_2\left[\frac{3}{2} - \frac{15}{64}\,k^2\,a^2\left(1 - \frac{4}{5}\,p\right)\right] = \frac{1}{2\,i}\,A\,k\,a \\ -\,C_1\,k^2\,a^2 &+ C_2\left[12 - \frac{3}{2}\,k^2\,a^2\right] = +\,i\,A\,k^3 a^3. \end{aligned}$$

In the second equation C_1 may again be replaced by its first approximation (18), and the second order approximation of C_1 is found from the first equation. Thus one obtains

$$C_2 = \frac{i}{24} A\, k^3 a^3,$$

(31)

$$C_1 = \frac{1}{2i} A\, k\, a \left\{1 + \frac{k^2 a^2}{8} (3 - 2p)\right\}$$

as the improved eq. (18). For the calculations leading to these results the author is indebted to Dr. E. Ruch.

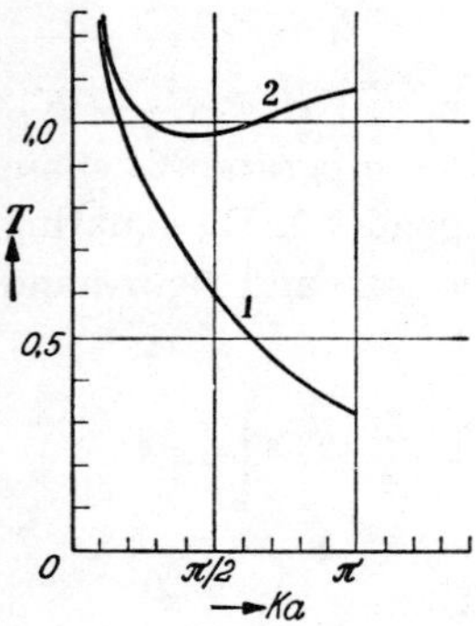

Fig. 82.
The transmission factor T as a function of the ratio of slit width to wavelength in the first and second order approximations.

Figure 82 illustrates the so-called transmission factor T in the first and second order approximations for the polarization $\mathsf{H} = H_y$. The transmission factor is defined as the ratio of the energy which light of finite wavelength actually carries through the slit to the energy which would pass through the slit in the limiting case of geometrical optics ($\lambda \to 0$). In both cases T is measured by the energy flow across a half-cylinder of infinite radius centered at the center line of the slit. One obtains in the first and second order approximations [corresponding to eq. (22) for C_0, eq. (29) for C_1 and C_0, respectively]

$$T_1 = \frac{1}{4\, k\, a\, |p|^2}, \quad T_2 = \frac{1}{4\, k\, a\, |p|^2} \left(1 + \frac{1}{4} (k\, a)^2\right). \tag{31 a}$$

Curve 1 is valid only for extremely narrow slits ($k\, a < 1/4$). For larger values of $k\, a$ curve 2 separates from 1 and exhibits a tendency to approach the value of geometrical optics ($T = 1$); curve 2 can be checked by comparing it with the work of P. M. Morse and J. Rubenstein[1] in which the problem of the slit is treated numerically and graphically by the theory of Mathieu functions with the help of tables of these functions. Our curve 2 agrees with the corresponding curve of the above authors sufficiently well for $k\, a < 2$. J. W. Miles[2] obtained about the same results as Morse and Rubenstein by means of a variational method. K. Schwarzschild[3] devised an approximation starting from the opposite limiting case, that of our solution for the half-

[1] Phys. Rev. **54**, p. 895, 1938; see in particular the top curve denoted by 90° in fig. **4**.
[2] Phys. Rev. **75**, p. 695, 1949.
[3] Mathem. Ann. **55**, p. 177, 1902.

plane, by a method of alternating successive approximations. Despite repeated attempts, the author has been unable to apply the method of Sec. 38 directly to the problem of the slit. But it should be noted again that the basic hypothesis (13) as to the form of the boundary values (which was also used by Levine-Schwinger) was prescribed by the method of Sec. 38.

40. The Resolving Power of Optical Instruments

The purpose of all spectroscopic apparatus is to obtain an increased resolving power. In spectroscopy "resolution" means the separation of two closely neighboring spectral lines. In the case of the microscope one is interested in distinct images of the structure of very fine tissues, while with a telescope one wishes to separate double stars, star clusters, discover new satellites, etc.

A. The Resolving Power of Line Gratings

According to Lord Rayleigh two spectral lines 1 and 2 can be considered resolved if the principal maximum of the diffraction pattern of 2 (wavelength $\lambda + \delta\lambda$) coincides with the first zero of the diffraction pattern of 1 (wavelength λ). The density of blackened grains on the photographic plate corresponds to a superposition of the intensity contours of 1 and 2. This sum of the two intensities has a depression between the two principal maxima which is sufficient to enable the eye to see the separate lines 1 and 2; see fig. 83 (we shall discuss fig. 83a later on). We shall now show that the ratio $\delta\lambda/\lambda$ measured in this way has a fixed value depending only on the nature and method of use of the grating. The resolving power is defined as the reciprocal of this ratio. Two lines are considered resolved if their $\lambda/\delta\lambda$ is less than the resolving power as defined.

We refer back to eq. (32.5). The zeros of the diffraction pattern are obtained by setting the numerator equal to zero. Hence, the value of $N\Delta/2$ at the first zero exceeds the value of $N\Delta/2$ at the principal maximum by π. At the principal maximum

$$\Delta = 2\pi h, \qquad \text{hence} \qquad N\Delta/2 = N\pi h,$$

where h is the order of the grating spectrum in which the observation is being made. Therefore the value of $N\Delta/2$ at the first zero is

$$N\pi h + \pi = N\pi \frac{d}{\lambda} (\alpha - \alpha_0). \tag{1}$$

The right-hand side of this equation follows from (32.1) and determines the angular deflection $\alpha - \alpha_0$ of the point in the spectrum under consideration which is the first zero of the diffraction pattern of the line 1. Now it is required that the principal maximum of the line 2 shall fall in the same direction $\alpha - \alpha_0$, which means that

$$N \pi h = N \pi \frac{d}{\lambda + \delta\lambda} (\alpha - \alpha_0). \tag{2}$$

Dividing the left- and right-hand sides of (1) and (2) by each other one obtains

$$\frac{N h + 1}{N h} = \frac{\lambda + \delta\lambda}{\lambda}$$

and hence

$$\frac{\lambda}{\delta\lambda} = N h. \tag{3}$$

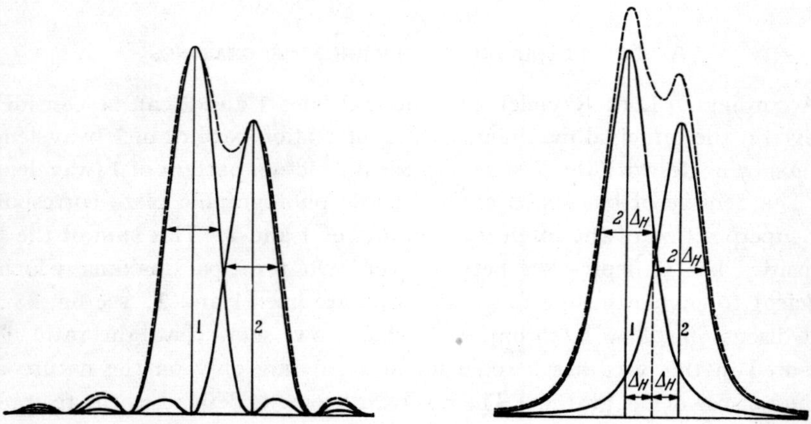

Fig. 83 and 83 a.
Rayleigh's criterion for the resolution of two spectral lines. Fig. 83 a illustrates an almost equivalent criterion.

In the second order spectrum, $h = 2$, the resolving power is twice that in the first order spectrum, a result which is used a great deal by spectroscopists. The resolving power depends only on the total number of grating lines N and not on the line spacing d. The close spacing of lines in the Rowland gratings is needed in order to put a sufficient number of lines within the width of the incident light bundle. A closer spacing of lines also increases the dispersion, that is, the angular separation of different spectral lines, but the spacing has nothing to do with the sharpness of the lines, i. e. with the resolving power.

This is the usual formulation of the theory of resolution for gratings. It is, however, valid only for spectra of *low orders* such as those which are used with Rowland gratings. The *greatest possible resolution* is attained in the spectrum of the highest possible order h_{max}, that is, when one observes at a very small grazing angle with respect to the grating surface. This order was denoted by h_{cr} on p. 182. For it $\alpha \sim 1$, and according to eq. (32.2) $h_{max} \sim d/\lambda$ for perpendicular incidence. From (3) follows therefore

$$\frac{\lambda}{\delta\lambda} = h_{max} N \sim \frac{N d}{\lambda}. \tag{4}$$

Hence, the *maximum* resolving power of the grating depends on the *total width Nd and not on the number of lines N* or, more precisely, it depends on the *path difference* Nd/λ between the rays coming from the first and last grating lines. In what follows we shall recognize this fact to contain the most general formulation of resolving power, a formulation which is valid for all spectral apparatus.

When observed at a *grazing angle*, a grating with 10 lines spaced 1 cm apart resolves just as well as a Rowland grating with 100,000 lines which are 1 μ apart. While in the case of the latter one might observe in the second order spectrum, with the former the twenty-thousandth order would have to be observed.

But observations of spectra of high orders have the serious disadvantage that with increasing order, spectra of *neighboring* orders overlap more and more. To show this we express the wavelength domain which can be observed without overlapping, say $D\lambda$, in terms of the wavelength the order number h, and the angle of deflection, or rather of its cosine, α:

$$(\alpha - \alpha_0)\, d = \lambda h = (\lambda + D\lambda)(h-1).$$

From this follows

$$\frac{D\lambda}{\lambda} = \frac{1}{h-1} \sim \frac{1}{h}.$$

This $D\lambda$ is also the wavelength interval $\delta\lambda$ between neighboring lines which is just measurable without overlapping. Hence in the twenty-thousandth order it is just barely possible to observe the structure of one narrow multiplet, and all other light must be removed by *pre-decomposition* in a prism spectrograph.

The grating with few lines has another and more serious disadvantage; the amplitudes produced by the 10 line and 100,000 line gratings are in the ratio of $1 : 10^4$; hence the intensity obtained from the former is only one 10^8 th of that produced by the latter. Moreover, the ruling of the 10 line grating would have to be just as precise as that of the 100,000 line grating and would therefore be no simpler to manufacture.

B. Echelon gratings and interference spectroscopy

In our method of treatment the grating lines play the role of secondary *light sources* which are excited by the incident light wave and which, owing to their positions, have fixed phase differences with respect to one another. It is possible to obtain the same intensity with a smaller number of secondary light sources, provided that these sources act as *directional radiators* which send most of their energy into a spectrum of high order. In other words, the function $f(\alpha)$ in eq. (32.3) must have a pronounced maximum in the direction of the desired high order spectrum. A very narrow slit or a finely ruled line will not do this. Instead, one must use a sequence of narrow parallel mirrors or narrow prisms which are placed so that by geometrical optics they reflect or refract the light in the desired direction. The almost insurmountable difficulties of the manufacture of such a grating were overcome by Michelson in an elegant manner. He stacked glass plates on top of one another so as to form a series of steps, see Fig. 84a. In making a grating of this type the

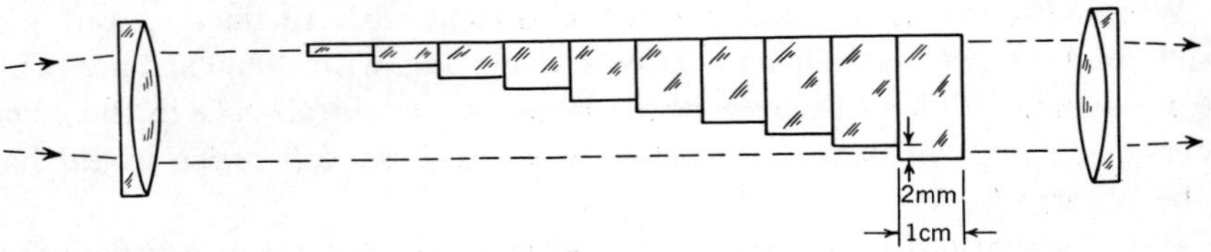

Fig. 84a. Ten-element echelon consisting of glass plates 1 cm thick, offset by 2 mm steps.

plates are cut from a single plane parallel plate, the thickness of which is everywhere constant to within a fraction of a wavelength. The steps are about 2 mm wide and perhaps 1 cm high; they are the grating elements of a "phase grating". With the help of a slit and collimator lens the grating is illuminated in a direction perpendicular to the surfaces of the glass plates, and the spectrum is observed through a telescope in the same direction. Thus the light rays form a *very small angle with the surface of the step grating*. If all grating elements except one are covered, one observes the very bright image of the slit which is, however, widened by diffraction and looks like the diffraction pattern of a slit 2 mm wide. When all the grating elements are uncovered, the image of the slit contracts to the image of the spectral line in one or two orders. As with all gratings, the resolving power of this "echelon grating" is given by the difference between the phase of the first and last ray, which is

(5) $$(n-1)\,N\,d/\lambda; \qquad \begin{array}{l} n = \text{index of refraction of the glass} \\ N = \text{number of steps.} \end{array}$$

In an echelon strips of the wave surface lying side by side interfere with one another. Therefore the wave surface must be made coherent throughout the entire extent of the grating by means of a collimator. As in the case of the line grating, the positions of the spectral lines depend on the direction of the incident light. If this direction is changed, then the phase differences between our secondary sources are changed; thus the positions of the sharp interference fringes are shifted. A wide collimator slit acts like the sum of many adjacent narrow slits; hence a wide slit causes the spectral lines to spread out to the width of the slit's image. This is the qualitative geometrical interpretation of the blurring of interference lines caused by insufficient coherence (see, for example, fig. 2).

This effect does not exist in "interference spectroscopy", by which we mean the spectroscopes of Perot-Fabry and Lummer. With these spectroscopes the positions of the interference fringes depend *only* on the wavelength and the thickness of the plate. The phase difference between interfering waves (though not their intensity) is *independent* of the position of the light source; in other words, the source may be extended without disturbing the interference, provided only that the source is sufficiently intense.

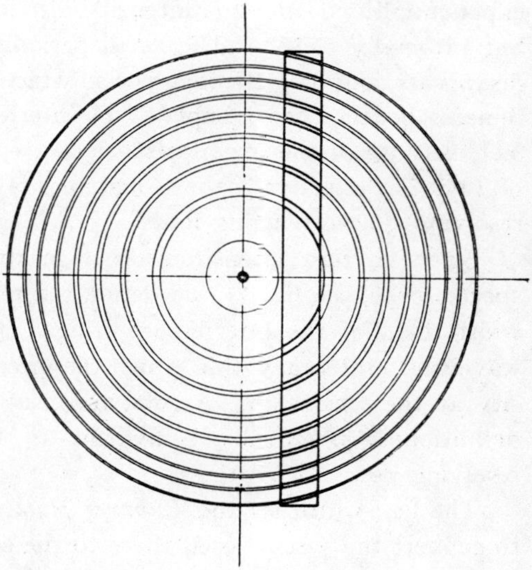

Fig. 84b.
Section from the field of view of a Perot-Fabry air plate.

As in the case of gratings the phase difference produced by interference spectroscopes changes as the angle of observation varies. To each angle between the wave normal and the plate surface corresponds a definite phase difference and therefore a definite wavelength. Thus, in the case of the Perot-Fabry interferometer the wave normals of a given spectral line lie on a narrow cone about the normal to the plate surface. This cone projects as a circle in the camera or on the retina. The different orders form concentric circles which are, however, visible only within the image of the extended source (all other wave normals are not excited). A prism spectroscope may be used for the necessary pre-decomposition. The interference spectroscope is placed between the prism and the telescope. The system of rings is bounded by the image of the slit which is made as wide as possible without causing disturbances due to neighboring spectral lines; see fig. 84b which is a schematic drawing of a Perot-Fabry spectrogram. In the case of the Lummer plate the wave normals of the spectral lines lie on a set of very wide cones about the normal to the plate and therefore appear on the photographic plate as a set of very flat hyperbolas which are almost straight lines.

While in gratings the amplitudes of the various interfering rays are equal, the amplitudes of the interfering rays of an interference spectroscope decrease exponentially. Hence the intensity distribution is no longer given by (32.5), but rather by (7.33). The small periodicity belonging to $N\Delta$ in eq. (32.5) disappears, and only the long period which is determined by the Δ' of eq. (32.1) remains because the number N of interfering rays is, so to speak, infinite. Yet, because of the exponential decrease in amplitude, only a finite number of rays is "effective", the remaining rays being too weak; therefore the resolving power remains finite.

Since no zero intensities occur in the fringe system of an interference spectroscope, see fig. 11, we define the resolving power in terms of the half, width $2\Delta_H$ of the interference fringe; that is, the resolving power is that wavelength interval within which the intensity is greater than half the intensity at the maximum. A comparison of figures 83 and 83a shows that this definition is practically equivalent to Lord Rayleigh's definition of the resolving power of gratings.

The half-width for the *Lummer plate* was computed in (7.28 a). In order to convert the φ-scale used there to the scale of wavelengths λ, we note that according to the definition (7.18a) φ is proportional to k, hence inversely proportional to λ. Therefore

$$\frac{d\varphi}{\varphi} = -\frac{d\lambda}{\lambda}. \tag{6}$$

If we substitute for $d\lambda$ the wavelength difference $\delta\lambda$ between the two spectral lines 1 and 2, then according to fig. 83a we must use for $d\varphi$ the half-width $2\,|\Delta\varphi| = 2\,(1-r)$ [eq. (7.28 a)], and we must substitute for φ the phase $2\pi z$ at the intensity maximum (see Sec. 7). Thus we obtain from (6) (the negative sign is immaterial)

$$\frac{\lambda}{\delta\lambda} = \frac{\pi}{1-r}\,z. \tag{6 a}$$

z represents the very high order of the interference fringe and corresponds to the order number $h = 1, 2, 3, \ldots$ of gratings. A comparison of (6 a) and (3) shows that the number of lines N of a grating is to be compared with the expression $\pi/(1-r)$. The orders of magnitude of the two factors contributing to the resolving power are thus interchanged for the two types of spectroscopes:

for a grating: N is very large, h is moderately large

for a plate: $\pi/(1-r)$ is moderately large, z is very large.

In order to complete the numerical comparison we recall the meaning of z in (7.28 a) and (7.18 a). If we disregard all insignificant factors in these formulae, then z is twice the plate thickness divided by the wavelength; hence for a 1 cm plate $z \sim 4 \times 10^4$. If $r \sim 0.9$, then $\pi/(1-r) \sim 30$. According to (6 a) the resolving power of such a Lummer plate is then about $30 \times 4 \times 10^4 \sim 10^6$. According to (3) this is the number of lines N in a grating with the same resolving power in the first order ($h = 1$). This means that if such a grating had 1000 lines per mm it would have to be 1 meter wide!

For the Perot-Fabry etalon one obtains similarly from the half-width (7.34) the resolving power

$$\frac{\lambda}{\delta\lambda} = \frac{1}{2}(1+g)\,z. \tag{7}$$

z is again the order number of the interference fringe and is therefore a very large number. The first factor, on the other hand, is only a moderately large number since the required light intensity limits the amount of silvering that can be applied to the surfaces. If we estimate g to be 9, then the first factor becomes 5. z being twice the spacing of the plates divided by the wavelength becomes 2×10^5 if we assume the plates to be 5 cm apart. The product of these two factors is 10^6, the same as for the Lummer plate considered above. *The resolving power of both plates exceeds that of the Rowland grating.* Because of its greater simplicity of operation, the Perot-Fabry etalon seems superior to the Lummer plate.

41. The Prism. Basic Theory of Resolving Power

We shall assume that the collimator lens provides completely parallel and monochromatic light. The telescope and collimator lenses will be assumed to be larger than the projections of the prism in the directions of the incident and refracted rays. Then the size of the ray bundle is limited by the size of the face of the prism through which the light emerges, see fig. 85. This surface is a rectangle which is perpendicular to the plane of the drawing and makes an oblique angle with the direction of the emerging ray. Using the notation of Sec. 36 A, the height of the rectangle is $2B$ and its width $2A$. This width equals the side 2–3 of our prism cross section and is also equal to 1–3. The height $2B$ of the rectangle cannot be shown in the figure and is also immaterial to the following considerations. The Fraunhofer diffraction pattern of this rectangle appears in the focal plane of the telescope. If the prism is illuminated through a slit which is parallel to the refracting edge, then the intensities of the patterns due to the elements of the slit add up along the direction of the height of the rectangle $2B$. The intensity distribution in the direction of the

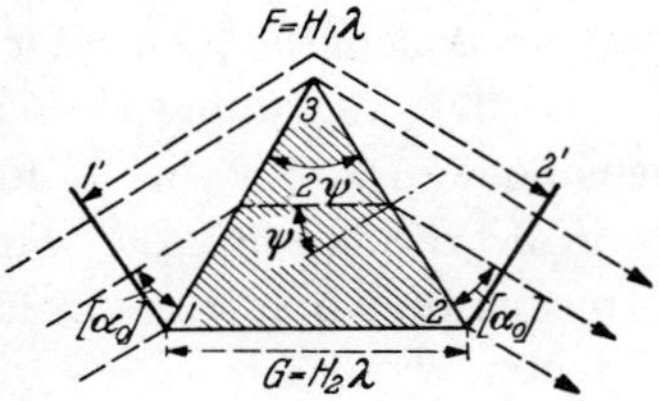

Fig. 85. Cross section of an isosceles prism, parallel to its base, with symmetrical light path. The emerging rays are drawn in the sense of geometrical optics without consideration for the diffraction at the edges of the prism. $[\alpha_0]$ means $\cos^{-1}\alpha_0$ for both the incident and emerging rays.

width $2A$ of the rectangle is given in Sec. 36 by eqs. (1), (2), (3). We note that this distribution almost agrees with that of a grating of width $Nd = 2A$, because in the vicinity of the principal maximum the factor $\sin \Delta/2$ in the grating formula (32.5) can be replaced by $\Delta/2$. The remaining calculations are similar to those in Sec. 40 for gratings. The positions of the first zeros to the right and left of the principal maximum are given by

$$(1) \qquad 2\pi A\,(\alpha_{1,2} - \alpha_0)/\lambda = \pm\pi.$$

α_0 is now the direction cosine of the emerging ray with respect to the surface of emergence of the prism; $\alpha_{1,2}$ are the direction cosines for the first zeros to the right and left of the ray α_0.

Let us now consider a second ray whose wavelength differs by $\delta\lambda$ from that of the previous ray. Because of dispersion this new ray will have a different index of refraction n' and a different direction of emergence α_0'. We want to know the value of $\delta\lambda$ for which the principal maximum of the second ray (direction α_0') will coincide with one of the two zeros α_1, α_2. α_0 and α_0' are determined by the law of refraction.[1] From exercise (III.2) we know that for symmetric path in a prism of refracting angle 2ψ

$$(2) \qquad \alpha_0 = n \sin \psi.$$

This relation holds for both refracting surfaces 1–3 and 2–3. The ray with wavelength $\lambda + \delta\lambda$ and with the same incident direction is no longer refracted

[1] Note the changed notation: The refracting angle which previously was φ is now $2\,\psi$; previously α, β, α', β' were the angles of incidence and refraction, now α_0, α_0', α are the direction cosines of the emerging (or diffracted) rays.

exactly symmetrically. It forms a small angle ε with the symmetry line so that on the surface 1–3

(2 a) $$\alpha_0 = n' \sin(\psi + \varepsilon)$$

and at the surface 2–3

(2 b) $$\alpha_0' = n' \sin(\psi - \varepsilon).$$

From (2 a) and (2 b) one finds

$$\alpha_0 + \alpha_0' = 2\, n' \sin\psi \cos\varepsilon;$$

subtracting from this equation twice the expression (2) for α_0 and calling the change in the index of refraction resulting from dispersion $\frac{dn}{d\lambda}\,\delta\lambda$, we obtain, neglecting terms of order ε^2

$$\alpha_0' - \alpha_0 = 2\,(n' \cos\varepsilon - n) \sin\psi \sim 2\,\frac{dn}{d\lambda}\,\delta\lambda \sin\psi.$$

According to Rayleigh's criterion for the resolving power, this difference must now agree with the value of $\alpha_{1,2} - \alpha_0$ as given by (1). It follows that

$$\frac{\lambda}{2A} = 2\,\frac{dn}{d\lambda}\,\delta\lambda \sin\psi$$

or

(3) $$\frac{\lambda}{\delta\lambda} = 4A\,\frac{dn}{d\lambda}\,\sin\psi = G\,\frac{dn}{d\lambda}.$$

$G = 4A \sin\psi$ (see figure) is the *base line* of our prism cross section. Only this base line and the *dispersion* of the glass $dn/d\lambda$ affect the resolving power. The larger the refracting angle $2\,\psi$, the smaller may we make the height of the triangle and the diameter of the lenses without decreasing the resolving power. As we approach the limiting angle of total reflection $\sin\psi = 1/n$, the resolving power becomes

(3 a) $$\frac{\lambda}{\delta\lambda} \sim 2\,A\,\frac{2}{n}\,\frac{dn}{d\lambda}.$$

The contribution $2A$ of the width of the prism surface is analogous to the contribution Nd of the width of a grating. The length $l = n/2\; d\lambda/dn$ now takes the place of the wavelength λ occurring in eq .(40.4) for gratings. For green light ($\lambda = 0.5\,\mu$) and heavy flint glass one has $n = 1.77$, $dn/d\lambda = 0.23\,\mu^{-1}$, $l = 3.84\,\mu$. Thus l is about eight times the corresponding wavelength $\lambda = 0.5\,\mu$. Therefore, with equal surfaces of emergence a prism attains only 1/8 the resolution attained by a comparable grating; but the prism is free of the superposition of the spectra of higher orders. This superposition, as well as the zeroth order spectrum, causes a considerable intensity loss in gratings. Therefore a prism spectrograph produces greater intensities than a comparable grating. Theoretically a prism could even have better resolution than a grating if it were possible to approach a characteristic frequency of the prismatic material sufficiently closely; for in the vicinity of such a frequency $dn/d\lambda$ becomes very large. Unfortunately, the strong absorption in the vicinity of a characteristic frequency prevents the utilization of such a region. An indication of this increased resolving power is already apparent in the violet ($\lambda = 0.41\,\mu$) where for the above-mentioned glass $l = 1.8\,\mu$; the resolution is here almost twice as good as in the green. If, instead of glass, quartz or rock salt is used, l is diminished still more, until in the far ultraviolet a rock salt prism becomes as good as a grating.

A. General considerations regarding resolving power

Let us compare the two following limiting light rays: one ray which when going from slit to crosshairs of the spectrograph passes through the vertex 3, and on the other hand, one which goes along the base 1 – 2. It suffices to measure the light path lengths from the wave surface 11′ which passes through the front edge of the prism to the wave surface 22′ which passes through the rear edge. We may limit our considerations to these portions of the rays because all rays from the slit to 11′ have the same path lengths and the same is true for all rays from 22′ to the cross hairs and therefore these portions of the paths do not contribute to the path difference between the rays. The extreme ray paths between the wave surfaces 11′ and 22′ are denoted by F and G in fig. 85. Their lengths measured in wavelengths shall be H_1 and H_2, respectively, and their difference shall be H. For the rays drawn in the figure which belong to the wavelength λ the value of H is, of course, zero because all rays have the same optical path length between two wave surfaces:

$$H = n\frac{G}{\lambda} - \frac{F}{\lambda} = 0. \tag{4}$$

The same holds true for the ray paths belonging to the wavelength $\lambda + \delta\lambda$ (not drawn in the figure) which terminate on another wave surface which is inclined with respect to 22′. But if we consider the path difference H for the changed wavelength along the *original* ray paths instead of along the changed paths (the geometrical paths F and G are kept fixed), then by varying λ we obtain from (4):

$$\delta H = \left(\frac{dn}{d\lambda}\frac{G}{\lambda} - \frac{nG}{\lambda^2} + \frac{F}{\lambda^2}\right)\delta\lambda = \frac{dn}{d\lambda}G\frac{\delta\lambda}{\lambda}. \tag{4 a}$$

Hence the above expression (3) for the resolving power is equivalent to the statement

$$\delta H = 1. \tag{5}$$

These considerations, first of all, explain the presence of the factor G in eq. (3) which may have seemed surprising at first glance: the larger G is, the easier it becomes for the dispersive power of the glass to produce the optical path difference between neighboring wavelengths which is necessary for resolution. Furthermore, the reasoning which led us to the criterion (5) can now be visualized and also *generalized* in the following way: $\delta H = 1$ means that the two extreme rays F and G, which arrive simultaneously at 2 and 2′ if they have the wavelength λ, have an optical path difference of exactly *one* wavelength if their wavelength is $\lambda + \delta\lambda$; and this path difference is a linear function of position along the original wave surface 22′. This means that in the focal plane of the telescope the rays with wavelength $\lambda + \delta\lambda$ are extinguished where the rays with wavelength λ which have equal phases along 22′ produce their diffraction maximum and vice versa. In other words, *the criterion* (5) *is equivalent to Rayleigh's criterion.*

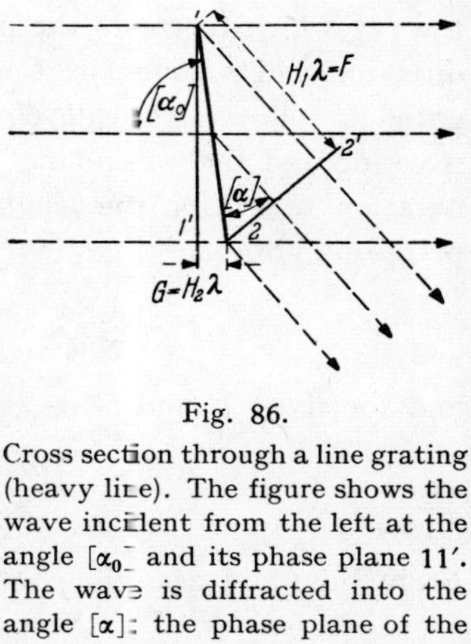

Fig. 86.

Cross section through a line grating (heavy line). The figure shows the wave incident from the left at the angle $[\alpha_0]$ and its phase plane 11′. The wave is diffracted into the angle $[\alpha]$; the phase plane of the diffracted wave is 22′.

B. Applications to Gratings and Interference Spectroscopes

The applicability of the above criterion to line gratings will be tested by means of fig. 86. 12 represents the trace of the grating on the plane of the drawing. For a plane wave incident from the left, 1 is the extreme left grating line and 2 is the extreme right grating line. 11′ is again the plane of constant phase for the incident wave and 22′ the plane of constant phase for the diffracted wave under consideration. $[\alpha_0]$ and $[\alpha]$ are the direction cosines of these planes with respect to the plane of the grating. The optical path lengths of the extreme left- and right-light paths between the two phase planes are

$$F = 12' = \alpha Nd, \qquad G = 1'2 = \alpha_0 Nd$$

where Nd is the width of the grating. From this it follows that

$$H = H_2 - H_1 = \frac{G - F}{\lambda} = \frac{Nd}{\lambda}(\alpha_0 - \alpha)$$

and (the directions α, α_0 are being kept fixed!)

$$\delta H = \frac{dH}{d\lambda}\,\delta\lambda = \frac{Nd}{\lambda^2}\,(\alpha-\alpha_0)\,\delta\lambda\,.$$

But according to the basic grating formula (32.2): $\alpha - \alpha_0 = h\,\lambda/d$ (h = order of the spectrum) and therefore

$$\delta H = Nh\,\frac{\delta\lambda}{\lambda}\,.$$

From this and our criterion (5) follows indeed the resolving power of the grating

$$\frac{\lambda}{\delta\lambda} = Nh \qquad \text{as in (40.3).} \tag{6}$$

Regarding interference spectroscopy a few words about the Perot-Fabry plate will suffice. We shall consider the ray which passes only once through the plate and then emerges as the "first ray". It follows the path F. We call the ray which traverses the plate $2\,p+1$ times, $p+1$ times in the forward direction and p times in the reverse direction, the "last ray" and the path it follows we call G. The number p depends on the amount of silvering and the weakening of the light caused by it. With our former notation of $z\lambda$ for the length of one forward and back path through the plate one obtains

$$F = \frac{1}{2}\,z\,\lambda, \qquad G = \left(p+\frac{1}{2}\right) z\,\lambda, \qquad H = p\,z$$

and for fixed F and G

$$\frac{\delta z}{z} = -\frac{\delta\lambda}{\lambda}\,, \qquad |\delta H| = p\,|\delta z| = p\,z\,\frac{\delta\lambda}{\lambda}\,.$$

From this, and according to (5), the resolving power becomes:

$$\frac{\lambda}{\delta\lambda} = p\,z. \tag{7}$$

This result is to be compared with eq. (40.7) where the quantity $g+1$ takes the place of our present $2p$. According to the discussion following eq. (7.29), g is a measure of the conductivity and the thickness of the silver layer, hence g is a measure of the reflecting power of the surface. On the other hand $2p$ is the number of reflections which can be observed without excessive weakening of the light [the notation is the same as in eqs. (7.20) and (7.21)]. Thus g and $2p$ mean qualitatively the same thing. Therefore, our present statement (7) agrees qualitatively with the statement (7) in Sec. 40.

42. The Telescope and the Eye. Michelson's Measurements of the Sizes of Fixed Stars

Let us assume that a telescope is directed at a pair of stars 1, 2 in such a way that the axis of the telescope points to 1. Then 1 produces in the focal plane a diffraction pattern of the type shown in fig. 68. According to eq. (36.11) the position of the first diffraction minimum is given by

$$s_1 = 0.61 \frac{\lambda}{a}, \tag{1}$$

where a is the radius of the objective, λ is a mean wavelength of the star's light, and s is defined as in (36.7) as the angle between the ray under observation and the direction of the principal maximum. The number 0.61 corresponds to the first root of the Bessel function J_1 and is approximately equal to 5/8; see (36.11 a).

If we agree that star 2 is clearly distinguishable from star 1 (either visually or on the photographic plate) only if its principal maximum is further away from the principal maximum of 1 than the first minimum of 1, then we see that the formula (1) also contains a *measure of the resolving power of the telescope.* The smaller s_1 is, the larger is the resolving power. Therefore we shall define the resolving power of the telescope as the *reciprocal* of the smallest resolved angular distance between 1 and 2 as determined by (1); that is, as the dimensionless number

$$\frac{1}{s_1} = \frac{a}{0.61\,\lambda}. \tag{2}$$

From this we conclude: *the resolving power is proportional to the size of the objective.* This fact is the reason for the giant telescopes on Mt. Wilson and for the large mirror at the Palomar observatory; in the case of the latter, $2\,a = 200$ in. ~ 5 meters! We note further that the resolution is somewhat better at the short wavelength end of the spectrum than at the long wavelength end.

In the *eye* the pupil takes the place of the rim of the objective; its diameter $2\,a$ varies between 1 mm and 8 mm depending on the brightness. It follows that for a medium value of λ equal to 5×10^{-4} mm

$$10^{-3} > \frac{\lambda}{a} > 1.2 \times 10^{-4},$$

thus

$$6 \cdot 10^{-4} > s_1 > 0.7 \times 10^{-4}$$

or in degrees instead of radians

$$2' > s_1 > 15''.$$

Therefore, quite aside from the cellular structure of the retina, diffraction imposes an upper limit on the resolving power of the eye. With strong illumination (small pupil) only differences in direction that are of the order of magnitude of one minute or more of arc can be perceived.

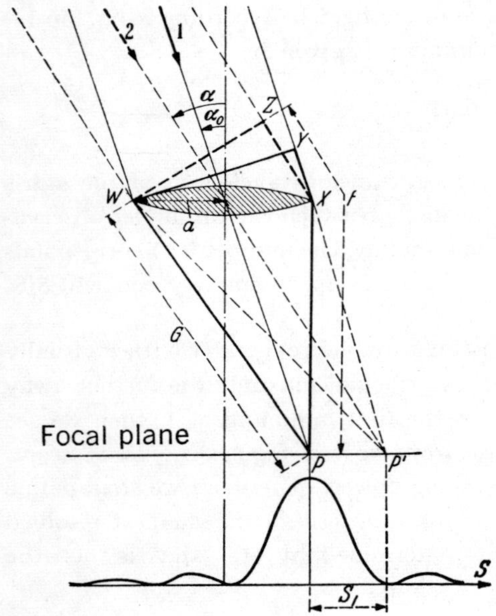

Fig. 87.
Diffraction of the light from a binary star 1, 2 when Rayleigh's criterion is fulfilled. Construction for the calculation of the difference $F-G$ between limiting rays.

Equation (2) can also be understood from the general point of view of optical path differences as formulated in eq. (41.5). Returning to the telescope we consider the two "limiting rays" which pass through diametrically opposite points at the rim of the objective. In fig. 87 these rays are drawn as full lines for star 1 and as dotted lines for star 2. P is the image of 1 in the focal plane; P' is the image of 2. The significant path lengths of the limiting rays from star 1 are

$$G = W P, \qquad F = Y X + X P,$$

where, of course, $F = G$ because, being the image of 1, P is that point where all the light from 1 arrives with the same phase. Therefore

$$(3) \quad Y X + X P - W P = 0.$$

But we are interested in the light paths of the rays from the star 2 to P which are

$$G = W P, \qquad F = Z X + X P$$

and so taking (3) into account

$$\text{(3 a)} \quad \begin{aligned} F - G &= Z X + X P - W P \\ &= Z X - Y X + (Y X + X P - W P) = Z X - Y X. \end{aligned}$$

The right triangles WZX and WYX show that

$$Z X = 2\, a \sin \alpha; \qquad Y X = 2\, a \sin \alpha_0.$$

Therefore, by (3 a)

$$\text{(3 b)} \quad F - G = 2\, a\, (\sin \alpha - \sin \alpha_0)$$

and

(3 c) $$H = \frac{F-G}{\lambda} = \frac{2\,a}{\lambda}\,(\sin\alpha - \sin\alpha_0).$$

This H-value must now be varied by varying the position of the object, that is by varying α_0 and not by varying λ as was done in the case of spectroscopic apparatus. This variation yields

(3 d) $$|\delta H| = \frac{2\,a}{\lambda}\,\delta\sin\alpha_0.$$

According to our condition $|\delta H| = 1$ two objects would be resolved or not resolved depending on whether

$$\delta\sin\alpha_0 \gtrless \frac{\lambda}{2\,a}.$$

The resolving power is therefore

(4) $$\frac{1}{\delta\sin\alpha_0} = \frac{2\,a}{\lambda}.$$

The difference between this and our definition (2) for the resolving power is trivial because it consists only of a numerical factor of $2 \times 0.61 = 1.22$. (The same trivial factor would appear in the resolving powers of spectroscopic devices such as gratings, prisms, etc. if the diffraction opening were bounded by a circle instead of by a rectangle. As an alternative we could replace the condition $\delta H = 1$ by $\delta H = 1.22$.) The intensity curve for the light of star 1 drawn at the bottom of fig. 87 indicates that our construction using limiting rays is equivalent to Rayleigh's condition: the image of star 2 coincides with the first minimum of the diffraction pattern of star 1.

The situation is different when "resolution" does not require (almost separate) images of the two stars but only some indication of whether or not a binary star is under observation at all. In that case it is possible to attain a much larger path difference of the limiting rays without having to require precision to within a wavelength for all rays over the entire surface of an objective. This leads to an arrangement of mirrors which had already been proposed by Fizeau but was first constructed successfully by Michelson; see fig. 88.

The two outer mirrors S, S' are a distance $b + b'$ of several meters apart; the spacing of the two inner mirrors lies within the diameter $2\,a$ of an ordinary telescopic objective, i. e. is of the order of a few inches.

Let us first consider only the component 1 of the twin stars and the light reaching the telescope from that star by way of S, s. The resulting diffraction pattern is determined by the cross section of the light bundle (which, in turn, is determined by the sizes of the mirrors S, s and by the diameter $2\,a$ of the objective). This diffraction pattern consists of a system of rings of the type described in Sec. 36 C. The same is true for the light reaching the telescope from the component 1 by way of S', s'. Since this light originates from the same source 1 as the previously considered light, the amplitudes of both the central spot and the system of rings would be doubled provided the arrangement of mirrors were perfectly symmetrical. Actually, the two mirrors S, S' are never exactly symmetrical with respect to the axis of the telescope and are never inclined at precisely 45° to that axis. Therefore there is also present a system of equidistant rectilinear interference fringes of the type which is already known to us from the Michelson experiment of Sec. 14 (these fringes are lines of equal optical path difference). Because of the lack of symmetry of the arrangement, we have denoted the distance $S'\,s'$ in the figure by b' to distinguish it from $S\,s = b$. The position of the system of fringes depends on the quantity

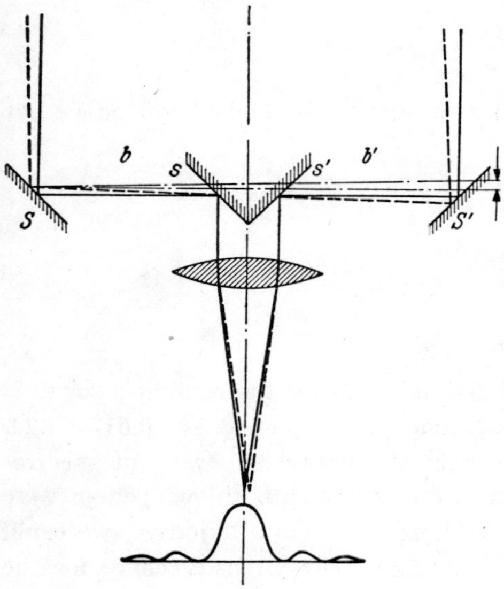

Fig. 88.
Michelson's Mirror Experiment.

$$(b - b')/\lambda.$$

Depending on whether this ratio (for a given λ and for a given position in the ray bundle between $S\,s$ and $S'\,s'$) is an integer or half-integer, we have a bright or a dark fringe.

Let us now turn to component 2 of the binary star. It yields a diffraction pattern of the same type as that produced by component 1. There are again a central spot, diffraction rings, and linear interference fringes. The central spots and rings of the two component stars coincide because it is assumed that the cross section of the ray bundle, or the diameter $2\,a$ of the objective, is not sufficiently large to resolve the double star. The central spot and diffraction rings of 2 are superimposed on those of 1 (of course they add intensity-wise because of the difference of the light sources). But the position of the system of rectilinear fringes from component 2 differs from that due to 1. The path difference responsible for these fringes is affected not only by the positions of the mirrors but also by the direction of the incident light. As in (3 d) the difference in the directions of incidence of the light of components 1 and 2 produces a path difference of

$$\frac{b+b'}{\lambda}\,\delta \sin \alpha_0. \tag{5}$$

Hence in general the two systems of fringes do not coincide but are displaced from one another by the amount given in (5). The mirrors S, S' are mounted on a rigid support and can be moved parallel to themselves so that the total distance $B = b + b'$ between the mirrors can be changed. In this way the shift between the fringe systems is also changed. Let us assume that at a given value of $B = B_n$ corresponding to a path difference of n wavelengths, the two systems of fringes coincide. Then

$$\frac{B_n}{\lambda}\,\delta \sin \alpha_0 = n. \tag{5 a}$$

If the mirrors are now shifted until the next coincidence of fringes is observed at $B = B_{n+1}$, then

$$\frac{B_{n+1}}{\lambda}\,\delta \sin \alpha_0 = n + 1. \tag{5 b}$$

Subtracting these two equations and letting $\Delta B = B_{n+1} - B_n$, one obtains

$$\frac{\Delta B}{\lambda}\,\delta \sin \alpha_0 = 1$$

and thus

$$\delta \sin \alpha_0 = \lambda / \Delta B. \tag{6}$$

ΔB can be measured precisely; a mean wavelength must, of course, be substituted for λ. *Owing to the enlarged scale B of the interference phenomenon, the existence of a binary star can be ascertained and the angular distance between the two components can be measured even if the resolving power of the telescope being used is insufficient.*

The same method can be applied to measure the size of a *single fixed star of exceptional size* which in the mirror arrangement behaves no longer like a point source but rather like a small disc. The stars which can be measured by this method are the so-called red giants (low temperature, hence the red color, but nevertheless great brightness because of the enormous luminous surface); see examples below. Such a small disc can be thought of as divided into a left, a right and a middle third, and the two outer thirds can be treated like a binary system; the light from the middle third will weaken the contrasts of the interference fringes belonging to the left and right thirds, but they will not be extinguished. A study of the fringe coincidences as described by eq. (6) leads to an estimate of the angular distance between the edges of the star. Michelson obtained the following values in seconds of arc:

Betelgeuse	0.047″,
Antares	0.040″,
Arcturus	0.022″.

Since the distances of these stars from the solar system (their parallaxes) are known from other measurements, their linear diameters can be calculated. These diameters turn out to be of the order of 10^8 km, which is about one hundred times the diameter of our sun and about equal to the diameter of the earth's orbit.

43. The Microscope

Helmholtz[1] treated the resolution of the microscope in the same way as that of the telescope. For an object let us take two luminous points a distance d apart and located in the lower focal plane F_1 of the objective; see fig. 89. The objective produces an image of these points at infinity because the spherical waves emitted by the two points leave the objective as plane waves forming two bundles of parallel rays, the directions of which differ by an angle α. If the medium on both sides of the objective is air, then this angle is the same as the angle between the two central rays from the objects to the optical center of the objective. The value of this angle is d/f, where f is the lower focal length of the objective. If a medium of greater index of refraction $n > 1$ occupies the space between the object and the objective (immersion in oil), then for small angles of incidence the law of refraction gives

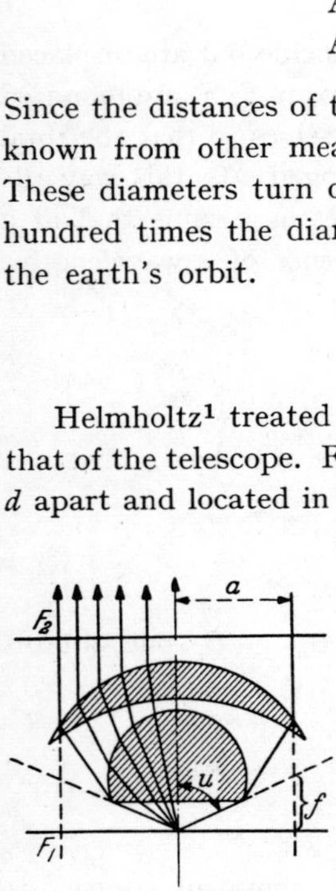

Fig. 89.

Paths of rays in a microscope. F_1 and F_2 are the front and rear focal planes.

$$\alpha = n\,d/f. \tag{1}$$

We assume that the objective is, from the point of view of geometrical optics, a perfectly corrected system of lenses. The rim of the objective and all other diaphragms are projected as geometrical images (real or virtual) by the succeeding lenses. *The smallest of these images is called the "exit pupil"*

[1] Die theoretische Grenze für die Leistungsfähigkeit der Mikroskope, Ann. d. Physik, 1874. Fraunhofer stated much earlier (Bayerische Akademie June 14, 1823) that the limit of effectiveness of a microscope depends on diffraction.

of the lens system and it limits the size of our ray bundles. In general the exit pupil is the virtual image of the rim of the front lens. Let the radius of the exit pupil be a.

$$A = \frac{a\,n}{f} \tag{2}$$

is called the "numerical aperture"; from the elementary geometrical optics of lens systems (sine condition, Sec. 48) it follows that $a = f \sin u$, where $2\,u$ is the angle of opening of the object ray cone. Hence, the definition (2) of A becomes

$$A = n \sin u. \tag{2 a}$$

Each of our two ray bundles produces at infinity the Fraunhofer diffraction image of the exit pupil as described by eq. (36.9). Hence the microscope reproduces each of our luminous points in the form of a diffraction pattern (central field plus diffraction rings). In order to render these patterns observable at a finite distance, the eyepiece contains a converging lens on whose focal plane the diffraction patterns are reproduced. This image is observed through the eyepiece which acts as a magnifying glass.

This projection of the image upon a finite plane is of no concern to the theoretical investigation of resolution. Instead, the original diffraction pattern at infinity can be treated directly. As in the case of the telescope, eq. (42.1), we are then led to the result $n\,d/f = 0.61\,\lambda/a$, which, according to the definition (2) of A, we can also write in the form

$$d = 0.61\,\frac{\lambda}{A}. \tag{3}$$

Two luminous points are resolved only if the distance between them is greater than that given by (3).

Recalling the meaning (2 *a*) of A, we note that while the resolving power of the telescope depends on the *size of the objective,* the resolving power of the microscope depends on the *angle u subtended by the objective at the specimen.*

A. Abbe's theory of the microscope

With Helmholtz's theory the question of the resolving power of the microscope is essentially settled. What remains to be explained is the remarkable effect which the manner of illumination (bright or dark field) has on the resolution of different tissue structures even though it does not influence the resolution of two-point objects. This is where Abbe's theory is of importance. Abbe regarded the object as a diffraction grating (amplitude or phase grating). Because of the thinness of the specimens observed and the low depth of focus of powerful objectives, these gratings can be considered plane and their extent as to depth can be neglected. If we illuminate the object with coherent light parallel to the axis of the microscope, then plane waves emerge from the object in the directions of the grating spectra of various orders. Those orders which are emitted inside the angle of opening u are collected in the form of a Fraunhofer diffraction pattern on the upper focal plane F_2 of the objective which is located above the objective and very close to it. This pattern is easily observed with the eye piece removed. But the rays continue, and at infinity or on the focal plane of the collector lens they combine into a more or less faithful image of the object grating.

If the spectra on the upper focal plane of the objective are stopped down even further than this is done by the exit pupil, or if an objective of smaller aperture is used, the image becomes less distinct. If with oblique incidence, for instance, at least two spectra are present, then one only sees a sinusoidal structure without other details. If only *one* spectrum is admitted, the image disappears in a uniformly illuminated surface. It is also possible to simulate a structure that does not exist. If, for instance, the first order spectrum is stopped but the second order spectrum is allowed to pass, a grating with twice the actual number of lines appears. For the correct grating period to be just visible, at least both first order spectra must appear at the edge of the exit pupil of the objective; this means that the first order spectra may emerge from the objective at angles of at most $\pm u$. From the formula for the grating spectrum (32.2) we obtain for air and perpendicular incidence[1]

$$\sin u \geqq \alpha = \frac{\lambda}{d}. \tag{4}$$

If, on the other hand, the object and the first lens of the objective are embedded in a medium of index of refraction n (immersion), then λ must be replaced by the smaller wavelength $\lambda' = \lambda/n$; the grating spectra then crowd closer together, which explains the significance of immersion from the point of view of Abbe's theory. The condition (4) becomes now

$$\sin u \geqq \alpha = \frac{\lambda'}{d} = \frac{\lambda}{n\,d}. \tag{4 a}$$

By (2 a) this means

$$A = n \sin u \geqq \frac{\lambda}{d}. \tag{5}$$

[1] α and α_0 continue to stand for direction cosines; see footnote 1, p. **303**.

We can now explain the advantage of *oblique* illumination. Let us assume, for instance, that the zeroth order spectrum falls on one edge of the exit pupil and that the first order spectrum falls on the opposite edge; this situation still suffices to show the existence of a structure and to reproduce its correct grating constant but no other details. The angle between the two spectra can now be twice as large as before or the spacing of grating lines can be half that of before. In place of eqs. (4 a) and (5) we obtain

$$2 \sin u = \alpha - \alpha_0 = \frac{\lambda'}{d} = \frac{\lambda}{n\,d}, \tag{6}$$

$$A = n \sin u = \frac{1}{2}\frac{\lambda}{d}, \qquad \text{hence} \qquad d = 0.5\frac{\lambda}{A}, \tag{6 a}$$

which involves only the small improvement in the ratio 0.5 : 0.61 over (3). If, instead of the spectra of zeroth and first order, the first and second order spectra or spectra of two still higher orders are used, then one speaks of "dark field illumination". The direct light (zeroth order spectrum) does not enter the objective and with no object in place the field of vision remains dark. For $n \sim 1.6$, $\sin u \sim 1$, $A \sim 1.6$ a numerical estimate yields according to eq. (6 a)

$$d \sim \frac{\lambda}{3}.$$

Smaller spacings can be resolved only by the use of shorter wavelengths: an ultraviolet microscope equipped with quartz and fluorite lenses which is effective down to $\lambda = 0.2\,\mu$ or an electron microscope. In the case of the latter the use of hard cathode rays makes the resolving power theoretically almost infinite.

These considerations are valid not only for the one-dimensional gratings which we have had in mind so far, but also for arbitrary plane structures which, according to the Fourier theorem for two dimensions, can always be considered as a superposition of cross gratings. The structure of the object and the structure of the diffraction image in the objective focal plane are "reciprocally" related to one another; one is the "Fourier-transform" of the other; see Vol. VI, eq. (4.13). This "reciprocity" means that the diffraction image of the second of the above structures again yields the structure of the original object. Also in the case of two-dimensional structures any loss of diffraction spectra resulting from stopping down reduces the similarity between the final image and the structure of the object.

B. Significance of phase gratings in microscopy

As an example, we shall consider the "laminary profile" of fig. 70. Like any pure phase grating, it is completely invisible if the image is perfect, for neither the retina nor the photographic plate can perceive phase differences. We would like to be able to see this grating as a set of bright and dark fringes, that is, as an *amplitude grating*. To accomplish this it suffices, according to the concluding remarks in Sec. 36 D, to shift the phase of the zeroth order spectrum by $\pi/2$ with respect to the phases of the spectra of higher orders. F. Zernicke[1] produced this phase shift in the following way: a glass plate is placed in the focal plane of the objective. A thin layer of transparent material is attached to the center of this plate where for axis-parallel illumination the spectrum of zeroth order appears. While the spectra of higher orders are unaffected, the phase of the zeroth order spectrum is changed by an amount which depends on the thickness of the thin layer and its index of refraction relative to the surrounding medium. In order for the phase change to be $\pi/2$, the path difference must be $\lambda/4$ and the thickness

$$d = \frac{\lambda}{4} \Big/ (n-1).$$

This is a real "quarter wave plate"; the layer is less than 1 μ thick in contrast to the "quarter wave plate" of crystal optics, see Sec. 30 B, where the thickness was determined by the small difference $n_2 - n_1$ between the indices of refraction in the two principal directions of oscillation rather than by the much larger difference $n-1$ between the indices of refraction of the plate and its surroundings.

This *phase contrast method* of Zernicke is perhaps the most sensitive way of making very weak phase structures visible. Previously microscopists had been obliged to use more or less oblique illumination. The resulting loss of several spectra caused blurred images of the object. The Zernicke method, on the other hand, fully utilizes the tissue structure and makes it visible to the eye in the same way as an ideal staining method would do.

C. Luminous and illuminated objects

Because of the great successes of Abbe's theory, it was thought for a long time that Helmholtz's theory applied only to *luminous sources*, while in the treatment of *illuminated objects* only Abbe's theory was believed to hold[2].

[1] Z. f. techn. Physik, Vol. 11, 1935.

[2] Such as in the review article by O. Lummer and F. Reiche: Die Lehre von der Bildentstehung im Mikroskop von Ernst Abbe, 1910.

However, Laue[1] proved by means of a simple hypothetical experiment that the image of any small luminous object, for instance a glowing Wollaston wire, is perfectly complementary to the image of the same object when illuminated by an external source. If the wire is in a cavity of constant temperature, then according to the laws of radiation it is entirely invisible. The radiation emitted by the wire and that originating from the walls of the cavity and absorbed and re-emitted by the wire combine to produce the same radiation density as that of the background. The same holds for observations with a microscope: the images resulting from self-luminous objects and those resulting from homogeneous illumination of equal brightness of the same objects must structurally complement one another entirely. The resolution of neighboring objects must be the same in both cases.

To be sure, the illumination is not homogeneous in the case of a microscope. With the usual arrangement of a lower illuminating mirror, the illumination is merely as uniform as possible within the aperture. However, the illumination from above is missing, but this cannot make much difference as long as the object is not (or only slightly) reflecting. We may think of this illumination from above as added or we may just omit it. The rays which do not enter the aperture are equally insignificant. Therefore Laue was justified in applying the law of black-body radiation to the microscope.

The superiority of Abbe's point of view becomes apparent only with oblique illumination of the type used in dark field observations. For, Helmholtz's theory assumes that all points of the object radiate uniformly in all directions, and this is generally not true for the elements of a tissue structure.

44. On Young's Interpretation of Diffraction

Even before Fresnel, Thomas Young[2] attempted to find a wave-theoretical explanation for the diffraction phenomena which had been discovered by Grimaldi. Young assumed that the incident light undergoes "a kind of reflection" at the edges of the diffraction opening and he explained the diffraction fringes in terms of the principle of interference which he had discovered as caused by the interaction between these edge rays and the incident light rays. In this way he achieved a qualitative understanding of the diffraction pattern of the slit in particular. However, Fresnel in his prize

[1] M. von Laue, Zur Theorie der optischen Abbildung, Ann. d. Phys. Lpz. **43**, 1914.
[2] Phil. Trans. Roy. Soc. London **20**, 1802.

essay of 1818 showed that Young's assumption did not suffice for a quantitative explanation, and as a result Young's theory was forgotten for a long time.

In this connection we wish to recall our treatment of the half-plane in Sec. 38. The light entering the geometrical shadow is a cylindrical wave which appears to originate on the edge of the screen; the diffraction fringes in the illuminated region were calculated from the interaction of this cylindrical wave with the incident light. The cylindrical wave does not, of course, radiate uniformly in all directions; rather its intensity depends in a definite way on the angle of diffraction. Furthermore, the edge of the screen is not an actual light source with infinite amplitude but only appears as such to a sufficiently distant observer because of a representation of the light field which is valid only asymptotically at large distances. From this we see the following: if we wish to talk of a reflection of the incident light as Thomas Young did, then the "kind of reflection" is very specialized and must be defined precisely.

The question arises whether and in what way Young's interpretation may be extended to arbitrary diffraction screens. This question was answered conclusively by A. Rubinowicz[1].

A. Reformulation of Kirchhoff's solution of the problem of diffraction

We shall make the following assumptions:

1. The screen shall be arbitrarily bounded. We apply the Kirchhoff formula (34.4); we cannot use the simplified representation (34.6) in terms of the Green's function of the half-space because (even in the case of a plane screen) we will have to integrate not only over the diffraction opening but also over the surface of a cone.

2. Let the light source be a luminous point at a finite distance as in (34.4 b). We shall change the notation from r' (distance of a point on the surface of integration from the light source) to ρ. We shall not carry out the specialization to an incident *plane* wave ($\rho \to \infty$) because this would complicate rather than simplify the presentation.

3. We shall regard the light field as *scalar*, just as Kirchhoff did. Thus we shall actually discuss the diffraction problem of *acoustics* rather than the vectorial problem of *optics*. However, this will suffice to point out the essential features of Young's interpretation.

---

[1] Ann. d. Phys. Lpz. **53**, 1917 and **73**, 1924.

4. Upon applying eqs. (34.4 b) in which we can set $A = 1$, the integrand in Kirchhoff's formula (34.4) becomes

$$J = \frac{e^{ikr}}{r} \frac{\partial}{\partial n} \frac{e^{ik\rho}}{\rho} - \frac{e^{ik\rho}}{\rho} \frac{\partial}{\partial n} \frac{e^{ikr}}{r}, \tag{1}$$

where r is the distance of the point of integration from the point of observation P. First we perform the integration in the manner of Kirchhoff over a surface σ which in some way spans the diffraction opening. This surface together with the diffraction screen S will separate a region of space containing the light source P' from a portion which contains P. Equation (34.4) then reads

$$4\pi v_P = \int_\sigma J \, d\sigma. \tag{2}$$

The following calculations serve only to transform this Kirchhoff formula and will not improve it or change its essence.

We adopt the following point of view: the surface σ over which (2) is to be integrated is entirely arbitrary; its choice is limited only by the requirement that σ shall pass through the curve s which forms the boundary of the diffraction opening. Hence (2) depends only on s and not on σ. Therefore it must be possible to transform the surface integral $\int d\sigma$ into a line integral $\int ds$. To accomplish this we construct the cone formed by the rays emitted by P' and passing through the boundary of the diffraction aperture, see fig. 90. We call the surface of this cone f and its surface elements df. We now consider the space bounded by σ and f and apply Kirchhoff's eq. (34.4) to this region. The boundary values to be used on f are

$$v = \frac{e^{ik\rho}}{\rho} \quad \text{and} \quad \frac{\partial v}{\partial n} = \frac{\partial}{\partial n} \frac{e^{ik\rho}}{\rho} = 0,$$

the latter because dn is perpendicular to $d\rho$ and $e^{ik\rho}/\rho$ is only a function of ρ. The integrand of (1) simplifies to

$$J' = -\frac{e^{ik\rho}}{\rho} \frac{\partial}{\partial n} \frac{e^{ikr}}{r}. \tag{3}$$

For this region bounded by σ and f we obtain in place of eq. (2) a value differing from v_P:

$$4\pi v_P' = \int_\sigma J \, d\sigma + \int_f J' \, df. \tag{4}$$

However, we know the exact value of v_P'. If P lies inside our truncated ray cone, then

$$v_P' = \frac{e^{ik\rho}}{\rho}; \tag{4 a}$$

if P lies outside this region and also outside the region which is directly illuminated by P', then

$$v_P' = 0. \tag{4 b}$$

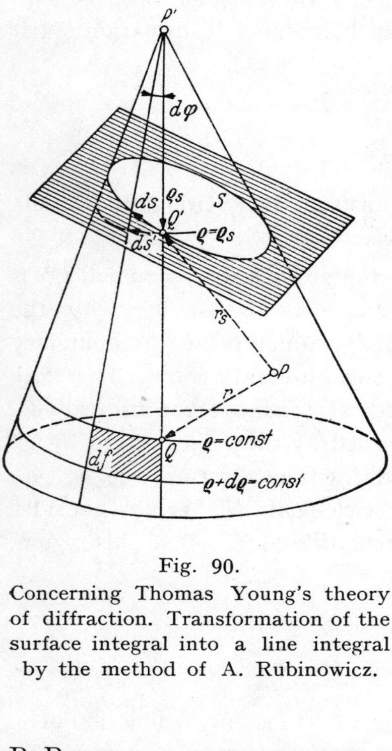

Fig. 90.
Concerning Thomas Young's theory of diffraction. Transformation of the surface integral into a line integral by the method of A. Rubinowicz.

Indeed we have incorporated into (4) the exact boundary values on both σ and f of the solution $u = e^{ik\rho}/\rho$ of the wave equation $\Delta u + k^2 u = 0$. Since by Green's theorem Kirchhoff's equation follows rigorously from this wave equation when the exact boundary values are known, v_P' agrees exactly with the solution $u = e^{ik\rho}/\rho$ inside the truncated cone and vanishes outside it. If we substitute (4 a, b) and (2) in (4), then for interior points

$$v_P = \frac{e^{ik\rho}}{\rho} - \frac{1}{4\pi}\int_f J'\,df, \tag{5 a}$$

and for exterior points

$$v_P = -\frac{1}{4\pi}\int_f J'\,df. \tag{5 b}$$

This completes the first step in transforming the Kirchhoff integral (2).

B. Reduction of the surface integral over the cone to a line integral over the boundary of the diffraction opening. Sharpening of Young's theory

In fig. 90 we have drawn two neighboring generators of the cone f which enclose an angle $d\varphi$. In the lower part of the figure we have drawn the intersections of the spheres ρ = constant and $\rho + d\rho$ = constant with the cone surface. The shaded surface element df in the figure is then

$$df = \rho\, d\rho\, d\varphi.$$

We now replace $d\varphi$ by the element of arc ds' which the sphere $\rho = \rho_s$ intercepts at the point Q' on the diffraction edge. Next we express ds' in terms of the line element ds of the boundary curve:

$$ds' = \rho_s \, d\varphi = ds \cos(ds', ds) = ds \sin(\rho_s, ds).$$

Thus we obtain

(6) $$df = \frac{\rho}{\rho_s} \sin(\rho_s, ds) \, d\rho \, ds.$$

We then evaluate the differential quotient occurring in (3) at the point Q on df

(7) $$\frac{\partial}{\partial n} \frac{e^{ikr}}{r} = \frac{\partial}{\partial r} \frac{e^{ikr}}{r} \cos(n, r) = \left(\frac{ik}{r} - \frac{1}{r^2}\right) e^{ikr} \cos(n, r),$$

where n is the direction perpendicular to df at the point Q and r is again the distance of the element of integration from the point of observation P. Besides this distance r, we have also indicated in the figure the distance r_s of the point of observation from Q', that is, from the vicinity of the boundary element ds. It is seen then that

(8) $$r \cos(n, r) = r_s \cos(n, r_s);$$

indeed, the left- and right-hand sides of this equation are both equal to the shortest distance from the point of observation to the surface of the cone; it should be noted here that the normal n to the cone surface at Q is parallel to the normal to the cone at Q'.

From (3), (6), (7) and (8) we finally obtain

(9) $$\frac{1}{4\pi} \int_f J' \, df = \frac{1}{4\pi} \int_s ds \sin(\rho_s, ds) \cos(n, r_s) \frac{r_s}{\rho_s} \int_{\rho_s}^{\infty} e^{ik(\rho + r)} \left(\frac{ik}{r^2} - \frac{1}{r^3}\right) d\rho.$$

The integrand of the first integral on the right-hand side consists of factors which depend only on the boundary curve s. The second integrand contains all the factors which are functions of the element df and of ρ; among these is also r because of the following relation which is derived from the triangle $Q \, P \, Q'$:

(10) $$r^2 = r_s^2 + (\rho - \rho_s)^2 + 2 r_s (\rho - \rho_s) \cos(r_s, \rho_s).$$

From this there follows by differentiation with respect to ρ, keeping ρ_s and r_s fixed (shifting Q with unchanged positions of P and Q')

$$r \frac{dr}{d\rho} = \rho - \rho_s + r_s \cos(r_s, \rho_s).$$

Hence also

$$r\left(1+\frac{dr}{d\rho}\right)=r+\rho-\rho_s+r_s\cos(r_s,\rho_s). \tag{10 a}$$

We now claim that the integrand of the ρ-integral in (9) is a perfect differential quotient, specifically, that

$$e^{ik(\rho+r)}\left(\frac{ik}{r^2}-\frac{1}{r^3}\right)=\frac{d}{d\rho}\left\{e^{ik(\rho+r)}\Big/ r[\;]\right\}, \tag{11}$$

where [ ] symbolizes the right-hand side of (10 a). For, upon carrying out the differentiation with respect to ρ on the right-hand side of (11), one obtains the following three terms, which can immediately be simplified by applying eq. (10 a):

$$\text{(a)}\qquad ik\left(1+\frac{dr}{d\rho}\right)e^{ik(\rho+r)}\Big/ r[\;]=\frac{ik}{r^2}e^{ik(\rho+r)},$$

$$\text{(b)}\qquad -e^{ik(\rho+r)}r^{-2}\frac{dr}{d\rho}\Big/[\;]=-\frac{1}{r^3}e^{ik(\rho+r)}\frac{dr/d\rho}{1+dr/d\rho},$$

$$\text{(c)}\qquad -e^{ik(\rho+r)}r^{-1}\left(\frac{dr}{d\rho}+1\right)\Big/[\;]^2=-\frac{1}{r^3}e^{ik(\rho+r)}\frac{1}{1+dr/d\rho}.$$

The sum (*a*) + (*b*) + (*c*) indeed equals the left-hand side of (11). Thus the value of the ρ-integral in eq. (9) becomes

$$\left\{e^{ik(\rho+r)}r^{-1}[\;]^{-1}\right\}_{\rho_s}^{\infty}=-\frac{e^{ik(\rho_s+r_s)}}{r_s^2(1+\cos(r_s,\rho_s))}. \tag{12}$$

As a result the right-hand side of (9) becomes a single integral over the edge of the diffraction opening:

$$\frac{1}{4\pi}\int_s ds\,\frac{e^{ik\rho_s}}{\rho_s}\,\frac{e^{ikr_s}}{r_s}\,\frac{\cos(n,r_s)}{1+\cos(r_s,\rho_s)}\sin(\rho_s,ds). \tag{13}$$

The first factor in the integrand represents the phase and amplitude of the wave incident on the edge; the second factor corresponds to the phase at the point of observation of the spherical wave reflected by the edge; the third factor determines the rather complicated angular dependence of the reflected wave. (r_s, ρ_s) is the angle of reflection at the edge; (n, r_s) is, so to speak, the angle of reflection at the surface of the cone; (ρ_s, ds) is the angle of incidence at the curve element of the edge.

Returning to eqs. (5 a, b), we can say in agreement with Thomas Young: according to eq. (5 a) the diffraction fringes in the illuminated region result from the interference of the incident light with a wave which is reflected by the edge; according to eq. (5 b) only this edge wave is present in the shadow region. We have improved Young's qualitative statement by defining the kind of reflection at the edge quantitatively in such a way that in both regions the resulting excitation v_P agrees exactly with that found by means of the Kirchhoff-Huygens theory. We have not, however, gone beyond the limits of validity of the Kirchhoff theory. Therefore the new formulae which have been adapted to Young's point of view are valid only if the wavelength is small compared to the diffraction opening and if the vectorial character of the electromagnetic problem does not come into play; see Sec. 46.

C. Discussion of the contour integral

Rubinowicz approximates the contour integral (13) by the method of *stationary phase* (the saddle-point method, simplified and adapted to a real domain): only those points on the boundary curve yield a substantial contribution to the integral at which the phase is stationary with respect to translation along the curve; the contributions of all other portions of the curve are small of higher order because of interference with the contributions of neighboring line elements. According to (13) the phase of the integrand on the boundary curve is

$$i\,k\,(\rho_s + r_s).$$

This remains constant under translation along the contour when

$$\frac{d\rho_s}{ds} = -\frac{dr_s}{ds}, \tag{14}$$

or, what is the same, when the "reflection condition"

$$\cos(\rho_s, ds) = -\cos(r_s, ds) \tag{14 a}$$

is satisfied. There are in general a finite number of points $s = s_1, s_2, \ldots$ on the curve which satisfy this condition. Each of these points radiates a substantial intensity to the point of observation P, and the line integral may be evaluated with sufficient accuracy as the sum of these radiations. The locus of points P which receive radiation from any one point s_ν on the edge is a circular half-cone with the apex at s_ν and with the axis ds. This has been proved experimentally by E. Maey[1] for the simple case of the half-plane, for which there exists only one such point s_1.

Thus Young's point of view, when analytically formulated, also leads to a quantitative explanation of diffraction phenomena — however, only within the same limits of applicability which already restrict Kirchhoff's method.

---

[1] Ann. d. Phys. (Lpz.) **49**, p. 93, 1893.

It is to be noted that the discontinuity of the diffraction field at the shadow boundary, which a comparison of (5 a) and (5b) seems to indicate, does not in fact exist. This apparent discontinuity is exactly compensated by a jump in the value of our line integral, which is due to the fact that the denominator $1 + \cos(r_s, \rho_s)$ in eq. (13) vanishes as one passes through the shadow boundary.

Finally, we rècall our consideration of "light fans" in Sec. 36 E, where these were explained from the point of view of Fresnel's zones. We now see that this phenomenon can be understood particularly well in terms of Young-Rubinowicz reflections. Each point s_ν on the diffraction edge radiates a conical light fan; when there are not merely discrete points s_ν but continuous sequences of such points, the light fans become particularly strong. This is the case for rectangles and more generally for polygonally bounded diffraction apertures, for then our cone surface f consists of plane portions which satisfy the reflection condition (14 a) along finite line segments. The specific intensity in these light fans is of the same order of magnitude as that of the incident light; the phenomenon of shadow formation thus disappears.

45. Diffraction Near Focal Points

From daily life we are well acquainted with the variously shaped caustics (focal lines) which appear within a teacup illuminated by a point source. These curves can be constructed by geometrical optics as the envelopes of pencils of rays. A more exact investigation of the vicinities of such lines leads to diffraction problems which have been treated particularly by Airy.

According to geometrical optics a focal point is an infinite concentration of rays. Wave optics resolves this (physically obviously inadmissible) singularity into a strong light concentration of finite amplitude and finite extent. In passing through a focal point a phase jump of magnitude π occurs. This jump has been studied experimentally by Gouy a d Sagnac, among others. In the case of a focal line of the type which results, for instance, from the convergence of the rays of a cylindrical wave, the phase j mp is $\pi/2$ instead of π. Rubinowicz[1] used his line integral (Sec. 44) as a starting point for the theoretical explanation of these phase jumps. He considered a ray bundle selected from a converging spherical or cylindrical wave by means of a diaphragm and treated its further course as a diffraction problem.

---

[1] A. Rubinowicz, Phys. Rev. **54**, 931, 1938; see also C. J. Bouwkamp, Physica **7**, 485, 1940.

We shall adopt a simpler method which is due to Debye[2]. He removed the diaphragm to infinity and in that way obtained a solution of the differential equation of optics which is valid in the whole space and exactly describes not only the phase jump but also the diffraction patterns in the vicinity of the focal point (or line). Debye's method is not limited to Kirchhoff's approximation but is based on the fundamentals of wave optics. His solution can claim the same degree of exactness as, for instance, our treatment of the problem of the straight edge in Sec. 38. There we had assumed as given the incident light at infinity in one half-space (plane wave) and had required that in the other half-space the radiation condition be fulfilled (hence no incident light). Correspondingly, Debye prescribes incident light (as a converging spherical wave) in one portion of infinity and requires that everywhere else at infinity no light shall be incident but shall only emerge.

A. The hypothesis of Debye

The expression

$$u = e^{-ikr\cos\Theta}, \quad \cos\Theta = \cos\vartheta\cos\vartheta_0 + \sin\vartheta\sin\vartheta_0\cos(\varphi-\varphi_0) \tag{1}$$

represents a plane wave coming from infinity, which is incident from the direction $\vartheta = \vartheta_0$, $\varphi = \varphi_0$ and which, after passing through the point $r = 0$, radiates toward infinity in the direction $\vartheta = \vartheta_0 + \pi$, $\varphi = \varphi_0$. As always, the time factor $\exp(-i\,\omega\,t)$ is to be thought of as added. $r\cos\Theta$ is a linear function of the coordinates

$$x = r\sin\vartheta\cos\varphi, \qquad y = r\sin\vartheta\sin\varphi, \qquad z = r\cos\vartheta$$

having the coefficients

$$\alpha = \sin\vartheta_0\cos\varphi_0, \qquad \beta = \sin\vartheta_0\sin\varphi_0, \qquad \gamma = \cos\vartheta_0,$$

the sum of the squares of which equals 1. Hence u satisfies the wave equation $\Delta u + k^2 u = 0$ in the whole x, y, z space including the point $r = 0$ which is not a singular point of u.

The same is true for the wave packet

$$U = \iint e^{-ikr\cos\Theta}\,d\Omega, \qquad d\Omega = \sin\vartheta_0\,d\vartheta_0\,d\varphi_0, \tag{2}$$

which represents *incident* waves only within the (arbitrarily defined) solid angle Ω. However this same expression (2) yields divergent waves for all ϑ, φ outside the solid angle Ω (and not merely in the solid angle which is diametrically opposite to Ω). This is due to the manner in which (2) was constructed as a superposition of the waves (1). Since it is an exact solution of the wave equation, U contains the answer to all questions regarding the behavior of the wave bundle in the vicinity of the focal point $r = 0$.

[2] Das Verhalten von Lichtwellen in der Nähe eines Brennpunktes oder einer Brennlinie. Ann. d. Phys. (Lpz.) **30**, 755, 1909.

It should be noted that no *boundary conditions* of any kind need to be satisfied. It is just this requirement, that boundary conditions be satisfied, which in other problems makes sclutions in closed form impossible. Thus in contrast to other diffraction problems, Debye's formulation of the problem of focal point diffraction involves a simple *summation method,* as is brought out clearly by the form of (2). To be sure, we have performed only a scalar summation in (2) and not a vector summation as required by the directional character of the electromagnetic light field. But Debye has shown that his expression can be applied without change to describe the rectangular components of the Hertz vector from which the vectorial optical field can be derived.

B. The diffraction field in the neighborhood of the focal point

First we shall show that the singularity of the light field at the focus which results from geometrical optics does not really exist, that according to wave optics the field is entirely regular. For the sake of brevity we shall make reference to several formulae from Vol. VI. According to eq. (VI. 22.35)

$$e^{-i\rho\cos\Theta} = \sum_{n=0}^{\infty} (2\,n+1)\,(-i)^n\,\psi_n(\rho)\,P_n(\cos\Theta). \tag{3}$$

where $\rho = k\,r$. The P_n are the Legendre polynomials

$$P_0(x) = 1, \qquad P_1(x) = x, \qquad P_2(x) = \frac{1}{2}(3\,x^2-1),\ \ldots,\ x = \cos\Theta; \tag{4}$$

according to Vol. VI, eq. (21.11) the ψ_n are the modified Bessel functions

$$\begin{aligned} \psi_n(\rho) &= \sqrt{\frac{\pi}{2\rho}}\,J_{n+\frac{1}{2}}(\rho) = \frac{\rho^n}{1\cdot 3\ldots(2\,n+1)}\left(1-\frac{\rho^2}{2\,(2\,n+3)}+\ldots\right) \\ \psi_0(\rho) &= \frac{\sin\rho}{\rho}, \qquad \psi_1(\rho) = \frac{\sin\rho-\rho\cos\rho}{\rho^2},\ \ldots \end{aligned} \tag{5}$$

If we neglect in (3) all powers of ρ higher than the second and perform the integration with respect to Ω indicated in (2), then we obtain, by using (4) and (5)

$$U = \left(1-\frac{\rho^2}{6}\right)\int d\Omega - i\,\rho\int\cos\Theta\,d\Omega - \frac{\rho^2}{3}\int P_2(\cos\Theta)\,d\Omega. \tag{6}$$

For convenience we define the solid angle Ω as the interior of a circular cone with its apex at the focus. Thus the integration is to be extended over

$$0 < \vartheta_0 < \alpha, \qquad -\pi < \varphi_0 < +\pi.$$

Then

$$\int d\Omega = 2\pi \int_0^\alpha \sin\vartheta_0\, d\vartheta_0 = 2\pi\,(1-\cos\alpha) = \Omega,$$

and recalling the meaning of $\cos\Theta$ from eq. (1) [the term containing $\cos(\varphi-\varphi_0)$ vanishes in the integration with respect to φ]

$$\int \cos\Theta\, d\Omega = 2\pi\cos\vartheta \int_0^\alpha \cos\vartheta_0 \sin\vartheta_0\, d\vartheta_0 = \pi\cos\vartheta\,(1-\cos^2\alpha)$$

$$= \frac{1}{2}\cos\vartheta\,(1+\cos\alpha)\,\Omega;$$

furthermore, using the addition theorem for spherical harmonics [Vol. VI, eq. (22.36)]

$$\int P_2(\cos\Theta)\, d\Omega = 2\pi\, P_2(\cos\vartheta) \int_0^\alpha P_2(\cos\vartheta_0 \sin\vartheta_0\, d\vartheta_0$$

$$= \pi\, P_2(\cos\vartheta) \int_0^\alpha (3\cos^2\vartheta_0 - 1)\sin\vartheta_0\, d\vartheta_0 = \pi\, P^2(\cos\vartheta)\cos\alpha\,(1-\cos^2\alpha)$$

$$= \frac{1}{2}\, P_2(\cos\vartheta)\cos\alpha\,(1+\cos\alpha)\,\Omega.$$

Substituting this in (6) one obtains

$$\text{(7)}\qquad \frac{U}{\Omega} = 1 - \frac{\rho^2}{6} - i\,\frac{\rho}{2}\cos\vartheta\,(1+\cos\alpha) - \frac{\rho^2}{6}\, P_2(\cos\vartheta)\cos\alpha\,(1+\cos\alpha).$$

For $\rho = 0$, the *finite* value

$$\text{(8)}\qquad U = \Omega$$

is obtained (for our particular normalization of the incident amplitude). Hence, *in contrast to geometrical optics there is no singularity*. Going on from U to determine $|U|^2$, one finds, consistently disregarding higher powers of ρ than the second,

$$\text{(9)}\qquad |U|^2/\Omega^2 = 1 - a_1\rho^2 + a_2\rho^2\cos^2\vartheta.$$

$$a_1 = \frac{1}{3} - \frac{1}{6}\cos\alpha\,(1+\cos\alpha),\quad a_2 = \frac{1}{4}(1-\cos^2\alpha),\quad a_1 - a_2 = \frac{1}{12}(1-\cos\alpha)^2.$$

We are interested in the region of large amplitude surrounding the origin which represents the wave-optical spreading of the focal point of geometrical optics. We consider the first surface of extinction $U = 0$ as the outer limit of this region. Calculating it from the approximation (9) and setting

$$\rho^2 = k^2(x^2+y^2+z^2), \qquad \rho\cos\vartheta = k\,z,$$

we obtain

$$k^2 a_1 (x^2 + y^2) + k^2 (a_1 - a_2) z^2 = 1. \tag{10}$$

This is the equation of an ellipsoid of revolution which is elongated in the direction of incidence. The smaller α is, the larger are the principal axes $1/k\sqrt{a_1}$, $1/k\sqrt{a_1 - a_2}$. This means that the focal region becomes more extended as the incident bundle of rays becomes narrower. It decreases in size with decreasing wavelength (increasing k).

C. Amplitude and phase along and near the axis of the light cone

On the axis of the light cone $\vartheta = 0$ in front of the focus and $\vartheta = \pi$ beyond the focus. Hence on the axis $\cos\Theta = \pm\cos\vartheta_0$. For these values the integration in (2) can be carried out by elementary methods and yields for $\vartheta = 0$

$$\frac{U}{2\pi} = \int_0^\alpha e^{-ikr\cos\vartheta_0} \sin\vartheta_0 \, d\vartheta_0 = \frac{e^{-ikr} - e^{-ikr\cos\alpha}}{-ikr}, \tag{11 a}$$

and for $\vartheta = \pi$

$$\frac{U}{2\pi} = \int_0^\alpha e^{+ikr\cos\vartheta_0} \sin\vartheta_0 \, d\vartheta_0 = \frac{e^{+ikr} - e^{+ikr\cos\alpha}}{+ikr}. \tag{11 b}$$

Both expressions agree for $r = 0$ with the value of U given by (8). They remind us of the elementary representation of the diffraction pattern along the central axis of a circular disc or opening which was derived in Sec. 35 C and D. Like the latter expressions, eqs. (11 a, b) are valid only on the axis of symmetry of the light bundle. We know from Sec. 35 C that the "Poisson spot" disappears only a short distance off the axis. The same is true of the interference patterns given by (11 a, b).

We shall prove this in a somewhat indirect way by differentiating eq. (2) with respect to the angle α. In the case of a circular light cone which we are now considering, α appears in (2) as an upper limit of integration. Thus the integration with respect to ϑ_0 is to be omitted and the integral with respect to φ_0 can be written in terms of the Bessel function J_0:

$$\frac{\partial U}{\partial \alpha} = e^{-ikr\cos\vartheta\cos\alpha} \int_0^{2\pi} e^{-ikr\sin\vartheta\sin\alpha\cos(\varphi-\varphi_0)} \sin\alpha \, d\varphi_0$$

$$= 2\pi \sin\alpha \, e^{-ikr\cos\vartheta\cos\alpha} J_0(kr\sin\vartheta\sin\alpha).$$

For all values of r which are physically of interest the argument of J_0 is a very large number except when $\vartheta = 0$ or π (or also when $\alpha = 0$ or π). We know, however, that J_0 vanishes for large real arguments; see, for instance Vol. VI, eq. (20.57). Therefore

$$\frac{\partial U}{\partial \alpha} = 0 \quad \text{for all } \vartheta \text{ except } \vartheta = 0 \quad \text{and} \quad \vartheta = \pi.$$

Except on these two half-rays U is, at infinity, independent of the angle α. From this we conclude that on the right-hand sides of (11 a) and (11 b) the second term in each of the numerators is to be neglected. Thus we obtain

$$\text{(12 a)} \qquad \frac{U}{2\pi} = \frac{e^{-ikr}}{-ikr} \quad \text{for} \quad 0 < \vartheta \leqq \pi/2,$$

$$\text{(12 b)} \qquad \frac{U}{2\pi} = \frac{e^{+ikr}}{+ikr} \quad \text{for} \quad \pi/2 \leqq \vartheta < \pi.$$

In both cases the interferences have disappeared; at large distances in front of and behind the focal point the light propagates as a spherical wave just as in geometrical optics.

We are mainly interested in the phase factors in eqs. (12 a, b), namely

$$-\frac{1}{i} = e^{+i\pi/2} \qquad \text{in (12 a) and} \qquad +\frac{1}{i} = e^{-i\pi/2} \qquad \text{in (12 b).}$$

The phase jumps by π at the focal point as stated at the beginning of this paragraph. Since this phase change is in no way connected with the amplitude pattern at the focus but refers to the state of the spherical wave at large distances from the focus, we are justified in considering it as a property of geometrical optics; see Rubinowicz, loc. cit.

D. The cylindrical wave and its phase jump

The two-dimensional analogue of the above phenomenon concerns a ray bundle propagating in the r, φ-plane

$$\text{(13)} \qquad U = \int_{-\alpha}^{+\alpha} e^{-ikr\cos(\varphi - \varphi_0)}\, d\varphi_0,$$

which, coming from infinity within the range $-\alpha < \varphi_0 < +\alpha$, proceeds toward the origin. In the language of three dimensions this origin is a geometrical focal line which is perpendicular to the r, φ-plane. We shall see again that wave-optically no singularity appears. For the proof we use formula (21.2 b) of Vol. VI:

$$\text{(14)} \qquad e^{-i\rho\cos\psi} = J_0(\rho) + 2\sum_{n=1}^{\infty} (-i)^n J_n(\rho) \cos n\psi.$$

The J are ordinary Bessel functions with integer indices. Neglecting higher powers of ρ, we approximate these Bessel functions by

$$J_0 = 1 - \left(\frac{\rho}{2}\right)^2, \qquad J_1 = \frac{\rho}{2}, \qquad J_2 = \frac{1}{2}\left(\frac{\rho}{2}\right)^2$$

and obtain from (13) and (14)

$$\frac{U}{2\,\alpha} = 1 - \frac{\rho^2}{4} - i\,\rho\cos\varphi\,\frac{\sin\alpha}{\alpha} - \frac{\rho^2}{4}\cos 2\,\varphi\,\frac{\sin 2\,\alpha}{2\,\alpha}\,. \tag{15}$$

At $\rho = 0$, U has (in contrast to geometrical optics) the *finite* value

$$U = 2\,\alpha. \tag{15 a}$$

In order to find the intensity distribution for small ρ, we calculate $|U|^2$, limiting ourselves consistently to quadratic terms in ρ. Using $k\,x = \rho\cos\varphi$, $k\,y = \rho\sin\varphi$, we find

$$|U|^2/4\,\alpha^2 = 1 - k^2\,a_1\,x^2 - k^2\,a_2\,y^2, \tag{16}$$

$$a_1 = \frac{1}{2}\left(1 + \frac{\sin 2\,\alpha}{2\,\alpha} - 2\,\frac{\sin^2\alpha}{\alpha^2}\right), \qquad a_2 = \frac{1}{2}\left(1 - \frac{\sin 2\,\alpha}{2\,\alpha}\right).$$

Setting $|U| = 0$ we obtain a measure of the size of the focal spot in the x, y-plane. According to (16), $|U| = 0$ on an ellipse whose major axis $1/k\sqrt{a_1}$ lies in the direction of incidence.

While in the three-dimensional case we obtained incident and radiating *spherical waves* (12 a, b) at large distances in front of and behind the focus, we now obtain for $k\,r \gg 1$ unperturbed convergent and divergent *cylindrical waves* in the region of the light bundles. These are represented by Hankel functions of the second and first kind, respectively,

$$H_0^2\,(k\,r) \cong \sqrt{\frac{2}{\pi\,k\,r}}\,e^{-i\,(k\,r - \pi/4)} \tag{17 a}$$

and

$$H_0^1\,(k\,r) \cong \sqrt{\frac{2}{\pi\,k\,r}}\,e^{+\,i\,(k\,r - \pi/4)}. \tag{17 b}$$

Here we have restricted ourselves to the asymptotic representations of Vol. VI, eqs. (19.55) and (19.56), which are the only ones of interest to us. From these we conclude that in passing through the focus the phase jumps from $\exp\,(+\,i\,\pi/4)$ in eq. (17 a) to $\exp\,(-\,i\,\pi/4)$ in eq. (17 b). *The phase jump in a cylindrical wave amounts to* $\pi/2$.

Actually the phase, like the amplitude, varies continuously in the vicinity of the focus; the "phase jump" appears only because in (12 a, b) and (17 a, b) we have compared the phases at points very far in front of and behind the focus.

46. The Huygens' Principle of the Electromagnetic Vector Problem

We have at various times pointed out the difference between the scalar acoustic and the vectorial optical problems. By limiting ourselves to a two-dimensional case in deriving the exact solutions of Sec. 38 and 39, we were able to reduce the vector problem to that of two scalar problems which described two different polarizations of the incident light. Huygens' principle is entirely *scalar* from the very beginning, so is the Fresnel-Kirchhoff application of Huygens' principle. We shall here, very briefly, discuss the *vectorial* formulation of this principle.

The problem is to calculate the vector fields E, H behind a diffraction opening σ in an opaque screen when the tangential components of E and H in the opening are given. We shall denote the latter, considered as vectors, by E_0 and H_0. If these were known *exactly*, then it would be possible to calculate E and H exactly. If, instead of the exact values, the unperturbed E, H-values of the incident wave are used, one obtains only a first approximation which is valid for small wavelengths, just as in the Fresnel-Kirchhoff treatment.

First, we shall state the formulation of the vectorial Huygens' principle which is due to W. Franz[1]:

$$(1) \qquad 4\pi\,\mathsf{E} = \operatorname{curl}\int\limits_{\sigma}[d\boldsymbol{\sigma}\times\mathsf{E}_0]\,\frac{e^{ikR}}{R} - \frac{1}{i\,\omega\,\varepsilon}\operatorname{curl}\operatorname{curl}\int\limits_{\sigma}[d\boldsymbol{\sigma}\times\mathsf{H}_0]\,\frac{e^{ikR}}{R},$$

$$(2) \qquad 4\pi\,\mathsf{H} = \operatorname{curl}\int\limits_{\sigma}[d\boldsymbol{\sigma}\times\mathsf{H}_0]\,\frac{e^{ikR}}{R} + \frac{1}{i\,\omega\,\varepsilon}\operatorname{curl}\operatorname{curl}\int\limits_{\sigma}[d\boldsymbol{\sigma}\times\mathsf{E}_0]\,\frac{e^{ikR}}{R}.$$

The following will explain the notation:

a) $d\boldsymbol{\sigma}$ is an element of area with which is associated the direction perpendicular to the surface σ; the vector product of $d\boldsymbol{\sigma}$ and E_0 is therefore a vector lying in the tangential plane of σ. The normal component of E_0 does not enter into the calculation of this vector product. The same holds for the vector product of $d\boldsymbol{\sigma}$ and H.

b) R is the distance from the point of observation x y, z to the point of integration ξ, η, ζ. E, H are functions of x, y, z; E_0, H_0 are functions of ξ, η, ζ.

---

[1] ZS. f. Naturforschung, Vol. **3a**, 500, 1948; we cannot here discuss the resulting corrections to the Kirchhoff diffraction calculations. See also Stratton and Chu, Phys. Rev. Vol. **56**, 99, 1939, as well as the book by J. A. Stratton, Electromagnetic Theory, Internat. Series in Pure and Appl. Physics, New York, 1941.

c) The eqs. (1) and (2) therefore represent superpositions of spherical waves radiating from the points ξ, η, ζ just as in the elementary Huygens' principle; these waves interact at the point x, y, z, but now they combine vectorially.

d) The operation curl is everywhere to be performed with respect to the coordinates of the point of observation x, y, z. If cartesian coordinates are chosen, the identity

$$\text{curl curl} = \text{grad div} - \Delta \tag{3}$$

may be used, as is done in the following.

e) In Franz's work the arrow of $d\sigma$ points toward the back of σ, that is, into the region where E and H are to be determined.

f) The values of ε, μ are everywhere constant and can be identified with the values in vacuum.

We now show that the E and H as defined in (1) and (2) satisfy Maxwell's equations. We write the latter for a purely periodic state of frequency ω:

$$\text{curl } \mathsf{E} = i\,\omega\,\mu\,\mathsf{H}, \qquad \text{curl } \mathsf{H} = -i\,\omega\,\varepsilon\,\mathsf{E}. \tag{4}$$

Next we form curl E from eq. (1). Then the first term on the right-hand side becomes identical with the second term of eq. (2) multiplied by $i\,\omega\,\mu$. Using cartesian coordinates temporarily and applying (3), the second term of curl E becomes

$$-\frac{1}{i\,\omega\,\varepsilon}\,\text{curl (grad div} - \Delta) \int\limits_\sigma [d\boldsymbol{\sigma} \times \mathsf{H}_0]\,\frac{e^{ikR}}{R}.$$

Curl grad vanishes and there remains

$$\frac{1}{i\,\omega\,\varepsilon}\,\text{curl} \int [d\boldsymbol{\sigma} \times \mathsf{H}_0]\,\Delta\,\frac{e^{ikR}}{R} = -\frac{k^2}{i\,\omega\,\varepsilon}\,\text{curl} \int [d\boldsymbol{\sigma} \times \mathsf{H}_0]\,\frac{e^{ikR}}{R}, \tag{5}$$

since $u = e^{ikR}/R$ satisfies the wave equation $\Delta\,u + k^2\,u = 0$. Moreover

$$k^2 = \omega^2/c^2 = \varepsilon\,\mu\,\omega^2, \qquad \text{hence} \qquad -k^2/i\,\omega\,\varepsilon = i\,\omega\,\mu.$$

Accordingly, (5) is identical with the first term on the right side of (2) multiplied by $i\,\omega\,\mu$. Thus we have proved that eqs. (1) and (2) satisfy the first Maxwell eq. (4). Quite analogously it can be verified that the second eq. (4) is also fulfilled.

But as Franz points out, this representation satisfies the boundary conditions

$$\mathsf{E} \to \mathsf{E}_0, \qquad \mathsf{H} \to \mathsf{H}_0 \qquad \text{as} \qquad x, y, z \to \sigma \tag{6}$$

only approximately.

In contrast to this we now show that the integral representation

$$2\pi\,\mathsf{E} = \operatorname{curl}\int_\sigma [d\boldsymbol{\sigma} \times \mathsf{E}_0]\,\frac{e^{ikR}}{R}, \tag{7}$$

$$2\pi\,\mathsf{H} = \operatorname{curl}\int_\sigma [d\boldsymbol{\sigma} \times \mathsf{H}_0]\,\frac{e^{ikR}}{R} \tag{8}$$

satisfies the boundary conditions (6) but not the differential eqs. (4).

We shall merely make this statement plausible by assuming that E_0, H_0 are differentiable functions of position everywhere in the opening (including its boundary!) and by restricting ourselves to the case of a *plane* screen and hence a *plane* opening σ. We choose as the origin of a cartesian ξ, η, ζ-coordinate system the point O on σ, which is the point that P is to approach. The ζ-axis is to be perpendicular to $d\boldsymbol{\sigma}$, and its positive direction is to coincide with the direction of $d\boldsymbol{\sigma}$. The x, y, z-system in which the point of observation P is defined shall be parallel to the ξ, η, ζ-system. With this choice of coordinate system the three components of the vector product $[d\boldsymbol{\sigma} \times \mathsf{E}_0]$ have the values

$$(-\,\mathsf{E}_{o\,y},\ \mathsf{E}_{o\,x},\ 0)\ d\xi\, d\eta. \tag{9}$$

We now form

$$\operatorname{curl}_x\left\{[d\boldsymbol{\sigma} \times \mathsf{E}_0]\,\frac{e^{ikR}}{R}\right\} = \left\{[d\boldsymbol{\sigma} \times \mathsf{E}_0]_z\,\frac{\partial}{\partial y} - [d\boldsymbol{\sigma} \times \mathsf{E}_0]_y\,\frac{\partial}{\partial z}\right\}\frac{e^{ikR}}{R}. \tag{9 a}$$

According to (9) this is equal to

$$= -\,d\xi\, d\eta\ \mathsf{E}_{0\,x}\,\frac{\partial}{\partial z}\,\frac{1}{R},$$

where, in this last expression, we have set $e^{ikR} \sim 1$ in the neighborhood of O. We now calculate the right-hand side of the x-component of (7) and obtain:

$$-\int\!\!\int d\xi\, d\eta\ \mathsf{E}_{0\,x}\,\frac{\partial}{\partial z}\,\frac{1}{R} = \int\!\!\int \mathsf{E}_{0\,x}\,\frac{z}{R}\,\frac{d\xi\, d\eta}{R^2}. \tag{10}$$

On the right, z/R is the cosine of the angle between the directions z and R; hence $d\xi\, d\eta\, z/R$ is the projection of $d\xi\, d\eta$ onto the sphere of radius R centered at P. Divided by R^2 this yields the solid angle which $d\xi\, d\eta$ subtends at P. As $P \to O$ this solid angle approaches 2π in the vicinity of O; in the more distant parts of the ξ, η-plane it goes to zero. Hence (10) becomes equal to $2\pi\ \mathsf{E}_{0x}$. In the same way the y-component of (7) may be checked.

The same calculation, applied to (8) shows that the boundary condition (6) is satisfied also for H. If one substitutes the expressions (7) and (8) into the differential eqs. (4), and if one performs suitable integrations by parts and applies eq. (34.20), the resulting surface integrals on both sides are found to be equal. But there also appear line integrals over the boundary of σ ("magnetic currents" as Stratton calls them, loc. cit.) which do not compensate one another.

The above remarks do not, of course, solve the problem posed in this paragraph; they only describe its general aspects. In any case, an exact solution of the diffraction problem could be obtained only if the exact boundary values $\mathbf{E}_0$ and $\mathbf{H}_0$ (or more correctly $\mathbf{E}_0$ *or* $\mathbf{H}_0$) were known. This is, of course, not the case. Rather, these boundary values can only be found simultaneously with the solution of the diffraction problem. Even in the case of the circular opening, the two auxiliary functions (potentials) which must be introduced to solve the problem are coupled in a complicated manner[1]. *The vectorial Huygens' principle is no magic wand for the solution of boundary value problems,* but it is of interest as a generalization of the time-honored idea of Christian Huygens.

47. Cerenkov Radiation

According to the theory of relativity a material body cannot possibly move with a velocity v greater then the speed of light c. However, we know that in a medium of index of refraction n, light propagates with the phase velocity $u = c/n < c$; see Sec. 2. Hard cathode rays and Compton electrons produced by very hard γ-rays can attain velocities in the range

$$u < v < c.$$

What happens in this velocity range?

We might expect to find phenomena which are known to us from the field of ballistics; see Vol. II, fig. 45 a, b: The projectile overtakes the pressure wave which it produces and thus causes a Mach cone with the characteristic angle $\sin\vartheta = c/v$ (c = velocity of sound). While this phenomenon can be derived only with some difficulty from the non-linear equations of aerodynamics, the corresponding electro-optical phenomenon follows simply and rigorously from Maxwell's equations.

The following discussion will adhere closely to an early paper by the author which was communicated to the Amsterdam Academy[2] by H. A. Lorentz; the only difference being that where that paper (which preceded the theory of relativity!) deals with velocities below and above that of light, $v < c$ and $v > c$, we must now substitute $v < u$ and $u < v < c$, respectively. For velocities below that of light the electron carries its own field with it and no

[1] J. Meixner, ZS. f. Naturforschung, Vol. **3a**, 506, 1948.

[2] Proc. Nov. 26, 1904, particularly p. 359 where older papers by Heaviside and Des Coudres are referred to; see also Göttinger Nachrichten 1905, p. 201.

energy is radiated; see Vol. III, Sec. 30 A. However for velocities above that of light the electron leaves its field behind in the shape of a Mach cone. The field radiates in directions perpendicular to the surface of the cone, and because of the nature of dispersion, this radiation consists mainly of visible light. Since the electron looses energy in the form of radiation, its velocity decreases rapidly to the value of the light velocity $v = u$. The radiated light is polarized so that the electric vector lies in the plane passing through the trajectory of the electron.

This radiation, then an unheard-of optical phenomenon, was first observed in 1934 by P. A. Cerenkov[1]. At first he used Compton electrons and later many different types of cathode rays. These observations were soon thereafter explained by Frank and Tamm[2] in the manner here indicated and were compared quantitatively with the theory.

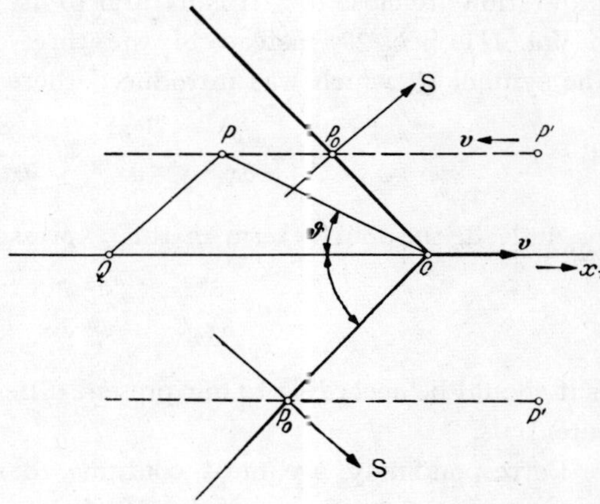

Fig. 91.
The Mach cone of the Cerenkov electron and its radiation vector **S**.

A. The field of the Cerenkov electron

Let us assume that an electron moves with a velocity v (to be thought of as constant) through a medium of refractive index $n > 1$ such that

$$u < v < c, \qquad u = c/n. \tag{1}$$

At the time $t = 0$ let the electron be at the point O, and let Q be the position of the electron at the earlier time $\tau < 0$. We are to find the field of the electron at an arbitrary point P at the time $t = 0$. Let

$$r = \text{distance}\, OP, \qquad \vartheta = \text{angle}\, QOP \tag{2}$$

(see fig. 91). Choosing the x-axis in the direction of motion of the electron, we denote the space-time coordinates of P by

$$x_1 = -r\cos\vartheta, \qquad x_2 = r\sin\vartheta\cos\varphi, \qquad x_3 = r\sin\vartheta\sin\varphi, \qquad x_4 = i\,c\,t = 0 \tag{3}$$

[1] Ac. Sci. USSR., **2**, 451, 1934, **3**, 414, 1936, **20** et seq. Phys. Rev. **52**, 378, 1937.
[2] C. R. Ac. Sci. USSR. **14**, 109, 1937; Tamm, Journ. of Sci. USSR. **1**, 409, 1939.

and those of Q by

(3a) $$\xi_1 = v\tau < 0, \qquad \xi_2 = \xi_3 = 0, \qquad \xi_4 = i c \tau.$$

As we saw in Sec. **2**, in a medium of refractive index n the light velocity c occurring in the wave equations for **E** and **H** is to be replaced by the phase velocity $u = c/n$. The same substitution must be made in the differential equations for the retarded potentials in Vol. III, Sec. 19. Since the velocities v in question are close to c, it is natural to use the four-dimensional potential Ω of Vol. III, Sec. 29 instead of the three-dimensional retarded potentials. The symbol $\Box$ which was introduced there must now be amended to mean

(4) $$\Box = \frac{\partial^2}{\partial x_1{}^2} + \frac{\partial^2}{\partial x_2{}^2} + \frac{\partial^2}{\partial x_3{}^2} + n^2 \frac{\partial^2}{\partial x_4{}^2};$$

for, indeed, the fourth term in this expression is

$$n^2 \frac{\partial^2}{\partial x_4{}^2} = -\frac{n^2}{c^2}\frac{\partial^2}{\partial t^2} = -\frac{1}{u^2}\frac{\partial^2}{\partial t^2},$$

as it should be according to our present differential equations for the retarded potentials.

Correspondingly, we must continue to replace c wherever it occurs by $u = c/n$. Thus the "current density four-vector Γ" introduced in Vol. III, eq. (28.16) is now defined as

(4 a) $$\Gamma = (\rho\, \mathbf{v}, i\, u\, \rho) = (\rho\, v, 0, 0, i\, u\, \rho),$$

where ρ is the charge density of the electron. The connection between Ω and the retarded potentials $\mathbf{A}$ and ψ which in vacuum was $\Omega = (\mathbf{A}, i\,\psi/c)$ by eq. (26.4), Vol. III is now to be changed to

(4 b) $$\Omega = (\mathbf{A}, i\,\psi/u).$$

With these changes in the definitions, the differential equation of our problem reads exactly as in Vol. III, Sec. **26**:

(4 c) $$\Box\, \Omega = -\mu_0\, \Gamma.$$

In order to integrate (4 c) we require, as in Vol. III, Sec. 29, a solution of the differential equation $\Box\, U = 0$ which is infinite at $P = Q$. This solution is now defined by

(5) $$U = \frac{1}{R^2}, \qquad R^2 = (\xi_1 - x_1)^2 + (\xi_2 - x_2)^2 + (\xi_3 - x_3)^2 + \frac{1}{n^2}(\xi_4 - x_4)^2.$$

The reader may convince himself that this function satisfies the differential equation $\Box\, U = 0$ in both sets of quadruple variables except at $P = Q$.

Applying Green's theorem, one obtains an expression identical with eq. (29.6), Vol. III except that the factor c occurring in ξ_4 must again be replaced by $u = c/n$. This results in the occurrence of the denominator n in the following equation:

$$4\pi^2 \Omega/\mu_0 = \int \Gamma \frac{d\xi_1\, d\xi_2\, d\xi_3\, d\xi_4}{n\, R^2}.$$

If the electron is considered to be a point and the integration with respect to ξ_1, ξ_2, ξ_3 is carried out, one obtains

(6) $$4\,\pi^2\,\Omega_1/\mu_0 = \frac{e\,v}{n}\int \frac{d\xi_4}{R^2}, \qquad \Omega_2 = \Omega_3 = 0,$$

$$4\,\pi^2\,\Omega_4/\mu_0 = \frac{i\,e\,u}{n}\int \frac{d\xi_4}{R^2}.$$

As shown in the text accompanying fig. 41, Vol. III, the integration with respect to ξ_4 must be performed by evaluating the contour integral around the negative-imaginary ξ_4-axis because the position of the electron is given only for $\tau < 0$. This integral is non-vanishing only if there are points on this semi-axis at which the denominator R^2 vanishes. According to (3), (3 a) and (5) the condition for this is

(7) $$R^2 = (v^2 - u^2)\,\tau^2 + 2\,v\,r\,\tau\cos\vartheta + r^2 = 0.$$

The roots $\tau_\pm$ of this quadratic equation are given by

(7 a) $$(v^2 - u^2)\,\tau_\pm = -\,v\,r\left(\cos\vartheta \pm \sqrt{\frac{u^2}{v^2} - \sin^2\vartheta}\right).$$

We see, therefore, that for

(8) $$\vartheta > \vartheta_M, \qquad \text{where } \vartheta_M = \text{Mach angle, } \sin\vartheta_M = \frac{u}{v},$$

no real roots of (7) exist. Therefore, in this region the integrals in (6) vanish and hence not only $\Omega_2 = \Omega_3 = 0$ but also $\Omega_1 = \Omega_4 = 0$. *The field is zero everywhere outside the Mach cone.*

On the other hand, for all points P within the Mach cone, (7 a) has *two real negative solutions* (also τ_- is negative because $v > u$); *both* points for which $R = 0$ lie on the negative imaginary ξ_4-axis and hence contribute to the integral. These contributions are, moreover, equal when the integration is performed in *opposite* directions around the two points[1]. *An electromagnetic field exists everywhere in the interior of the Mach cone.*

---

[1] If the directions of integration were the *same*, the sum of the two contributions would be zero. Thus we would not obtain an actual solution of the differential eq. (1). Therefore the path must be defined as a lemniscoid loop about the two points $\tau_\pm$; this is certainly permissible.

Before determining this field, we shall consider the connection between the times $\tau_\pm$ calculated in (7 a) and the so-called relaxation time of the retarded potentials, the "retardation" time of Vol. III, eq. (19.13 c)

$$\tau = r_{PQ}/c. \tag{8 a}$$

Here we have used r_{PQ} instead of the letter r of Vol. III in order to emphasize that what is meant is the distance between the position Q of the electron at time $t = \tau$ and the point P at which the field is being observed at the time $t = 0$. Using the notation of fig. 91 this distance is, according to the theorem of Pythagoras (r means here the distance PO):

$$r_{PQ} = \sqrt{v^2\tau^2 + r^2 + 2\,v\,r\,\tau\cos\vartheta}.$$

This, substituted in (8 a), where however c is replaced by u, yields indeed eq. (7). We note in this connection that *both* light points L and L' of fig. 41 of Vol. III are given by one and the same quadratic equation, namely our eq. (7); the only difference is that while the two points had different signs in Vol. III, they are now both negative. We also note that for $u = c$ the Mach angle (8) becomes imaginary and that is why there was *no* Mach cone in the problems in Vol. III.

We return now to the representation (6) of the four-vector potential. In order to evaluate the integrals involved we rewrite the expression (7) for R in the form

$$R^2 = (v^2 - u^2)\,(\tau - \tau_+)\,(\tau - \tau_-).$$

In evaluating the loop integral around $\tau = \tau_+$ we can replace the factor $\tau - \tau_-$ by $\tau_+ - \tau_-$. If we also write $d\xi_4 = i\,c\,d\tau$, we obtain

$$\oint \frac{d\xi_4}{R^2} = \frac{i\,c}{(v^2 - u^2)\,(\tau_+ - \tau_-)} \oint \frac{d\tau}{\tau - \tau_+} = \frac{-2\pi c}{(v^2 - u^2)\,(\tau_+ - \tau_-)}\,. \tag{9}$$

According to (7 a) the denominator of this fraction is equal to

$$-2\,v\,r\left(\frac{u^2}{v^2} - \sin^2\vartheta\right)^{1/2}$$

and therefore (9) becomes

$$\oint \frac{d\xi_4}{R^2} = \frac{\pi c}{v\,r}\left(\frac{u^2}{v^2} - \sin^2\vartheta\right)^{-1/2}.$$

If we carry out the same calculation for the zero $\tau = \tau_-$, then $\tau_- - \tau_+$ takes the place of $\tau_+ - \tau_-$ in the denominator of (9). Therefore, with opposite directions of integration the two contributions are equal as stated above. One obtains therefore from (6)

$$2\pi\,\Omega_1/\mu_0\,e\,u\,v = 2\pi\,\Omega_4/\mu_0\,i\,e\,u^2 = \frac{1}{v\,r}\left(\frac{u^2}{v^2} - \sin^2\vartheta\right)^{-1/2} \tag{10}$$

$$= (u^2 r^2 - v^2 r^2 \sin^2\vartheta)^{-1/2},$$

or written in terms of the coordinates x_1, x_2, x_3 of the point of observation

$$(10\text{ a})\quad 2\pi\,\Omega_1/\mu_0\, e\, u\, v = 2\pi\,\Omega_4/\mu_0\, i\, e\, u^2 = \{u^2\, x_1{}^2 - (v^2 - u^2)\,(x_2{}^2 + x_3{}^2)\}^{-1/2}.$$

The equipotential surfaces are hyperboloids; they take the place of the Heaviside ellipsoids of Vol. III, Sec. 242, footnote 1. The field is finite and continuous everywhere inside the Mach cone; the reason why according to (10) the field becomes infinite on the surface of the cone ($\sin\vartheta = u/v$) is that we have assumed the electron to be a point charge; for an electron of finite radius a the field would attain only a maximum of the order of magnitude $1/a$ [1].

We have derived the existence of the Mach cone from the phenomenological theory of a continuum. Physically, the cone, like the refractive index is caused by the molecular structure of matter. We need only consider that the electron moving with the velocity $v > u$ arrives at the molecules in its path sooner than the radiation emitted by the previously excited molecules. If dispersion is taken into account (see below), then the difference between the *phase* and *group velocity* affects the size of the Mach cone in an interesting way: The velocity of the wave fronts emitted by the previously excited molecules is not equal to the phase velocity u, but according to Sec. 22, it equals the group velocity $g < u$. Therefore the Mach angle is somewhat smaller than the value given by (8) because u must be replaced by g. A precise measurement of the angle of the Cerenkov waves (see fig. 92 below) should show this[2].

B. The radiation of the Cerenkov electron

So far we have considered only the instantaneous picture of the field produced by the electron which we were able to characterize arbitrarily by the time coordinate $t = 0$. Therefore the field has apparently been independent of t. But for an observer in the laboratory, the field obviously depends on t. If the observer is at the point P' in fig. 91, he sees nothing until the surface of the Mach cone reaches him. For convenience fig. 91 has been drawn as if the observer were travelling into the cone along the dotted straight line $P'\,P$

[1] See the above-mentioned paper in the Proc. of the Amsterdam Academy. There it will also be seen that the Mach cone is surrounded by a *marginal zone* of width 2 a in which the field decreases continuously from its maximum on the cone to zero on the outside.

[2] A full discussion of this statement has recently been given by H. Motz and L. I. Schiff, Am. Journ. Phys. **21** (1953), p. 258. This paper also shows that the discussion given in the following sections B.-D. cannot be accepted without modification. (Translator).

with a velocity opposite to v. At the point P_0 on the surface of the cone the observer perceives the maximum field (in our approximation, actually an infinite field). Also at all interior points P a field is observed which becomes weaker with increasing distance from P_0.

Mathematically, we obtain this time dependence simply by replacing in the above formulae

$$x_1 \quad \text{by} \quad x - v\,t.$$

That is, we consider the point O in fig. 91 not as fixed but as moving with time. In addition we shall write y, z instead of x_2, x_3 and express Ω in terms of the real potentials A, ψ in the manner of eq. (4 b). Then eq. (10 a) becomes

$$(11) \quad 2\pi\,\mathsf{A}_x/\mu_0\,e\,u\,v = 2\pi\,\psi/\mu_0\,e\,u^3 = \{u^2\,(x - v\,t)^2 - (v^2 - u^2)\,(y^2 + z^2\}^{-1/2}.$$

From this it is easy to establish the validity of the relationship

$$(12) \qquad \frac{\partial}{\partial t}\mathsf{A}_x = -\frac{v^2}{u^2}\frac{\partial}{\partial x}\psi.$$

We use this to calculate the electric field E from the relation in Vol. III, eq. (19.7) $\mathsf{E} = -\dot{\mathsf{A}} - \nabla\psi$ and obtain

$$
(13) \qquad
\begin{aligned}
\mathsf{E}_x &= \left(\frac{v^2}{u^2} - 1\right)\frac{\partial\psi}{\partial x} = \frac{\mu_0\,e\,u^3}{2\pi}\,(v^2 - u^2)\,(x - v\,t)\left\{\ \right\}^{-3/2}\\
\mathsf{E}_y &= \quad - \quad \frac{\partial\psi}{\partial y} = \frac{\mu_0\,e\,u^3}{2\pi}\,(v^2 - u^2)\,y\left\{\ \right\}^{-3/2},\\
\mathsf{E}_z &= \quad - \quad \frac{\partial\psi}{\partial z} = \frac{\mu_0\,e\,u^3}{2\pi}\,(v^2 - u^2)\,z\left\{\ \right\}^{-3/2},
\end{aligned}
$$

where $\{\ \}$ stands for the expression inside the curly brackets of eq. (11). Thus the electric field has the direction of the vector

$$x - v\,t,\ y,\ z$$

which points from the instantaneous position of the electron to the point of observation. *If, in particular, the latter has the position P_0 of fig. 91, then the direction of the electric field coincides with a generator of the Mach cone.*

To calculate H the relation of Vol. III, eq. (19.6) $\mathsf{B} = \text{curl}\,\mathsf{A}$ is used. According to (11) this yields besides $\mathsf{H}_x = 0$

$$
(14) \qquad
\begin{aligned}
\mathsf{H}_y &= \frac{1}{\mu_0}\frac{\partial\mathsf{A}_x}{\partial z} = \frac{e\,u\,v}{2\pi}\,(v^2 - u^2)\,z\left\{\ \right\}^{-3/2},\\
\mathsf{H}_z &= -\frac{1}{\mu_0}\frac{\partial\mathsf{A}_x}{\partial y} = -\frac{e\,u\,v}{2\,\pi}\,(v^2 - u^2)\,y\left\{\ \right\}^{-3/2}.
\end{aligned}
$$

The magnetic field lines therefore form circles around the trajectory of the electron. *At the point P_0 the direction of H is tangential to a circular cross-section of the Mach cone.*

The directions of $\mathbf{E}$ and $\mathbf{H}$ determine the direction of the ray $\mathbf{S} = \mathbf{E} \times \mathbf{H}$. *At the point P_0 the ray direction is perpendicular to the Mach cone. The radiated light is polarized*; the electric vector lies in the plane passing through the trajectory of the electron. Thus we have proved our initial statements about the character of the Cerenkov radiation. We still have to show that the spectrum lies principally in the *visible frequency range*. We note also that the radiation proceeds almost like a shock wave because the factor $\{\ \}^{-3}$ which occurs in the product of $\mathbf{E}$ and $\mathbf{H}$ is large only in the immediate vicinity of the surface of the Mach cone.

To produce Cerenkov radiation it is best to use a thin resin plate on which the electrons impinge perpendicularly. Thus only a small portion of the Mach cone appears. The emitted radiation fills a thin annular cone perpendicular to that portion; see fig. 92. The exposed area on a photographic plate placed behind the resin plate renders the annular trace of this cone visible.

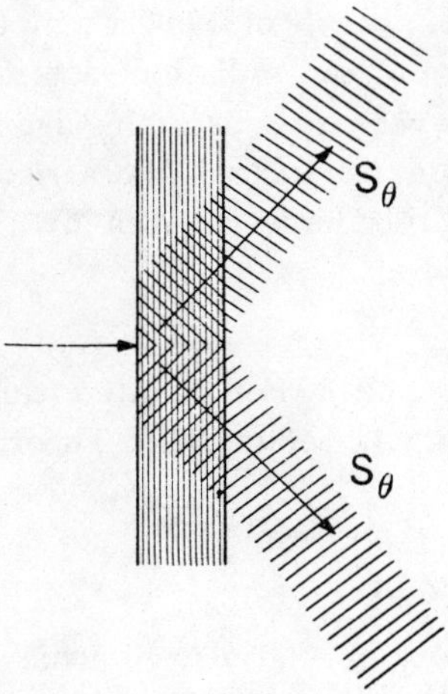

Fig. 92.
Observation of Cerenkov radiation behind a plate of large dielectric constant.

C. Cerenkov Radiation with Dispersion Taken into Account

So far we have carried out all calculations as if the index of refraction were a fixed number. Actually the refractive index is a function of frequency which rapidly approaches the value 1 in the far ultraviolet. But as n approaches 1, the interval $u < v < c$ in which the Cerenkov effect can take place shrinks to zero.

Therefore our previous treatment does not yet suffice for a quantitative analysis of experimental observations. In order to complete the calculations we would have to decompose the time dependence of the radiation field into its Fourier components in terms of ωt, and each of these components would have to be provided with its own $n(\omega)$. We would then find that only those Fourier components which lie in the visible spectrum make a noticeable contribution to the Cerenkov effect, while the ultraviolet spectrum is unable to excite Cerenkov radiation. Therefore we can state what has already been indicated in the introduction: *the Cerenkov effect makes electrons visible.*

However, owing to the singularity of the field on the Mach cone, the Fourier analysis involves certain formal difficulties. These can be avoided if, instead of a point-electron, an electron of finite extent (radius a) is used (see footnote 1 on p. 341), for then the singularity is smoothed out into a transition zone with finite field strength. We cannot present these somewhat involved calculations here. Tamm (loc. cit.) overcame the above-mentioned difficulties by using so-called δ-functions[1]; but these calculations, too, are complicated.

It is interesting to note that because in quantum mechanics one is forced to decompose the electromagnetic field into its Fourier components from the very start, the quantum-mechanical treatment of the Cerenkov effect[2] leads directly to a representation of the field in which dispersion can be taken into account, and therefore the visible character of the Cerenkov light is directly put in evidence.

Finally we wish to mention that in the author's work of 1904 a simple relation was derived for the force F which, because of the radiation, acts to decelerate the electron. If c is replaced by the phase velocity u, assuming again a constant refractive index, and if the electron is treated like a rigid sphere with uniform spatial charge distribution, this force is given by

$$4\pi\,|\mathsf{F}| = \frac{1}{4}\,\frac{e^2}{\varepsilon_0\,a^2}\left(1 - \frac{u^2}{v^2}\right). \tag{15}$$

An equivalent expression was derived by Tamm (loc. cit.). For an interesting application of eq. (15) to ballistics see F. Klein and A. Sommerfeld, Theorie des Kreisels, IV, p. 925, Leipzig, 1910.

D. A final critical remark

We have used the formalism of the special theory of relativity to achieve a simple integration of the wave equation for the case of the Cerenkov electron. But in this formalism we have everywhere replaced the light velocity c by the phase velocity u. This change must also be made in the equation $\operatorname{Div}\Omega = 0$, Vol. III, eq. (26.7) which therefore becomes

$$\operatorname{div}\mathsf{A} + \frac{1}{i\,u}\,\frac{\partial}{\partial t}\,\frac{i\,\psi}{u} = \operatorname{div}\mathsf{A} + \psi/u^2 = 0. \tag{16}$$

This form of the equation was implicit in the above calculations.

However, we must note that the four-component quantities Ω, Γ which we have used are not really four-vectors; they are not relativistic covariants. Rather, they are based specifically on the system of the stationary dielectric medium in which the electron is moving. In order to change from this reference system to another (such as the rest system of the electron, for instance), one should not use the usual Lorentz transformation but rather Minkowski's electrodynamics of moving media (Vol. III, chap. IV).

[1] Similarly in G. Beck, Phys. Rev. **74**, 795, 1948.

[2] K. M. Watson and J. M. Jauch, Phys. Rev. **75**, 1249, 1949.

48. Supplement on Geometrical Optics. Curved Light Rays, Sine Condition, Lens Formulae, Rainbow

In Sec. 35 A we based geometrical optics on the existence of the *eikonal*, that is, a system of surfaces

$$S\,(x, y, z) = \text{constant}$$

the orthogonal trajectories of which are *rays*. According to (35.3) the equation of the eikonal is

$$D\,(S) = (\operatorname{grad} S \cdot \operatorname{grad} S) = n^2, \qquad (n = \text{refractive index}).$$

The unit vector[1] in the ray direction, hence the normal to the eikonal surface passing through the point in question, is given by

$$\mathbf{s} = \frac{\operatorname{grad} S}{\sqrt{D\,(S)}} = \frac{1}{n} \operatorname{grad} S. \tag{1}$$

From this follows

$$\operatorname{curl}\,(n\mathbf{s}) = 0. \tag{2}$$

This condition is equivalent to the existence of the eikonal. *All ray bundles (straight or curvilinear) realized in geometrical optics are normals to surfaces* and are distinguished from more general systems of curves in that they satisfy the condition (2).

Parallel rays and rays diverging from a luminous point source, which are the types of rays usually considered, obviously have the property of being *normals to surfaces.* A theorem which had already been stated by Malus says that this property is preserved under arbitrary reflections and under refractions in arbitrary lens systems. This theorem is self-evident to us because of the existence of the eikonal before and after every reflection or refraction, and it is also expressed by (2).

The following integral requirement is equivalent to the differential condition (2):

$$\oint n\,(\mathbf{s}_x\,dx + \mathbf{s}_y\,dy + \mathbf{s}_z\,dz) = 0. \tag{2 a}$$

We can also abbreviate this as follows

$$\oint n\,\mathbf{s} \cdot \vec{ds} = 0. \tag{2 b}$$

As a consequence

$$\int_1^2 n\,\mathbf{s} \cdot \vec{ds} = S_2 - S_1. \tag{2 c}$$

---

[1] On its use is based the work by Sommerfeld and Iris Runge cited on p. 201. It will also simplify and clarify the following presentation.

That is, the line integral from a point 1 in the field to a point 2 is independent of the path and is equal to the difference between the values of the eikonal at the two points. Later we shall call the starting point 1 of the integral the "object" and the end point 2 the "image".

A. The curvature of light rays

At the point P under consideration we construct the osculating plane to the curved light ray. Next we construct the tangent unit vectors $\mathbf{s}$ at P and $\mathbf{s}'$ at a neighboring point P' lying on the same ray. The curvature of the ray is defined as the angle between these two vectors divided by the distance $P\,P' = ds$. Since $|\mathbf{s}| = 1$ the above angle equals the vector difference $\mathbf{s}' - \mathbf{s}$ which we call $d\mathbf{s}$. Hence we define the curvature as

$$\vec{K} = \frac{d\mathbf{s}}{ds}. \tag{3}$$

The absolute value of this vector gives the magnitude of the curvature; the direction of $d\mathbf{s}$ in (3) gives the position of the radius of curvature in the osculating plane.

For clarity we temporarily use cartesian coordinates[1] to transform the right-hand side of (3):

$$\frac{d\mathbf{s}}{ds} = \frac{\partial \mathbf{s}}{\partial x}\frac{dx}{ds} + \frac{\partial \mathbf{s}}{\partial y}\frac{dy}{ds} + \frac{\partial \mathbf{s}}{\partial z}\frac{dz}{ds}.$$

The factors $dx/ds\ldots$ are nothing else but the components of the unit vector $\mathbf{s}$. Hence we have

$$\frac{d\mathbf{s}}{ds} = \frac{\partial \mathbf{s}}{\partial x} s_x + \frac{\partial \mathbf{s}}{\partial y} s_y + \frac{\partial \mathbf{s}}{\partial z} s_z. \tag{3 a}$$

In addition, it follows from $|\mathbf{s}|^2 = 1$ for every direction of the gradient:

$$0 = \frac{1}{2}\,\mathbf{grad}\,|\mathbf{s}|^2 = s_x\,\mathbf{grad}\,s_x + s_y\,\mathbf{grad}\,s_y + s_z\,\mathbf{grad}\,s_z. \tag{3 b}$$

Subtracting this from (3 a) one obtains

$$\frac{d\mathbf{s}}{ds} = s_x\left(\frac{\partial \mathbf{s}}{\partial x} - \mathbf{grad}\,s_x\right) + s_y\left(\frac{\partial \mathbf{s}}{\partial y} - \mathbf{grad}\,s_y\right) + s_z\left(\frac{\partial \mathbf{s}}{\partial z} - \mathbf{grad}\,s_z\right). \tag{3 c}$$

The x-component of this vector equation (put $\mathbf{s} = s_x$, $\mathbf{grad} = \partial/\partial x$) reads

$$\frac{ds_x}{ds} = s_y\left(\frac{\partial s_x}{\partial y} - \frac{\partial s_y}{\partial x}\right) + s_z\left(\frac{\partial s_x}{\partial z} - \frac{\partial s_z}{\partial x}\right)$$
$$= -\,s_y\,\mathbf{curl}_z\,\mathbf{s} + s_z\,\mathbf{curl}_y\,\mathbf{s} = (\mathbf{curl}\,\mathbf{s} \times \mathbf{s})_x,$$

[1] Otherwise we would have to use tensor calculus which we wish to avoid.

and corresponding results hold for the y- and z-components of (3 c). Combin' these into a single vector relation we have

$$\frac{d\mathbf{s}}{ds} = \mathbf{curl\, s} \times \mathbf{s}. \tag{4}$$

This is true for every unit vector **s** and not merely for the surface normals of optics.

We now use the fundamental eq. (2) which characterizes the light vector **s** and which may be written in the form

$$n\, \mathbf{curl\, s} - \mathbf{s} \times \mathbf{grad}\, n = 0$$

or

$$\mathbf{curl\, s} = \frac{1}{n} \mathbf{s} \times \mathbf{grad}\, n. \tag{4 a}$$

We substitute this in (4) and form the absolute values of both sides. We note that according to (4 a) **curl s** is perpendicular to **s** and that therefore the absolute value of the vector product in (4) is equal to the product of $|\mathbf{curl\, s}|$ and $|\mathbf{s}|$. Thus we obtain

$$\left|\frac{d\mathbf{s}}{ds}\right| = \frac{1}{n} |\mathbf{s} \times \mathbf{grad}\, n|. \tag{5}$$

According to (3) this is equal to the curvature

$$K = \frac{1}{n} |\mathbf{s} \times \mathbf{grad}\, n| = \frac{1}{n} |\mathbf{grad}\, n| \sin \alpha, \qquad \alpha = \text{angle}\, (\mathbf{s}, \mathbf{grad}\, n). \tag{6}$$

For the direction of the radius of curvature we obtain from (4) and (4 a):

$$\vec{K} = \frac{1}{n} [\mathbf{s} \times \mathbf{grad}\, n] \times \mathbf{s},$$

or, using a well-known vector theorem,

$$n \vec{K} = \mathbf{grad}\, n - \mathbf{s}\, (\mathbf{s} \cdot \mathbf{grad}\, n). \tag{6 a}$$

From this we see that the principal normal $\vec{K}$, the tangent **s**, and the gradient of n all lie in one plane, namely the osculating plane. Or, to say it better, if we consider as given not the osculating plane but the gradient of n and the direction **s** of the light ray, then the osculating plane passes through **s** and the gradient of n. The principal normal $\vec{K}$ also lies in this plane. According to (4 a) the binormal has the direction of the (axial) vector **curl s**.

Equation (6) contains a theorem which was already used in 35 A: *in a homogeneous medium* (n = *constant*) *the light rays are straight lines* ($K = 0$).

As an example of an optically inhomogeneous medium, let us consider the earth's atmosphere at sunset. The refractive index n of the air decreases with altitude; therefore the gradient of n points toward the center of the earth. The plane of the drawing in fig. 93 is the osculating plane of the curved light ray shown; this plane is perpendicular to the earth's surface. The direction of $\vec{K}$ is essentially given by the first term on the right-hand side of (6 a) because the second term is directed almost horizontally. Hence the light ray is curved concavely toward the earth. In a more elementary way of speaking the light ray is "refracted" in the direction of increasing air density. It follows that the setting sun appears *elevated* as indicated by the dash-dot tangent at the point P in the figure.

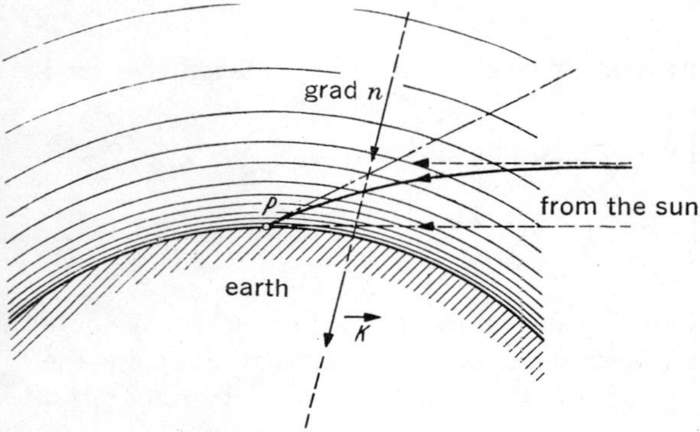

Fig. 93.
Curvature of the sun's rays in the atmosphere of the earth.

In the present example of a medium with parallel strata, the equation of the curved ray can be written explicitly directly from the law of refraction in the form

$$n \sin \alpha = \text{const.} \tag{6 b}$$

It is easy to show (by logarithmic differentiation and evaluation of $K = d\alpha/ds$) that this equation agrees with the generally valid eq. (6).

B. Abbe's sine condition

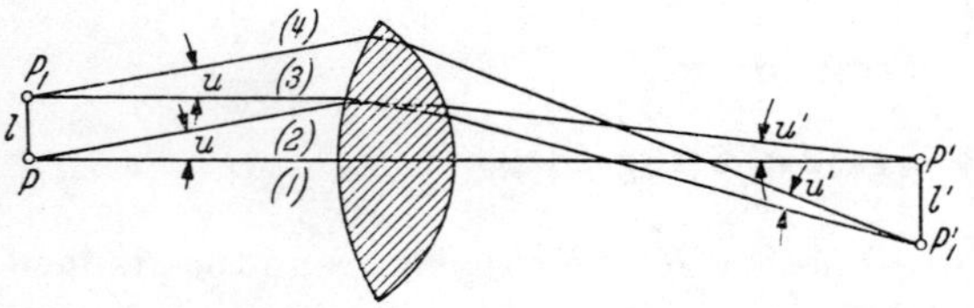

Fig. 94.
On the sine condition

Let us consider a system of lenses which is symmetrical about an axis $P\,P'$ and which, by means of bundles of rays of *finite* opening, maps the points on the plane $P\,P_1$ in the object space as (almost) faultless images onto the plane $P'\,P_1'$ in the image space. Let u and u' be the angular apertures of the bundles; let n and n' be the indices of refraction in front of and behind the system of lenses; $n > n'$ implies immersion.

Let (1) be the axial ray from P to P' and (2) the limiting ray of the bundle emitted by P (because of the rotational symmetry only one meridian plane need be considered, see fig. 94). The object point P_1 is at a distance l from P; the rays which are emitted by P_1 and are parallel to (1) and (2) shall be called (3) and (4). Let the distance of the image P_1' from P' be l'. We are not concerned with the (possibly curved) ray paths inside the lens system; outside the lenses the rays are straight.

According to eq. (2 c) two line integrals $\int n\,\mathbf{s}\cdot\overrightarrow{ds}$ between the same beginning and end points are equal to one another and to the difference between the respective values of the eikonal

$$(7) \qquad \begin{aligned}(1) &= (2) = S(x', 0) - S(x, 0),\\ (3) &= (4) = S(x', l') - S(x, l),\end{aligned}$$

where x and x' are the abscissae of object and image measured along the central axis. We wish to find the difference (3) – (1) and the equal difference (4) – (2).

From (7) there follows for sufficiently small l' and l

$$(8) \qquad \begin{aligned}(3) - (1) &= S(x', l') - S(x', 0) - \{S(x, l) - S(x, 0)\}\\ &= l' \frac{\partial}{\partial y'} S(x', y') - l \frac{\partial}{\partial y} S(x, y) \quad \text{with} \quad y' = y = 0.\end{aligned}$$

The two differential quotients, originating as they do from a Taylor expansion, are to be evaluated on the path 1. The first derivative is to be evaluated in the image space at the point P', the second in the object space at the point P. On the other hand, because S is the line integral of $n\,\mathbf{s}$, these two differential quotients are equal to

$$(8\text{ a}) \qquad n' s_y \text{ in } P' \qquad \text{and} \qquad n\, s_y \text{ in } P, \text{ respectively}$$

Since they are evaluated on the path (1) where $s_y = 0$ they are both zero. Therefore

$$(8\text{ b}) \qquad (3) - (1) = 0.$$

If we calculate (4) – (2) by the same method, we again obtain from the Taylor expansion the right-hand side of (8) where, however, the derivatives with respect to y are to be evaluated on the path (2) in the image and object space, respectively; on the other hand, by the definition of S as a line integral, these derivatives are again given by the values (8 a) but are now taken for the path (2). These are, see figure,

$$n' s_y = n' \sin u' \qquad \text{and } n\, s_y = n \sin u, \text{ respectively.}$$

Hence one obtains

$$(9) \qquad (4) - (2) = l' n' \sin u' - l\, n \sin u.$$

But according to (7), (4) – (2) = (3) – (1) and since (3) – (1) = 0 by (8 b), also (4) – (2) = 0. This is *Abbe's sine condition* which we had anticipated on p. 307:

$$n' \, l' \sin u' = n \, l \sin u. \tag{10}$$

As was shown by Straubel[1], this theorem is contained in a general reciprocity theorem of geometrical optics.

A similar theorem can be derived by considering two points P and P_1 which are *axially* displaced with respect to one another instead of being displaced *transversely* as before. The assumption now is that light bundles of finite opening project these points onto two image points P', P_1' which are also axially displaced with respect to one another. The only difference in this arrangement is that in all the above formulae the derivatives of the eikonal are to be taken with respect to x and x' instead of y and y'. The result is

$$\begin{aligned}(3) - (1) &= -n\,l + n'\,l', \\ (4) - (2) &= -n\,l \cos u + n'\,l' \cos u',\end{aligned}$$

which yields the relation

$$n'\,l'\,(1 - \cos u') = n\,l\,(1 - \cos u). \tag{11}$$

The incompatibility of this expression with the sine condition (10) makes it clear that no optical system can simultaneously produce sharp images of transversally and axially neighboring points by means of ray bundles with wide angles of opening.

C. On the structure of rectilinear ray bundles

Ray bundles in homogeneous media and consisting of straight lines are, of course, of special interest. Kummer investigated the most general, not necessarily normal, bundles of straight lines. We shall limit ourselves solely to bundles which are possible in optics, that is, to those which for n = constant satisfy the condition [eq. (2)]

$$\mathbf{curl\,s} = 0. \tag{12}$$

We call one ray the central ray and consider only those rays which deviate a very slight distance from the central ray. That is, we consider an infinitesimally thin bundle. We construct a plane E perpendicular to the central ray and mark the points at which the rays of our infinitesimally small bundle intersect the plane. We do the same with a parallel plane E' which is located at the (small) distance δ from E. The corresponding points on the planes E and E' are related to one another by an affine transformation.

[1] R. Straubel, Physikal. ZS. **4**, 114, 1902.

According to the fundamental theorem of the kinematics of continuous media (see Vol. II, Sec. 1, specialized to a two-dimensional continuum), this transformation is composed of a deformation in two mutually perpendicular directions (symmetrical transformation coefficients) and a rotation about the axis perpendicular to these directions (antisymmetric transformation). A small circle drawn about the intersection of the central ray on the plane E is transformed into an ellipse by the deformation, and this ellipse is rotated by the rotational transformation through an angle given by $\frac{1}{2}$ **curl s**. This rotation is analogous to the angular velocity in hydrodynamics. Our condition (12) applied to the central ray states that the rotational component of the transformation vanishes and that therefore the principal axes of the deformation ellipses are parallel for all planes E, E'. These axes lie in two fixed mutually perpendicular planes which are *the symmetry planes of the structure of the ray bundle.*

As such they must contain the two degenerate cases of the deformation ellipse in which one of the two principal axes shrinks to zero. These degenerate ellipses are called the *focal lines of the ray bundle. They are perpendicular to one another and to the central ray* (*Sturm's theorem*). The points where the focal lines intersect the central ray are called the *focal points* of the ray bundle. The distance between the two focal points is called the *astigmatic difference d.* (In the case of general Kummer rays the two focal lines are not necessarily perpendicular; non-optical ray complexes have no symmetry planes but, since **curl s** $\neq 0$, they have a positive or negative sense of rotation.)

From p. 208 we know that to every optical ray bundle there corresponds a system of parallel surfaces, the surfaces $S =$ constant. In the simplest case of the spherical wave (and its special case, the plane wave) these surfaces are concentric spheres (or parallel planes). In this case the positions of the symmetry planes and of the perpendicular focal lines are indeterminate. The two focal points coincide, forming only *one* focal point; d becomes zero and the ray bundle converges to the focal point. This situation is characterized as *anastigmatism.*

D. On the lens formula

In high school one learns the lens formula

$$\frac{1}{a} + \frac{1}{b} = \frac{1}{f}, \tag{13}$$

$a =$ distance from object to lens
$b =$ distance from lens to image
$f =$ principal focal length of the lens; $1/f = (n-1)(1/R_1 + 1/R_2)$.

We shall show that this formula can be proved by the above method without recourse to trigonometry, to the law of refraction or to special graphical constructions.

The object point P emits spherical waves. According to (35.6) their eikonal for $n = 1$ (air) and the origin at P is

$$S = \sqrt{x^2 + y^2 + z^2}.$$

We consider the anastigmatic bundle which falls perpendicularly on the lens and place the z-axis along the central ray of this bundle. Near the lens ($z = a + \zeta$) we have, neglecting higher powers of x, y and ζ,

$$S = z\left(1 + \frac{1}{2}\frac{x^2 + y^2}{z^2}\right) = a + \zeta + \frac{x^2 + y^2}{2\,a}. \tag{14}$$

We now continue the bundle into the interior of the lens. Let the front surface of the lens be a sphere of radius R_1 with its center at $x = y = 0$, $z = a + R_1$. Its equation is therefore

$$x^2 + y^2 + (z - a - R_1)^2 = R_1{}^2. \tag{14 a}$$

The point where the central ray intersects this sphere shall be called T_1.

We assume that the perpendicularly incident bundle remains anastigmatic after being refracted. (This is not the case for oblique incidence.) The proof that this assumption is correct lies in the fact that it permits us to satisfy the boundary condition (continuity of S in passing through the spherical surface R_1). We think of the refracted bundle as rectilinearly extended backward through the boundary surface to its point of convergence Q_1. The distance $Q_1\,T_1$ shall be ρ_1. The eikonal of the ray bundle emerging from Q_1 is analogous to (14), where however a is replaced by ρ_1:

$$S = n\left(\rho_1 + \zeta + \frac{x^2 + y^2}{2\,\rho_1}\right) + S_0. \tag{14 b}$$

The factor n takes into account the fact that this ray bundle must be thought of as propagating in glass not only inside the lens but also outside it, since the bundle was constructed by a rectilinear extension across the boundary surface. The term S_0 is the integration constant available to us in integrating the differential equation of the eikonal.

We require that the eqs. (14) and (14 b) connect continuously with each other on the sphere (14 a) not only at the point T_1 on the central ray ($\zeta = 0$, $x = y = 0$) but also at points in its vicinity. If, as before, we set $z = a + \zeta$, these latter points are, according to (14 a), characterized by

$$\zeta = \frac{x^2 + y^2}{2\,R_1}. \tag{14 c}$$

Substituting this in (14) and (14 b) one obtains

$$\text{(15)}\qquad S = \begin{cases} a + \dfrac{x^2+y^2}{2\,R_1} + \dfrac{x^2+y^2}{2\,a}, \\ n\,\rho_1 + S_0 + n\,\dfrac{x^2+y^2}{2\,R_1} + n\,\dfrac{x^2+y^2}{2\,\rho_1}. \end{cases}$$

The available constant S_0 is chosen such that the constant terms in both lines are equal. A comparison of the variable terms yields

$$\frac{x^2+y^2}{2\,R_1} + \frac{x^2+y^2}{2\,a} = n\,\frac{x^2+y^2}{2\,R_1} + n\,\frac{x^2-y^2}{2\,\rho_1},$$

and hence

$$\text{(16)}\qquad \frac{n-1}{R_1} = \frac{1}{a} - \frac{n}{\rho_1}.$$

Next we consider a situation where the image point P' is the source of spherical waves which are refracted at the rear lens surface of radius R_2. The point where the central ray intersects this spherical surface shall be T_2. If R_1, a, ρ_1 are replaced by R_2, b, ρ_2, then eq. (15) is again valid. ρ_2 is now the distance between T_2 and the convergence point Q_2 of the bundle refracted at the rear surface of the lens. Equation (16) becomes then

$$\text{(16 a)}\qquad \frac{n-1}{R_2} = \frac{1}{b} - \frac{n}{\rho_2}.$$

But since P' is to be the image point of P, the rays emerging from P' must coincide inside the lens with the previously considered rays emitted by P. Hence Q_2 and Q_1 must coincide, and if we write henceforth $Q_1 = Q_2 = Q$, we obtain

$$\text{(16 b)}\qquad Q\,T_2 = Q\,T_1 + T_1\,T_2.$$

From it follows for $T_1\,T_2 = d =$ thickness of the lens and with the same meanings of ρ_1 and ρ_2, that

$$\text{(17)}\qquad -\rho_2 = \rho_1 + d.$$

A negative sign appears on the left because with respect to the rear surface of the lens the convergence point Q has the opposite position from that with respect to the front surface.

For a "thin" lens ($d \ll \rho_{1\,2}$) one has $\rho_2 = -\rho_1$. *The sum of the two eqs.* (16) *and* (16 a) *yields directly the lens formula* (13) *which was to be proved.* In order for the latter to be valid, it is not at all necessary for d to be small in any absolute sense. Our derivation requires only that d be small compared to the radii of curvature R_1, R_2 (or, what is approximately the same, with respect to ρ_1, ρ_2).

In order to progress somewhat beyond the usual school curriculum we shall also find the generalization for a lens of finite thickness.

For this purpose we calculate from (16)

$$\frac{\rho_1}{n} = 1 \Big/ \left(\frac{1}{a} - \frac{n-1}{R_1}\right)$$

and from (16 a) and (17)

$$\frac{-d-\rho_1}{n} = 1 \Big/ \left(\frac{1}{b} - \frac{n-1}{R_2}\right).$$

Adding these two equations as we previously added eqs. (16) *and* (16 a), *we obtain as a generalization of* (13)

$$\frac{d}{n} = 1 \Big/ \left(\frac{n-1}{R_1} - \frac{1}{a}\right) + 1 \Big/ \left(\frac{n-1}{R_2} - \frac{1}{b}\right). \tag{18}$$

We have restricted ourselves to the case of a single lens (same n in front and back). The result for a system of lenses is, of course, less simple than eq. (18). Furthermore, we have considered only the perpendicularly incident bundle. As was remarked above, an obliquely incident bundle does not remain anastigmatic after refraction; the resulting astigmatic ray structure has two focal lines, one of which lies in the plane of incidence and the other in the perpendicular plane. We wish to remark that the astigmatic bundle is decidedly the *normal case*: only with special axially symmetric arrangements is the anastigmatic character of the incident plane or spherical wave preserved.

As was noted above at eq. (13), it has not been necessary to make explicit use of the law of refraction in this entire section C. This is because the law of refraction by its very derivation (see beginning of Sec. 3) guarantees the equality of the phases of the incident and refracted waves. On the other hand, the eikonal, see Sec. 35 A, represents nothing but the phase of the wave; thus our requirement that the eikonal be continuous at the boundary surfaces of the lens represents an entirely valid substitute for the law of refraction. Indeed, the law of refraction can be obtained directly from our present eq. (2 b). It is only necessary to apply the latter to a shallow rectangle which surrounds and cuts across an element of surface of the refractive medium. Then (2 b) reads

$$n_1 s_1 = n_2 s_2, \tag{18 a}$$

where s_1 and s_2 are the tangential components of our unit vector and hence $s_1 = \sin\alpha$, $s_2 = \sin\beta$. Thus eq. (18 a) is indeed Snell's law of refraction.

E. Production of Curved Light Rays by Diffusion and a Remark on the Theory of the Rainbow

By means of a diffusion experiment it is possible to construct an optically inhomogeneous medium whose index of refraction varies in space in a simple and easily represented manner. A narrow tall parallel-walled glass tank is filled in its lower half, $x < 0$, with glycerine and in its upper half, $x > 0$, with water. While initially ($t = 0$) there exists a sharp plane of separation between the two media ($x = 0$), this separation becomes less and less distinct owing to diffusion. At $x = 0$ the concentration u always retains the value $u = 1/2$ which corresponds to a perfect mixture. At this point also the concentration curve $u(x)$ has an inflection tangent whose slope gradually increases with time. (Curves 1 and 2 in the figure.) At large positive values of x the curve $u(x)$ approaches the value 0 while at large negative values of x it approaches 1. As $t \to \infty$ (perfect mixture if the very small effect of gravity is neglected) $u \to 1/2$ (the line ∞ in the figure).

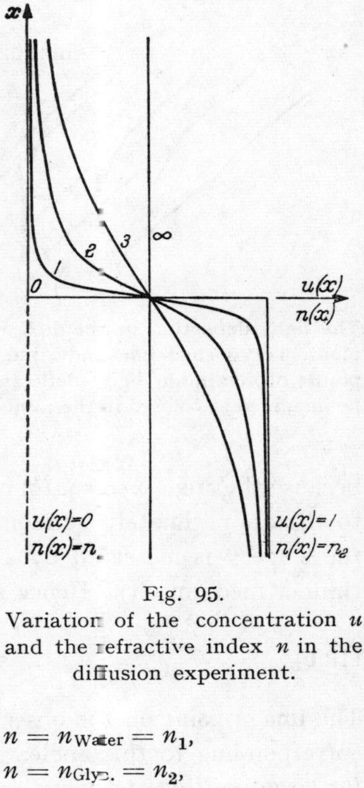

Fig. 95.
Variation of the concentration u and the refractive index n in the diffusion experiment.

We may assume that the refractive index n behaves qualitatively like u, so that fig. 95 may also be used to represent n. Then we need only to read the figure with the following new designations:

$$\text{the straight line } u = 0 \text{ becomes } n = n_{\text{Water}} = n_1,$$
$$\text{the straight line } u = 1 \text{ becomes } n = n_{\text{Glyc.}} = n_2,$$
$$\text{the straight line } u = \frac{1}{2} \text{ becomes } n = \frac{1}{2}(n_1 + n_2).$$

The significant point of this $n(x)$ diagram is the inflection point

$$(19)\quad x = 0, \qquad n(0) = \frac{n_1 + n_2}{2} \qquad \text{where } n'(0) = \text{maximum of } n'(x).$$

At this point the optical inhomogeneity of the medium, as represented by the gradient of n, is greatest. The inhomogeneity vanishes only at $t = \infty$.

Let us now paste tinfoil on the front surface of the tank so that this surface becomes opaque except for a narrow slit which is inclined 45° to the horizontal. Let us also illuminate the rear surface with perpendicularly incident light (arc lamp with collimator). One might think that a *rectilinear* image of the 45° slit should appear on an observation screen placed in front of the slit because each incident light ray at its position x passes through a horizontal layer with constant $n = n(x)$. But this is not so; the light ray is curved because $n'(x) \neq 0$, and because of the direction of the gradient of n, the center of curvature of the ray lies in the lower part of the trough; the center lies higher where the gradient of n is greater. Since the tank is very narrow, the curved light path is very short, and we may approximate it by a circular arc with a horizontal tangent at the rear wall of the trough and a downward sloping tangent at the front wall. If d is the inside width of the trough and R the radius of the circular path, the slope of the tangent at the front wall is $\gamma = d/R$. According to eq. (6)

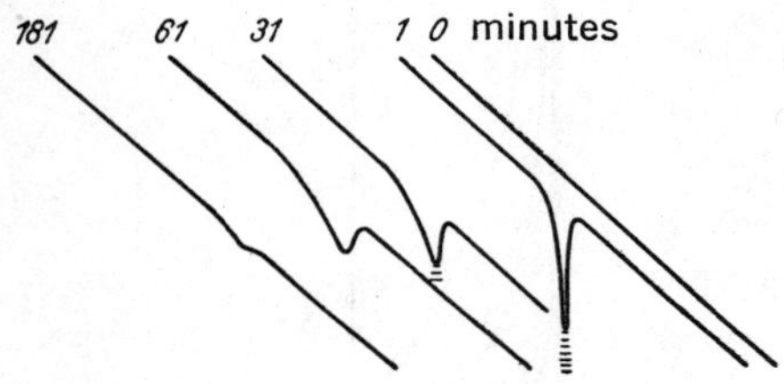

Fig. 96.

The light deflection in the diffusion experiment. The dashed lines indicated below the points of maximum light deflection should be imagined as colored in the rainbow colors.

$$\frac{1}{R} = \frac{1}{n}\left|\frac{dn}{dx}\right|, \tag{19 a}$$

because the angle α in eq. (6) is equal to $\pi/2$ (the gradient of n is perpendicular to the approximately horizontal light path). On emerging from the trough the slope γ is increased by a factor n (refraction at the interface with the thinner medium air). Hence the slope of the emerging light rays is given by

$$\gamma' = n\gamma = d\left|\frac{dn}{dx}\right|. \tag{19 b}$$

The image point on the observation screen is deflected downward a distance corresponding to this angle. The deflection is greatest at the level $x = 0$; for large positive and negative values of x the deflection goes to zero so that the upper and lower ends of the image curve form a straight line inclined at 45°.

Figure 96 shows a sketch of the time variation of the image curve which was prepared during a seminar demonstration by H. Ott, who was at that time an assistant at our institute. In this figure 1 is the shape of the curve immediately after the start of the diffusion process; 31 is its appearance 30 minutes later, 61 is its appearance another 30 minutes later and 181 is the curve two hours later when the mixing process is almost complete. This experiment was first performed by O. Wiener[1] and is described by

[1] O. Wiener, Ann. d. Phys. (Lpz.) **49**, 105, 1893.

Kohlrausch in "Praktische Physik" as a method of measuring the diffusion constant k. In order to explain this application we must digress very briefly into the theory of diffusion.

In the one-dimensional case the differential equation of diffusion is

$$\frac{\partial u}{\partial t} = k\frac{\partial^2 u}{\partial x^2}. \tag{20}$$

If the following substitution, which suggests itself for dimensional reasons, is made

$$u = f(\xi), \qquad \xi = \frac{x}{\sqrt{k\,t}}, \tag{20 a}$$

then

$$\frac{\partial u}{\partial t} = -\frac{\xi}{2\,t}f'(\xi), \qquad \frac{\partial^2 u}{\partial x^2} = \frac{1}{k\,t}f''(\xi)$$

and by (20)

$$f''(\xi) = -\frac{\xi}{2}f'(\xi).$$

It follows that

$$f'(\xi) = A\,e^{-\xi^2/4},$$

where A is a constant of integration. Hence from (20 a)

$$\frac{\partial u}{\partial x} = \frac{1}{\sqrt{k\,t}}\,A\,e^{-x^2/4kt}, \tag{20 b}$$

$$\frac{\partial^2 u}{\partial x^2} = -\frac{x}{2\,(k\,t)^{3/2}}\,A\,e^{-x^2/4kt}. \tag{20 c}$$

As a consequence of (20 b) u can be represented by the Gaussian error integral, a representation which is similar to that derived in Vol. VI, eq. (13.19) for an analogous diffusion problem. For, if B is a second constant of integration,

$$\left\{\begin{aligned} u &= \frac{1}{\sqrt{k\,t}}\,A\int_0^x e^{-x^2/4kt}\,dx + B = \sqrt{\pi}\,A\,\Phi\,(x/2\sqrt{k\,t}) + B, \\ \Phi\,(p) &= \frac{2}{\sqrt{\pi}}\int_0^p e^{-\alpha^2}\,d\alpha. \end{aligned}\right. \tag{20 d}$$

Rewriting the above equations in terms of the refractive index n in place of u, we conclude from the initial conditions for n

$$\begin{aligned} n &= n_1 \quad \text{for} \quad t = 0 \quad \text{and} \quad x > 0, \\ n &= n_2 \quad \text{for} \quad t = 0 \quad \text{and} \quad x < 0 \end{aligned}$$

and from (20 d) that

$$n_1 = \sqrt{\pi}\, A + B, \qquad n_2 = -\sqrt{\pi}\, A + B,$$

$$A = \frac{n_1 - n_2}{2\sqrt{\pi}}, \qquad B = \frac{n_1 + n_2}{2}. \tag{21}$$

From this it follows according to (20 d) that for $x = 0$ and arbitrary t

$$n(0) = \frac{n_1 + n_2}{2}, \tag{21 a}$$

which agrees with our assertion (19). Next we calculate from (20 b) and (21), again at $x = 0$,

$$\frac{\partial n}{\partial x} = \frac{n_1 - n_2}{2\sqrt{\pi k t}}. \tag{21 b}$$

From this the maximum angle of emergence γ' of the light ray is found by (19 b) to be

$$\gamma' = \frac{(n_2 - n_1)\, d}{2\sqrt{\pi k t}}, \tag{21 c}$$

which determines the maximum deflection of the light ray on the curves in fig. 96. This deflection decreases with time as $1/\sqrt{t}$. Lastly, we conclude from (20 c) that

$$\frac{\partial^2 n}{\partial x^2} = \frac{x}{4\,(k\,t)^{3/2}}\, \frac{n_2 - n_1}{\sqrt{\pi}}\, e^{-x^2/4kt} \quad \text{for } x = 0 \text{ and all } t. \tag{21 d}$$

The curve of the index of refraction plotted as a function of x has a permanent inflection point at $x = 0$.

Wiener used eq. (21 c) to calculate the diffusion constant k from the measured deflections γ'. In order to obtain a sharp deflection curve with a well-defined maximum deviation, he used *monochromatic* light. However we are interested in a phenomenon which appears when *white* light is used: the lowest point of the deflection curve, particularly of curve 1 in fig. 96, is decomposed into a horizontally narrow but vertically well-separated spectrum, which is indicated in the figure by seven horizontal lines (the "seven" colors of the rainbow starting with red at the top and ending with violet at the bottom). This spectrum rapidly contracts as the diffusion process proceeds in time and is just barely recognizable in curve 2 of fig. 96. At a small distance from the point of maximum deflection the spectrum is quite indistinct even in curve 1. All other points on curves 1, 2, etc. appear white. Let us interpret this phenomenon on the basis of the wave theory.

By (21 c) the angle γ' depends on wavelength, because of the dispersion of n_1 and n_2. Therefore the various colors are deflected in different directions and appear on the observation screen in sequence from red at the top to violet at the bottom. We can no longer restrict ourselves to the layer at $x = 0$ as before, but in order to obtain a finite color intensity on the screen we must also consider the neighboring layers. If we denote by $n\,(x, \lambda)$ the dispersion of the ray passing through the layer at x, then not only $n\,(0, \lambda)$ but also the behavior of $n\,(x, \lambda)$ in the vicinity of $x = 0$ is of importance. n is *stationary* at $x = 0$ and *only* there. The curve $n\,(x, \lambda)$ has, according to (21 d), an *inflection point* there. This means that in the immediate vicinity of $x = 0$ the deflection angle γ' [see eq. (19 b)] is the same as *at* $x = 0$. Here the color effect is amplified while at points $x \neq 0$ it is extinguished owing to superposition. Only at the very beginning of the diffusion process does the resulting spectrum show considerable extension. The denominator $\sqrt{k\,t}$ is (21 c) causes not only the deflection but also the dispersion to decrease rapidly. The latter is already considerably reduced in curve 2 of fig. 96 as compared to curve 1.

We have treated this example in some detail in order to illustrate the essential idea of the *theory of the rainbow*. In order for the rays which are reflected, refracted and dispersed in the water droplets to enter the eye as *parallel rays* with sufficient intensity, the wave front (or rather its trace on the plane of incidence) must have an *inflection point*. (Since we are here dealing with geometrical optics, we should really say "eikonal surface" instead of "wave front".) This necessary and sufficient condition fixes the radius of the principal rainbow (single reflection in the interior of the droplet) at about 41° and that of the secondary rainbow (double reflection) at about 51°20′. It is therefore clear that the rays contributing to the rainbow have an *extremal deviation* compared to all the other diverging rays which emerge from the droplets; for, indeed, the deviation of the inflection tangent perpendicular to the rainbow is an extremum compared to that of all other tangents. (In our diffusion experiment this was shown by the fact that the color spectrum appeared only at the lowest point of the deflection curve.) From this extremal position of the effective rays it follows that for the principal rainbow the diverging rays form a smaller angle with the incident rays of the sun than do the parallel rays and that therefore the diverging rays reach the eye from below the rainbow; it further follows that for the secondary rainbow the diverging rays form a larger angle with the incident radiation than do the parallel rays and that therefore the diverging rays reach the eye from above the bow; therefore the zone between the two rainbows appears darker than the regions below the principal and above the secondary rainbow.

We have emphasized before (p. 179) that, strictly speaking, the rainbow represents a difficult diffraction problem whose character changes from one case to the next depending on the sizes of the drops. As for the wave-theoretical treatment of the rainbow, we shall limit ourselves to a result which is directly connected with the existence of the inflection point. If we wish to investigate the propagation of a cylindrical wave surface, for example by the saddle-point method, and represent the trace of the wave surface by $y = S(x)$, then we must look for the point $x = x_0$ where $S'(x_0) = 0$. To do this we develop $S(x)$ in a Taylor series about this point:

$$y = S(x_0) + \frac{(x - x_0)^2}{2} S''(x_0) + \frac{(x - x_0)^3}{3!} S'''(x_0) + \dots .$$

The radiation is in general determined by the *curvature* of the trace of this wave surface as given by $S''(x_0)$. Since we are concerned with the approximation of an integral of the form

$$\int \exp(i\,k\,y)\,dx,$$

we are led to a *Fresnel* integral

$$\int \exp(i\,k\,a\,\tau^2)\,d\tau, \quad \tau = x - x_0, \quad a = \frac{1}{2} S''(x_0).$$

But if the trace has an inflection point $S''(x_0) = 0$, that is, if its curvature vanishes, this approximation breaks down; in place of the Fresnel integral one is then led to an Airy integral of the form

$$\int \exp(i\,k\,a\,\tau^3)\,d\tau, \quad \tau = x - x_0, \quad a = \frac{1}{6} S'''(x_0).$$

This could also be called a "rainbow integral", for it is characteristic of the quantitative investigation of the color distribution in the rainbow and of all similarly degenerate wave problems. One of these latter problems is discussed in detail in connection with the asymptotic representations of the Hankel functions in Vol. VI, Sec. 21 D; see in particular the final remarks in that section. Here we have only been interested in showing that the principal features of that most impressive of celestial phenomena, the rainbow, can be understood at least qualitatively in terms of geometrical optics and that, moreover, geometrical optics provides us with a hint for the quantitative treatment of the problem.

49. On the Nature of White Light. Photon Theory and Complementarity

In the historical chart of Sec. 1 we have listed an important achievement of Lord Rayleigh in the field of optics: his theory of natural white light as a *completely random process*. We shall first quote a remark made by Rayleigh[1] which concerns the opposite of white light, that is to say, regular monochromatic waves.

[1] Phil. Mag. **50**, 135, 1900, Sci. Papers, Vol. IV, p. 486.

"To suppose, as is sometimes done in optical speculations, that a train of simple waves may begin at a given epoch, continue for a certain time ..... and ultimately cease, is a contradiction in terms. A like contradiction is involved if we speak of unpolarized light as homogeneous, really homogeneous light being necessarily polarized."

Regarding the last point in this quotation we refer to our Sec. 2 where the elliptic polarization of an ideal monochromatic and plane wave was proved to be a mathematical consequence of Maxwell's equations. The first sentence quoted has been taken into account in Sec. 22 where a wave train of the type described by Rayleigh was decomposed into its Fourier spectrum and was therefore treated not as monochromatic but rather as polychromatic.

Concerning the problem of white light we again quote a remark[2] by Lord Rayleigh which refers to the then recent discovery by Röntgen: "The conclusion of Stokes and J. J. Thompson 'that the Röntgen rays are not waves of very short wavelength, but impulses', surprises me. From the fact of their being highly condensed impulses, I should conclude on the contrary that they *are* waves of short wavelength... What then becomes of Fourier's theorem...?"

"Is it contended that previous to resolution (whether merely theoretical, or practically effected by the spectroscope) the vibrations of ordinary (e. g. white) light are regular, and thus distinguished from disturbances made up of impulses? This view... has been shown to be untenable by Gouy, Schuster, and the present writer. A curve representative of white light, if it were drawn upon paper, would show no sequences of similar waves."

We wish to add that Emil Wiechert, independently of Stokes, advanced the same hypothesis regarding the nature of Röntgen rays and that the author cooperated with him on several papers at the beginning of this century. In contrast to these theories, Planck, in connection with his discovery of the quantum of action (see p. 9), recognized the necessity of assuming the complete phase independence of "natural light". Laue's discovery subsequently put the individual Fourier components of the continuous Röntgen spectrum into direct evidence and proved the distinction between pulsed and wave radiation to be meaningless, in complete agreement with Rayleigh.

We use the words *white light* and *natural light* synonymously. We perceive the sun's natural light as white, i. e. as lacking all spectral colors, because the eye is *adapted* to the sun; that is to say, because our eye and the associated physiological-psychological vision apparatus has in its evolution-

[2] Röntgen Rays and Ordinary Light. Nature 57, 607, 1898 and Sci. Papers, Vol. IV, p. 353.

ary development adapted itself to the spectrum of the sun. If we lived in the vicinity of a red giant, we would presumably perceive its red color as the normal white. As is well known, Goethe abhorred the theory that white light is a mixture of the seven colors of the rainbow (he was certainly correct in regard to the white-*sensation* which he had primarily in mind). But the rainbow should have convinced him that white light is decomposed into colors by a spectral apparatus (in this case water droplets). In this decomposition the periodicity originates not from the primary sun light but from the frequency-sensitive spectral apparatus.

Gouy (1886) was probably the first to ascertain that a line grating diffracts a single "plane" pulse[1] in the same manner as a plane wave. An obliquely incident primary pulse (or a random sequence of mutually independent pulses) impinges on the various grating lines at rhythmically equal time intervals. Thereupon the grating lines emit a staggered sequence of secondary pulses[2]; one may refer to fig. 37, bearing in mind, however, that it describes continuous waves, while we are now dealing with a discrete sequence of secondary cylindrical excitations. At a sufficiently large distance from the grating and in any given direction these pulses are spaced a constant distance

$$\lambda = d\,(\alpha - \alpha_0)$$

apart (α is the direction cosine of the incident pulse with respect to the plane of the grating). This is our former grating formula (32.2) for $h = 1$. However λ is now not the wavelength of a monochromatic sinusoid but is the distance between pulses. The white character of the light has thereby not been lost entirely. Our sequence of secondary pulses is still far from being monochromatic. The periodicity originates in the grating and not in the primary pulse. It differs for different directions of observation. The higher order spectra ($h > 1$) would be obtained if the secondary pulse were decomposed into pure sine waves in the manner of Fourier, that is, into a fundamental oscillation of wavelength λ and harmonics of wavelengths λ/h.

But what is the situation in the case of a *prism* which also produces monochromatic waves from white light without seeming to possess any periodic structure as the grating does? Gouy answers this question in the

[1] By "plane pulse" we mean the counterpart of a "plane wave", that is to say an electromagnetic disturbance which has an appreciable intensity only between two parallel infinite planes a small distance apart and which has a constant instantaneous value on every plane parallel to these two.

[2] By a "secondary pulse" we mean an excitation which has an appreciable intensity only between two coaxial cylinder surfaces centered about a grating line and which propagates outward in this cylindrical shape with the velocity of light.

following manner: a wide spectral line, i. e. a certain portion of the continuous spectrum, represents from a wave-optical viewpoint a *modulated* wave train (with beats between neighboring frequencies). These beats propagate with the *group velocity* and therefore lag behind the wave train which propagates with the *phase velocity*. This results in a periodic variation of the shape of the wave train: the wave train assumes the same shape only when the group has lagged behind the phase by one whole wavelength. The light path inside the prism increases continuously from the edge to the base. Those light paths which satisfy the above condition for any given wavelength define on the surface of emergence a sequence of equidistant parallel lines comparable to the grating structure. The resulting regularity of the emerging waves is thus due to the difference between group and phase velocity or, as we may also say, it is due to dispersion. Again the regularity originates in the prism and not in the incident (more or less white or colored) light.

With this concept of pulses we have extended the older, overly restricted conception of the wave theory in such a way that it comes closer to Einstein's hypothesis of the *light-quantum*. Indeed the mechanics of light quanta has haunted the science of optics from its very beginning.

What else is Fermat's principle of the shortest time of arrival but the principle of the shortest (geodetic) line in the mechanics of a force-free mass point? Both give the same result because in the force-free case the time of travel and the path length are proportional to one another. The same holds for the principle of least action because of the constancy of the kinetic energy, see Vol. I, eq. (37.1) and (37.2). Fermat's principle provides us with a truly popular exercise in the method of maxima and minima: given the starting and end points of the path and given the velocities in the first and second media, the incident particle of light travels along that path by which it reaches the end point in the second medium in the shortest possible time. The same is true if the end point also lies in the first medium and the accessory condition is added that the particle shall touch the surface separating media 1 and 2. Let us here use this principle to calculate the curvature of the trajectory in an inhomogeneous medium which we calculated in the preceding paragraph by means of the theory of the eikonal.

We start with the principle

$$\delta \int_{P_0}^{P} dt = 0, \tag{1}$$

where P_0 and P are the prescribed starting and end points of the path. Let us consider a stratified medium in which therefore the velocity u of the light particles (phase velocity of the light) is a given function of the coordinate x alone. We replace u by the velocity ratio $n(x) = c/u(x)$ where c is a standard velocity, the magnitude of which is of no importance here. (1) becomes then

$$\frac{1}{c}\,\delta \int_{P_0}^{P} n(x)\, ds = 0, \tag{2}$$

which we write in the form

$$\delta \int F(x, y')\, dx = 0, \qquad F(x, y') = n(x)\sqrt{1 + y'^2}.$$

The Lagrangian derivative for this variational problem reads

$$\frac{d}{dx}\frac{\partial F}{\partial y'} - \frac{\partial F}{\partial y} = 0$$

and hence, since F is independent of y

$$\frac{d}{dx}\frac{\partial F}{\partial y'} = n(x)\frac{d}{dx}\frac{y'}{\sqrt{1+y'^2}} + \frac{dn}{dx}\frac{y'}{\sqrt{1+y'^2}} = 0. \tag{3}$$

If α is the angle which the tangent to the curve $y = y(x)$ forms with the x-axis, then

$$\tan\alpha = y', \qquad \sin\alpha = \frac{y'}{\sqrt{1+y'^2}}, \qquad \cos\alpha\,\frac{d\alpha}{dx} = \frac{d}{dx}\frac{y'}{\sqrt{1+y'^2}}. \tag{4}$$

Therefore the last term in the double eq. (3) equals $\sin\alpha\, dn/dx$, the next to the last term equals $n(x)\cos\alpha\, d\alpha/dx$, and eq. (3) becomes

$$n(x)\cos\alpha\,\frac{d\alpha}{dx} + \sin\alpha\,\frac{dn}{dx} = 0. \tag{5}$$

But now

$$dx = \cos\alpha\, ds, \qquad \text{hence} \qquad \cos\alpha\,\frac{d\alpha}{dx} = \frac{d\alpha}{ds} = K, \tag{6}$$

where K is the curvature of the curve $y(x)$. Combining eqs. (5) and (6) we obtain

$$|K| = \frac{1}{n}\,|\mathbf{grad}\, n|\sin\alpha. \tag{7}$$

This is our eq. (48.6). *As Hamilton had already recognized, geometrical optics is identical with the ordinary mechanics of mass points* not only for homogeneous but also for inhomogeneous media.

However to pass from this primitive corpuscular theory of light to the modern "photon theory" required a bold step into quantum theory, a step which was taken by Einstein in 1905: the energy of the photon had to be set equal to $h\nu$ and its momentum to $h\nu/c$. Only in this way does one obtain the energy relations which are so drastically brought into evidence in the photoelectric effect, the Compton effect, and in the short wave limit of the continuous X-ray spectrum. Only in this way can classical optics be brought into harmony with atomic physics.

It is significant that Einstein made this advance in the quantum theory in the same year in which the theory of relativity was created. L. de Broglie[1] emphasizes that only relativistic mechanics satisfies the requirements of the photon theory. According to classical mechanics we would have for a photon of energy W, momentum g and velocity u

$$W = \frac{m}{2} u^2, \qquad g = m u, \qquad \text{hence} \qquad g = \sqrt{2 m W}.$$

If we substitute for the mass the value obtained from the generally valid relationship $W = m c^2$, then we obtain

$$g = \sqrt{2 \frac{W^2}{c^2}} = \sqrt{2} \frac{W}{c},$$

and not

$$g = \frac{W}{c} = \frac{h\nu}{c},$$

as required by the photon theory. A similar discrepancy by a factor 2 appears in the expression for the light pressure, depending upon whether it is calculated from classical mechanics or, as in Vol. III, Sec. 31, eq. (15), from relativistic electrodynamics.

The photon theory is a corpuscular theory of light such as the one which Newton envisioned. The wave theory of light has equal status with the photon theory. Which of the two will give the correct answer depends on the question that is posed in each particular experiment. Each completes the other — they are *complementary*. At the end of Chapter II we discussed the fact that the two theories do not contradict one another, and we mentioned the resulting far-reaching philosophical consequences. One is taught in school that the eye "sees the light waves". That is a myth. What our eye "sees" is the photoelectric processes taking place in the retina which, depending on the magnitude $h\nu$ of the incident light quanta, produce the varicolored world of our visual impressions. Certainly as far as our primary sensations go, there is no preference for a wave structure of light (however imbued we may be with it) over a quantum structure. It is fitting that we should conclude our volume on Optics by emphasizing once more that most remarkable and epistemologically most important result, the complementarity of wave and corpuscle.

---

[1] Rev. Mod. Physics Vol. 21, p. 345, 1949, on Einstein's seventieth birthday.

PROBLEMS

CHAPTER I.

I.1. Superposition of two parallel linear oscillations of equal frequency.

Let the two oscillations (in real notation) be

$$x_1 = a_1 \cos(\omega t + \alpha_1), \qquad x_2 = a_2 \cos(\omega t + \alpha_2) \tag{1}$$

By forming the vectorial (complex) sum of these oscillations, find the amplitude a and the phase α of the resulting oscillation

$$x = x_1 + x_2 = a \cos(\omega t + \alpha). \tag{2}$$

I.2 The curve described by the electric and magnetic vectors of a plane wave during one period.

In the ideal case (perfectly plane and perfectly monochromatic wave) this curve is an ellipse. Under what conditions does this ellipse degenerate into a circle or a straight line?

I.3. Concerning the surface charge on the boundary between I and II.

Show that the boundary surface must be free of charges in Sec. 3 B as well as in Sec. 3 A.

I.4 A check on figure 4.

Find the equations of the parabolas R_p and R_s as functions of α.

I.5 On the calculation of the reflective power r and the transmissivity d.

Confirm the energy theorem $r + d = 1$ for arbitrary material constants ε, μ.

I.6. Elliptic polarization of light through total reflection.

Starting with eq. (5.11), prove eq. (5.12) for the maximum phase difference $\gamma - \delta$ and the associated angle of incidence α_{max}.

I.7. The Perot-Fabry maxima considered as a resonance effect.

Following a suggestion by Kossel[1], investigate the electromagnetic *eigen-vibrations* of the Perot-Fabry air space between the silvered plates of the etalon. Consider only the case where the field depends only on the y-coordinate (perpen-

---

[1] W. Kossel, Ann. d. Phys. (Lpz.) 36, 1939; see remark at the bottom of p. 191 and top of p. 192 of that paper.

dicular to the plates) and where the silvering is so heavy that, for instance, the electric vector **E** oscillates everywhere parallel to the plates. Determine the frequency of this *free* oscillation and show that it agrees with the frequency of the maxima of the *forced* oscillation which is excited by a perpendicularly incident p-polarized wave.

I.8 Wiener's experiment with obliquely incident light.

Investigate the appearance of interference fringes for arbitrary angles of incidence α and for both cases of polarization.

Chapter III.

III.1. The reduced mass in the problem of intra-molecular oscillations.

For a molecule consisting of a positive ion of mass M_1 and a negative ion of mass M_2, prove the expression

$$\frac{1}{M} = \frac{1}{M_1} + \frac{1}{M_2},$$

which was used in (18.3). The ions are to be thought of as idealized mass points which attract one another with a central force. The same expression appeared in connection with the inelastic collision of two mass points in Vol. I, eq. (3.28 b).

III.2. The deflection angle δ of a prism.

Prove that the deflection angle is a minimum for symmetrical ray paths.

III.3 Direct vision and achromatic prisms.

For small prism angles and small angles of incidence calculate the deflection δ due to a double prism composed of two different glasses (refractive indices n_1, n_2; prism angles φ_1, φ_2); prism 1 is upright, prism 2 is upside down so that its edge adjoins the base surface of 1. Determine the ratio φ_2/φ_1 for $\delta = 0$ (direct vision prism) and for $d\delta/d\lambda = 0$ (achromatic prism).

III.4. Zeeman effect and Larmor precession.

Treat the motion of an electron in an arbitrary atomic field describable by a potential $V(r)$ a) when an additional homogeneous magnetic field B is present and b) when, instead, the motion is referred to a coordinate system which rotates with the angular velocity ω about the direction of B as an axis. Show that the motions a) and b) are the same when

$$\omega = \frac{1}{2}\frac{e}{m} B \quad \text{(Formula of the Larmor precession).}$$

Assume that the ordinary centrifugal force is negligible compared to the Coriolis force.

Chapter IV

IV.1. Geometrical derivation of the normal surface.

In Vol. II, solution to exercise I.6, the following two theorems were derived from the invariants of a tensor surface represented by an ellipsoid with principal axes a, b, c:

a) The sum of the inverse squares of any three mutually perpendicular semi-diameters is independent of the spatial orientation; hence, in particular, it is equal to

$$\frac{1}{a^2}+\frac{1}{b^2}+\frac{1}{c^2}.$$

b) The volume of any circumscribed parallelepiped of the ellipsoid is independent of its particular position or shape; hence, in particular, it is equal to

$$2a\cdot 2b\cdot 2c.$$

Apply these theorems to the index ellipsoid and to the construction described in Sec. 25, eqs. (12) to (19) and derive in this way the equation of the normal surface.

IV.2 Elementary geometrical derivation of the ray surface.

Apply the theorems a), b) of the previous exercise to Fresnel's ellipsoid and supplement the construction of Sec. 25 by a corresponding construction for the field vector $\mathbf{E}$. In this way obtain the equation of the ray surface.

IV.3. Proof of the approximate formula (31.9) for the phase difference due to the illumination of a crystal plate with converging light.

A sufficiently good approximation is obtained if in the exact formula for the phase difference between the two rays ABD and AC in fig. 47 one replaces β_1 and β_2 by a suitable mean angle β and if one treats both β_1 and β_2 as small quantities.

SOLUTIONS TO PROBLEMS

I.1. The usual method would be to set

$$x_1 = a_1 \cos \tau, \qquad x_2 = a_2 \cos(\tau + \delta), \qquad \tau = \omega t + \alpha_1, \qquad \delta = \alpha_2 - \alpha_1,$$
$$x_1 + x_2 = (a_1 + a_2 \cos \delta) \cos \tau - a_2 \sin \delta \sin \tau.$$

Comparison with eq. (2) in the exercise yields then

$$a_1 + a_2 \cos \delta = a \cos(\alpha - \alpha_1), \qquad a_2 \sin \delta = a \sin(\alpha - \alpha_1)$$

hence

(1) $$a^2 = (a_1 + a_2 \cos \delta)^2 + a_2^2 \sin^2 \delta = a_1^2 + a_2^2 + 2 a_1 a_2 \cos \delta,$$

(2) $$\tan(\alpha - \alpha_1) = \frac{a_2 \sin \delta}{a_1 + a_2 \cos \delta}.$$

It is much simpler, however, to proceed in the following manner: omitting the common factor $e^{i\omega t}$ one writes:

$$x_1 = a_1 e^{i\alpha_1}, \qquad x_2 = a_2 e^{i\alpha_2}, \qquad x = a e^{i\alpha},$$

where x_1, x_2 represent the vectors OP_1, OP_2 of lengths a_1, a_2 in the complex plane. Their sum is represented in fig. 97 by the diagonal OQ of the parallelogram formed from x_1 and x_2. By the theorem of Pythagoras the length a of the diagonal is given by

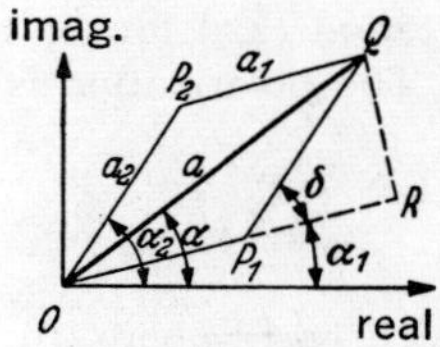

Fig. 97.

Superposition of two parallel oscillations of different phases in the complex plane.

(3) $$a^2 = a_1^2 + a_2^2 + 2 a_1 a_2 \cos \delta$$

as in eq. (1). The angle $\alpha - \alpha_1$ between OQ and OP is calculated from the right triangle OQR to be

(4) $$\tan(\alpha - \alpha_1) = \frac{a_2 \sin \delta}{a_1 + a_2 \cos \delta},$$

as in eq. (2).

For $a_2 = a_1$ and $\delta = \pi - \Delta$ it follows from (3) that

(5) $$a^2 = 2 a_1^2 (1 - \cos \Delta), \qquad a = 2 a_1 \sin \frac{\Delta}{2}.$$

We shall meet this formula again in an interference problem in Sec. 31.

I.2. Substituting in (2.1)

$$E_y = \eta, \qquad E_z = \zeta, \qquad A_y = A e^{-i\alpha}, \qquad A_z = B e^{-i\beta}$$

then in real notation;

$$\eta = A\cos(\tau - \alpha), \qquad \zeta = B\cos(\tau - \beta), \qquad \tau = k\,x - \omega\,t.$$

By eliminating $\sin\tau$ or $\cos\tau$ one finds

$$\cos\tau\sin(\beta - \alpha) = \frac{\eta}{A}\sin\beta - \frac{\zeta}{B}\sin\alpha,$$

$$\sin\tau\sin(\beta - \alpha) = -\frac{\eta}{A}\cos\beta + \frac{\zeta}{B}\cos\alpha$$

and by squaring and adding one obtains

$$(1) \qquad \left(\frac{\eta}{A}\right)^2 + \left(\frac{\zeta}{B}\right)^2 - 2\frac{\eta}{A}\frac{\zeta}{B}\cos\gamma = \sin^2\gamma, \qquad \gamma = \beta - \alpha.$$

This is the polar equation of an ellipse. The two principal axes are in general rotated with respect to the y- and z-axes; they coincide with the latter only when $\gamma = \pm\pi/2$.

If *in addition* $A = B$, then the ellipse becomes a circle, which corresponds to criterion (2.6) for circular polarization.

The polarization is linear when $\gamma = 0$ or π. For then (1) becomes

$$\left(\frac{\eta}{A} \mp \frac{\zeta}{B}\right)^2 = 0$$

which corresponds to criterion (2.6 a).

In view of (2.5) the same calculation yields for the magnetic vector $H_y = \eta$, $H_z = \zeta$

$$\eta = -B\sqrt{\frac{\varepsilon}{\mu}}\cos(\tau - \beta), \qquad \zeta = A\sqrt{\frac{\varepsilon}{\mu}}\cos(\tau - \alpha)$$

and hence in place of (1)

$$(2) \qquad \left(\frac{\eta}{B}\right)^2 + \left(\frac{\zeta}{A}\right)^2 + 2\frac{\eta}{B}\frac{\zeta}{A}\cos\gamma = \frac{\varepsilon}{\mu}\sin^2\gamma.$$

Thus the curve described by the magnetic vector becomes a circle or a straight line under the same conditions which hold for the electric vector.

I.3. The general proof of the non-existence of a surface charge rests upon the following considerations: from Maxwell's equations for non-conducting media it follows that $\operatorname{div}\mathbf{D} = \rho$ is independent of t. But since the field is assumed to be purely periodic in time, $\rho = f(x, y, z)$ is excluded and only $\rho = 0$ remains possible. The same statement holds for the surface divergence $\omega = D_n - D_{n'}$.

This can be formally verified for the case of Sec. 3 B in the following way: at $y = 0$ (see fig. 3 b)

$$E_y = \begin{cases} (A + C)\sin\alpha\, e^{i k_1 x \sin\alpha} & \text{in medium I.} \\ B\sin\beta\, e^{i k_2 x \sin\beta} & \text{in medium II.} \end{cases}$$

From this, the law of refraction, and the relation $\mathbf{D} = \varepsilon\,\mathbf{E}$ follows

$$D_y = \begin{Bmatrix} \varepsilon_1 (A + C) \sin\alpha \\ \varepsilon_2\, B \sin\beta \end{Bmatrix} e^{i k_1 x \sin\alpha}.$$

From eq. (3.14 a), the law of refraction and the definitions of m_{12} and n_{12} it is easily shown that these two values of D_y are equal.

I.4. From the law of refraction it follows that for small α in second approximation:

$$\beta\left(1 - \frac{\beta^2}{6} + \ldots\right) = \frac{\alpha}{n}\left(1 - \frac{\alpha^2}{6} + \ldots\right),$$

and hence, consistently neglecting higher powers of α,

$$\beta = \frac{\alpha}{n}\left(1 - \frac{\alpha^2}{6}\right)\left(1 + \frac{\alpha^2}{6\,n^2}\right) = \frac{\alpha}{n}\left(1 - \frac{n^2 - 1}{6\,n^2}\,\alpha^2\right),$$

(1) $$\alpha \pm \beta = \alpha\left\{1 \pm \frac{1}{n}\left(1 - \frac{n^2 - 1}{6\,n^2}\,\alpha^2\right)\right\}.$$

Thus one obtains to the same order of approximation

(2) $$\frac{\sin(\alpha - \beta)}{\sin(\alpha + \beta)} = \frac{n - 1 + \dfrac{\alpha^2}{6\,n^2}\left(n^2 - 1 - (n - 1)^3\right)}{n + 1 - \dfrac{\alpha^2}{6\,n^2}\left(n^2 - 1 + (n + 1)^3\right)}$$

$$= \frac{n - 1}{n + 1}\left(1 + \frac{\alpha^2}{6\,n^2}(2\,n + 4\,n)\right) = \frac{n - 1}{n + 1}\left(1 + \frac{\alpha^2}{n}\right),$$

which is given by (4.4) as the expression for R_p.

From (1) one obtains in the same approximation

(3) $$\frac{\cos(\alpha + \beta)}{\cos(\alpha - \beta)} = 1 - \frac{2\,\alpha^2}{n}$$

and the negative product of (2) and (3) is

$$R_s = -\frac{n - 1}{n + 1}\left(1 - \frac{\alpha^2}{n}\right).$$

as in (4.9).

I.5. In order to give a general proof (not only for the special case $\mu_2 = \mu_1$ considered in Sec. 4) we rely on eqs. (3.9) and (3.15). By (4.18) the equation to be proved becomes

$$\left|\frac{C}{A}\right|^2 + m\,\frac{\cos\beta}{\cos\alpha}\left|\frac{B}{A}\right|^2 = 1.$$

Dividing by $|B/A|^2$ it can be rewritten

$$\left|\frac{C}{B}\right|^2 + m\frac{\cos\beta}{\cos\alpha} = \left|\frac{A}{B}\right|^2.$$

According to (3.9) this equation becomes for *p-polarization*

$$\left(1 - m\frac{\cos\beta}{\cos\alpha}\right)^2 + 4\,m\frac{\cos\beta}{\cos\alpha} = \left(1 + m\frac{\cos\beta}{\cos\alpha}\right)^2$$

and according to (3.15) it becomes for *s-polarization*

$$\left(m - \frac{\cos\beta}{\cos\alpha}\right)^2 + 4\,m\frac{\cos\beta}{\cos\alpha} = \left(m + \frac{\cos\beta}{\cos\alpha}\right)^2.$$

Both equations are clearly identities.

I.6. By differentiating (5.11) with respect to α and setting the resulting differential quotient equal to zero one obtains [using the upper signs in the numerator and denominator of (5.11)]

$$(1)\qquad 0 = \frac{\cos(\alpha - i\beta')}{\sin(\alpha + i\beta')}\left(1 - i\frac{d\beta'}{d\alpha}\right) - \frac{\sin(\alpha - i\beta')}{\sin^2(\alpha + i\beta')}\cos(\alpha + i\beta')\left(1 + i\frac{d\beta'}{d\alpha}\right).$$

By differentiating the law of refraction $n\sin\alpha = \cos i\beta'$ one finds

$$\frac{d\beta'}{d\alpha} = \frac{in\cos\alpha}{\sin i\beta'}.$$

Substituting this in (1) one obtains

$$0 = \sin 2\,i\beta' + n\frac{\sin 2\,\alpha\cos\alpha}{\sin i\beta'}.$$

A second judicious application of the law of refraction yields

$$0 = 2\,n\frac{\sin\alpha}{\sin i\beta'}\{2 - (n^2 + 1)\sin^2\alpha\},$$

which contains the second formula (5.12) which was to be derived. Rewriting now (5.11) in the real notation and letting $\Delta = \gamma - \delta$ one writes

$$\frac{e^{i\Delta} - 1}{e^{i\Delta} + 1} = -\frac{\cos\alpha\sin i\beta'}{\sin\alpha\cos i\beta'}$$

or the identical expression

$$i\tan\frac{\Delta}{2} = -\cot\alpha\tan i\beta'.$$

From the above-derived value of $\sin \alpha_{max}$ and the law of refraction the two factors on the right-hand side are found to be equal to

$$\cot \alpha = \sqrt{\frac{n^2 - 1}{2}}, \qquad \tan i\beta' = \frac{i}{n}\sqrt{\frac{n^2 - 1}{2}}.$$

Thus also the first formula (5.12) has been proved.

I.7. If in the general trial solution of Sec. 7 A the coefficient A is set equal to zero, that is, if the continuous excitation supplied by the incident wave is omitted, then the solution of the problem of the plane parallel plate represents the free oscillations in the plate instead of the forced oscillations. The wave of amplitude C, which had previously been called the reflected wave, now represents, like the D-,wave the radiation emitted by the free oscillations into outer space. This radiation must clearly be present if the silver layer is not totally reflecting. Let the thickness of the air plate again be $2\,h$. The formulae of Sec. 7 are based upon the same p-polarization which is assumed in the exercise, but because of the given geometrical characteristics of the free oscillations (independence of x), one must now set $\alpha = \beta = \gamma = 0$. Furthermore, since all three media I, II and III are now air, $n = n_1 = 1$.

Therefore the four equations (7.30) and (7.31) simplify to

$$\text{(1)} \qquad -C\,e^{+ikh} - B\,e^{-ikh} + E\,e^{+ikh} = g\,C\,e^{+ikh} = g\,(B\,e^{-ikh} + E\,e^{+ikh}),$$

$$\text{(2)} \qquad -D\,e^{+ikh} - E\,e^{-ikh} + B\,e^{+ikh} = g\,D\,e^{+ikh} = g\,(E\,e^{-ikh} + B\,e^{+ikh}).$$

This way of writing the equations shows that the problem has become symmetrical in C and D as well as in B and E, which is due to the fact that the incident wave has been omitted. Therefore one can set (symmetrical type of solution)

$$\text{(3)} \qquad D = C, \qquad E = B$$

whereby (2) and (1) become identical so that only *one* double equation remains

$$\text{(4)} \qquad -C\,e^{ikh} + 2\,i\,B \sin k\,h = g\,C\,e^{ikh} = 2\,g\,B \cos k\,h.$$

By eliminating either B or C one obtains

$$\text{(5)} \qquad \tan k\,h = \frac{1 + g}{i}.$$

Alternately, one can set (antisymmetric type of solution)

$$\text{(3 a)} \qquad D = -C, \qquad E = -B.$$

Equations (1) and (2) again become identical except for the sign. In place of (4) and (5) one obtains

$$\text{(4 a)} \qquad -C\,e^{ikh} - 2\,B \cos k\,h = g\,C\,e^{ikh} = -2\,i\,g\,B \sin k\,h,$$

$$\text{(5 a)} \qquad \tan k\,h = \frac{1}{i\,(1 + g)}.$$

Setting

$$(6)\qquad \xi = k\,h, \qquad \tan k\,h = \frac{e^{i\xi} - e^{-i\xi}}{i\,(e^{i\xi} + e^{-i\xi})}, \qquad \beta = \frac{1}{1+g} \ll 1,$$

(5 a) yields

$$(7)\qquad e^{2i\xi} = \frac{1+\beta}{1-\beta} \sim 1 + 2\,\beta, \qquad \xi = \frac{1}{2\,i}\,(2\,\beta + 2\,m\pi\,i)$$

and (5) yields

$$(7\text{ a})\qquad e^{2i\xi} = -\frac{1+\beta}{1-\beta} \sim -(1 + 2\,\beta), \qquad \xi = \frac{1}{2\,i}\left(2\,\beta + (2\,m+1)\,\pi\,i\right).$$

From (7) and (7 a) follows, with ξ and β as defined in eq. (6)

$$(8)\qquad k\,h = m\,\pi - i/g,$$

$$(8\text{ a})\qquad k\,h = \left(m + \frac{1}{2}\right)\pi - i/g.$$

The symmetrical and antisymmetrical eigenvalues of $k\,h$ form an equidistant sequence with the spacing $\pi/2$ between neighboring values. This result is in complete agreement with the *forced* etalon-oscillations as represented in fig. 11.

According to (8) and (8 a) the damping constant $1/g$ is the same for all *free* oscillations. From this one concludes that the half-width is also the same for all *forced* oscillations as indicated in fig. 11. To prove this it is only necessary to compare the results for the simplest mechanical type of damped free oscillations and damped forced oscillations as derived in Vol. I, Sec. 19.

I.8. According to the general expression (3.1), the incident and reflected waves for arbitrary angle of incidence α are represented by

$$(1)\qquad \left.\begin{aligned} E_i &= A\; e^{ik(x\sin\alpha - y\cos\alpha)} \\ E_r &= C\; e^{ik(x\sin\alpha + y\cos\alpha)} \end{aligned}\right\} e^{-i\omega t}.$$

With *p-polarization* E is parallel to the z-axis and, because of the boundary condition at $y = 0$, $C = -A$ as in (8.3). Therefore

$$\operatorname{Re}(E_i + E_r) = 2\,A\cos(\omega\,t - k\,x\sin\alpha)\sin(k\,y\cos\alpha).$$

The locus of points of maximum electric field strength (maximum photographic effectiveness) is the system of parallel planes

$$(2)\qquad k\,y\cos\alpha = \left(m + \frac{1}{2}\right)\pi.$$

The spacing of these planes is larger than the spacing $\lambda/2$ for perpendicular incidence ($\alpha = 0$). In particular, for $\alpha = \pi/4$ it becomes equal to $\lambda/\sqrt{2}$.

With *s-polarization* H is in the z direction, and E has components in the x and y directions. From the boundary condition

$$E_{xi} + E_{xr} = 0 \qquad \text{for} \qquad y = 0$$

it follows that $C = -A$ (see fig. 3 b) and according to (1) for $y > 0$:

$$\text{Re}\,(E_i + E_r)_y = 2A \sin\alpha \sin(\omega t - k x \sin\alpha) \sin(k y \cos\alpha),$$
$$\text{Re}\,(E_i + E_r)_x = 2A \cos\alpha \cos(\omega t - k x \sin\alpha) \cos(k y \cos\alpha).$$

The time average of the sum of the squares of these components gives

$$J = 2A^2 \{\sin^2\alpha \sin^2(k y \cos\alpha) + \cos^2\alpha \cos^2(k y \cos\alpha)\}$$
$$= 2A^2 \{\cos^2\alpha - \cos 2\alpha \sin^2(k y \cos\alpha)\}.$$

Thus, for the angle of incidence $\alpha = \pi/4$ which was used by Wiener one has $J = A^2$; no fringe system results and the illumination is uniform. For other angles of incidence weak fringes appear superimposed on uniform brightness.

III.1. If the central force divided by the distance between the two mass points is denoted by f, then the equations of motion, written in terms of the cartesian coordinates x, y and x_1, y_1, are

$$M_1 \ddot{x}_1 = f(x_2 - x_1), \qquad M_1 \ddot{y}_1 = f(y_2 - y_1), \tag{1}$$
$$M_2 \ddot{x}_2 = f(x_1 - x_2), \qquad M_2 \ddot{y}_2 = f(y_1 - y_2).$$

Addition of the equations in each column yields the equations of motion of the center of mass; subtraction yields the equations of the relative motion of the two masses:

$$\ddot{\xi} = -\left(\frac{1}{M_1} + \frac{1}{M_2}\right) f\,\xi, \qquad \ddot{\eta} = -\left(\frac{1}{M_1} + \frac{1}{M_2}\right) f\,\eta, \qquad \begin{cases} \xi = x_1 - x_2, \\ \eta = y_1 - y_2. \end{cases} \tag{2}$$

If the definition of M is that given in the exercise, these equations describe the motion of a mass point M with coordinates ξ, η. If the binding force is quasi-elastic as we had generally assumed it to be in our dispersion calculations, then $f =$ const. and from (2) follows

$$\ddot{\xi} + \omega_0^2\,\xi = 0, \qquad \ddot{\eta} + \omega_0^2\,\eta = 0, \qquad \omega_0^2 = f/M. \tag{3}$$

Thus the motion is simply periodic with the frequency ω_0. The same is true in the case of Coulomb attraction (f proportional to r^{-3}) but not for arbitrary central forces.

III.2. Applied to the front and rear surfaces of the prism, the law of refraction requires that

$$\frac{\sin\alpha}{\sin\beta} = n, \qquad \frac{\sin\alpha'}{\sin\beta'} = n' \tag{1}$$

where $n' = 1/n$ if both surfaces border on air. From the sum of the angles in the triangle formed by the prism faces and the interior ray follows

$$\varphi = \beta + \alpha'. \tag{2}$$

At the front face the incident ray is deflected by $\delta_1 = \alpha - \beta$; the emerging ray is deflected by $\delta_2 = \beta' - \alpha'$ at the rear surface. The total deflection is therefore

$$\delta = \delta_1 + \delta_2 = \alpha - \beta + \beta' - \alpha'$$

and because of (2)

(3) $$\delta = \alpha + \beta' - \varphi.$$

Substitution in (1) yields

(4) $$\frac{\sin \alpha}{\sin (\varphi - \alpha')} = n, \qquad \frac{\sin \alpha'}{\sin (\delta + \varphi - \alpha)} = \frac{1}{n}.$$

Therefore, by eliminating α', δ can be represented as a function of α.

Differentiation of (4) with respect to α (before eliminating α') gives the following conditions for the minimum deflection $d\delta = 0$:

$$\cos \alpha \, d\alpha + n \cos (\varphi - \alpha') \, d\alpha' = 0,$$

$$n \cos \alpha' \, d\alpha' + \cos (\delta + \varphi - \alpha) \, d\alpha = 0.$$

These can be satisfied only if

(5) $$\begin{vmatrix} \cos \alpha & \cos (\varphi - \alpha') \\ \cos (\delta + \varphi - \alpha) & \cos \alpha' \end{vmatrix} = 0.$$

Equating the terms in the first and second column gives $\alpha = (\delta + \varphi)/2$, $\alpha' = \varphi/2$ and applying (2) and (3) yields $\beta = \alpha'$, $\beta' = \alpha$. Hence, if the ray is symmetrical with respect to the bisector plane of the prism angle, one obtains by substitution in one of the eqs. (1) an equation which is much used in the determination of n:

(6) $$n = \sin \frac{1}{2} (\delta + \varphi) \Big/ \sin \frac{1}{2} \varphi \, .$$

III.3. For small angles α, φ, α' it follows from the two eqs. (4) of the preceding solution by eliminating $\alpha + n \alpha'$ that

$$\delta = (n - 1) \varphi.$$

In order to be able to apply this result directly to the twin prisms, it is convenient to imagine prisms 1 and 2 as separated by a narrow air space. Thus, taking into account the opposite positions of the two prism edges, one obtains for the total deflection

(1) $$\delta = \delta_1 - \delta_2, \qquad \delta_1 = (n_1 - 1) \varphi_1, \qquad \delta_2 = (n_2 - 1) \varphi_2.$$

a) For a direct vision prism it is to be required that

(2) $$\delta = 0, \qquad \text{i. e.} \qquad (n_1 - 1) \varphi_1 - (n_2 - 1) \varphi_2 = 0, \qquad \frac{\varphi_2}{\varphi_1} = \frac{n_1 - 1}{n_2 - 1}.$$

Since n_1 and n_2 depend on the wavelength, this condition can only be satisfied for some average wavelength such as $\lambda = 0.590 \, \mu$.

b) For an achromatic prism one requires that

(3) $$\frac{d\delta}{d\lambda} = 0, \qquad \frac{dn_1}{d\lambda} \varphi_1 - \frac{dn_2}{d\lambda} \varphi_2 = 0, \qquad \frac{\varphi_2}{\varphi_1} = \frac{dn_1/d\lambda}{dn_2/d\lambda}.$$

Also this condition may in particular be satisfied for $\lambda = 0.590\,\mu$. The table below lists the refractive indices n_1 of light boron crown glass and n_2 of heavy flint glass for various wavelengths. For $\lambda = 0.590\,\mu$ eqs. (2) and (3) yield

$$\text{(4 a)} \quad \frac{\varphi_2}{\varphi_1} = \frac{0.5103}{0.7562}, \qquad \text{(4 b)} \quad \frac{\varphi_2}{\varphi_1} = \frac{4.18}{13.84}.$$

In the case a) the value of δ is small everywhere in the spectrum but it does depend strongly on color. In the case b) the (also for $\lambda = 0.590\,\mu$) non-vanishing deflection is quite independent of color (only at the violet end of the spectrum does the deflection decrease slightly). Once the angle φ_1 is arbitrarily chosen (though it must be small), the angle φ_2 is determined by (4 a, b).

Dispersion of crown glass (n_1) and of flint glass (n_2)

λ	n_1	n_2
0.761	1.5050	1.7390
0.656	1.5076	1.7473
0.590	1.5103	1.7562
0.486	1.5156	1.7792
0.397	1.5245	1.8403

A much more important problem is that of *achromatic lenses*. For these a condition similar to (3) must be satisfied.

III.4. In case a) the inertial force tending to deflect the electron from its orbit must be balanced by the force $-\partial V/\partial r$ of the atomic field and by the Lorentz force $\mathbf{K} = e\,\mathbf{v} \times \mathbf{B}$. We do not need to go into the shape of the orbit or the velocity variations along it. In case b) the ordinary centrifugal force

$$|\mathbf{Z}| = m\,\rho\,\omega^2 \qquad (\rho \text{ distance from the axis of rotation})$$

and the Coriolis force

$$\mathbf{C} = 2\,m\,\mathbf{v} \times \vec{\omega} \qquad (\mathbf{v} \text{ velocity relative to the rotating system})$$

take the place of $\mathbf{K}$ (see Vol. I, Sec. 29), while the force due to the atom $-\partial V/\partial r$ is the same as in a). It is to be assumed (see wording of problem) that $\mathbf{Z}$ is negligible compared to $\mathbf{C}$. Then one obtains equilibrium in case b) by setting $\mathbf{C} = \mathbf{K}$. This yields

$$2\,m\,\mathbf{v} \times \vec{\omega} = e\,\mathbf{v} \times \mathbf{B}; \qquad \omega = \frac{1}{2}\frac{e}{m}B$$

which proves *Larmor's theorem*.

In order to be able to neglect $\mathbf{Z}$ we must have

$$m\,\rho\,\omega^2 \ll 2\,m\,|\mathbf{v}|\,\omega.$$

This is equivalent to saying that the velocity $\rho\,\omega$ imparted to the electron when the magnetic field is switched on is small compared to the velocity $|\mathbf{v}|$ which the electron would have without the magnetic field. For practically attainable fields $\mathbf{B}$ this condition is always fulfilled.

We conclude therefore that the theory of the Zeeman effect developed in Sec. 21 remains valid when the quasi-elastic binding, there assumed, is replaced by a Coulomb field (hydrogen atom) or by an arbitrary atomic field $V(r)$. In particular the theorems on the normal Zeeman effect [$\Delta\omega = 0$ for longitudinal observation, $2\Delta\omega = \pm (e/m) B$ for transverse observation] are preserved because of the general validity of Larmor's theorem.

IV.1. The index ellipsoid [normalized in accordance with (25.12)]

(1) $$u_1^2 x_1^2 + u_2^2 x_2^2 + u_3^2 x_3^2 = C, \qquad C = 2 W_e/\mu_0$$

and the plane E perpendicular to the wave number vector $\mathbf{k}$

(2) $$k_1 x_1 + k_2 x_2 + k_3 x_3 = 0$$

intersect in an ellipse. As in (25.19) we denote the reciprocals of the principal axes of this ellipse by $u'/\sqrt{C}, u''/\sqrt{C}$ without, however, presupposing the former definitions of u', u'', as wave velocities to hold. We construct a third axis perpendicular to these two and call its length from the origin to the ellipsoid $OP = l$. The coordinates of P are $x_i = l\, k_i/k$. Substituting these in (1) one obtains

(3) $$\frac{C}{l^2} = \frac{1}{k^2} \sum u_i^2 k_i^2.$$

Then theorem a) gives the following relationship between the three axis lengths $\sqrt{C}/u'$, $\sqrt{C}/u''$, l:

$$\frac{1}{C}(u'^2 + u''^2) + \frac{1}{l^2} = \frac{1}{C}(u_1^2 + u_2^2 + u_3^2),$$

and hence by (3)

(4) $$u'^2 + u''^2 = \Sigma u_i^2 \left(1 - \frac{k_i^2}{k^2}\right).$$

In order to be able to apply theorem b) we must construct the plane E' which is tangent to the ellipsoid (1) and parallel to E. The equation of any arbitrary tangent plane with the point of tangency $\xi_1\, \xi_2\, \xi_3$ is

(5) $$\sum u_i^2 \xi_i (x_i - \xi_i) = 0.$$

If this plane is to be perpendicular to $\mathbf{K}$, then

(5 a) $$u_i^2 \xi_i = \rho k_i, \qquad (\rho = \text{constant of proportionality}).$$

Since, in particular, the point ξ must lie on the ellipsoid (1):

(6) $$\rho^2 \sum \frac{k_i^2}{u_i^2} = C.$$

Because of (5 a) and (1), eq. (5) becomes

(7) $$\rho \sum k_i x_i - C = 0.$$

By the rules of analytic geometry the distance of this plane E' from the center of the ellipsoid is

$$(8) \qquad p = \frac{C}{\rho k}, \qquad \text{and hence by (6)} \quad p = \frac{1}{k}\sqrt{C \Sigma k_i^2/u_i^2}.$$

E' and its diametrically opposite parallel plane E'', together with the planes which are tangent to the ellipsoid at the end points of the principal axes of the intersectional ellipse, form a circumscribed parallelepiped of the ellipsoid. Its volume is

$$2p \cdot \frac{2\sqrt{C}}{u'} \cdot \frac{2\sqrt{C}}{u''} = \frac{8}{k}\frac{C^{3/2}}{u'u''}\sqrt{C \Sigma k_i^2/u_i^2}.$$

According to theorem b) this volume equals the volume of the rectangular parallelepiped formed by the three principal axes of lengths $\sqrt{C}/u_i$ of the ellipsoid; that is, it is equal to $8\,C^{3/2}\,u_1\,u_2\,u_3$. It follows therefore that

$$(9) \qquad u'^2 u''^2 = \frac{u_1^2 u_2^2 u_3^2}{k^2} \sum \frac{k_i^2}{u_i^2}.$$

The two symmetric functions $u'^2 + u''^2$, eq. (4) and $u'^2 u''^2$, eq. (9) yield a quadratic equation in u^2, the roots of which are u'^2 and u''^2. It is easy to show that this equation agrees with the eq. (26.19 b) of the *normal surface* when in the latter the expressions (26.19) for the ξ_i are substituted.

IV.2. The equation (24.6 a) of the Fresnel ellipsoid is written in a form analogous to eq. (1) of the preceding problem

$$(1) \qquad \frac{x_1^2}{u_1^2} + \frac{x_2^2}{u_2^2} + \frac{x_3^2}{u_3^2} = C, \qquad C = 2\mu_0 W_e,$$

where the x_i now denote the components of $\mathbf{E}$. Since $\mathbf{E}$ is perpendicular to the ray vector $\mathbf{S}$, hence also to the parallel unit vector $\mathbf{s} = s_1, s_2, s_3$, one must now cut the ellipsoid with the plane

$$(2) \qquad s_1 x_1 + s_2 x_2 + s_3 x_3 = 0$$

and calculate the principal axes of the intersectional ellipse so formed. Except for the changed form of the subsidiary conditions (1) and (2), the extremum problem to be solved here is the same as that in Sec. 25.

Let the principal axis lengths of the intersectional ellipse be $\sqrt{C}\,v'$, $\sqrt{C}\,v''$: let l be the length from the origin to the ellipsoid of the axis perpendicular to these principal axes. Since its end point has the coordinates $\xi_i = l\,s_i$, eq. (1) yields

$$(3) \qquad \frac{C}{l^2} = \sum \frac{s_i^2}{u_i^2}.$$

Theorem a) gives the following relationship for the three axis lengths $\sqrt{C}\,v'$, $\sqrt{C}\,v''$, C:

$$\frac{1}{C}\left(\frac{1}{v'^2} + \frac{1}{v''^2}\right) + \frac{1}{l^2} = \frac{1}{C}\sum \frac{1}{u_i^2}$$

and hence by (3)

(4) $$\frac{1}{v'^2} + \frac{1}{v''^2} = \sum \frac{1 - s_i^2}{u_i^2}.$$

Theorem b) concerns the tangential planes E', E'' of the Fresnel ellipsoid which are parallel to our present intersectional ellipse. The equation of one of these planes is

(5) $$\sum \frac{\xi_i}{u_i^2}(x_i - \xi_i) = 0$$

or, see the preceding problem,

(6) $$\rho \sum s_i x_i - C = 0, \quad \text{where} \quad (7) \quad \rho^2 \sum u_i^2 s_i^2 = C.$$

It follows that the distance of the plane from the origin is

(8) $$p = \frac{C}{\rho} = \sqrt{C \Sigma s_i^2 u_i^2}$$

and the volume of the parallelepiped to be considered here is

$$2p \cdot 2\sqrt{C}\, v' \cdot 2\sqrt{C}\, v'' = 8\, C^{3/2} v' v'' \sqrt{\Sigma s_i^2 u_i^2}.$$

From theorem b) one obtains therefore

(9) $$\frac{1}{v'^2\, v''^2} = \frac{\Sigma s_i^2 u_i^2}{u_1^2 u_2^2 u_3^2}.$$

From (4) and (9) it follows that v'^2, v''^2 are the roots of a quadratic equation in v^2 which we may write

$$0 = \left(\frac{1}{v^2} - \frac{1}{v'^2}\right)\left(\frac{1}{v^2} - \frac{1}{v''^2}\right) = \frac{1}{v^4} - \frac{1}{v^2} \sum \frac{1 - s_i^2}{u_i^2} + \frac{1}{u_1^2 u_2^2 u_3^2} \sum s_i^2 u_i^2.$$

It is easy to verify that this equation is identical with the equation (26.13 b) of the *ray surface*

IV.3. From fig. 47 one finds that

(1) $$A C = d/\cos\beta_2, \qquad A B = d/\cos\beta_1, \qquad B D = B C \sin\alpha,$$

(1 a) $$B C = E C - E B = (\tan\beta_2 - \tan\beta_1)\, d$$

while from the law of refraction it follows that

(2) $$\sin\alpha = \sin\beta_1 \frac{k_1}{k} = \sin\beta_2 \frac{k_2}{k}.$$

Hence

$$\sin\alpha \tan\beta_1 = \frac{\sin^2\beta_1}{\cos\beta_1}\frac{k_1}{k}, \qquad \sin\alpha \tan\beta_2 = \frac{\sin^2\beta_2}{\cos\beta_2}\frac{k_2}{k},$$

where the wave number k refers to the surrounding air, and k_1 refers to the more strongly and k_2 to the less strongly refracted ray. Therefore by (1) and (1 a)

$$B\,D = \frac{d}{k}\left(\frac{\sin^2\beta_2}{\cos\beta_2}k_2 - \frac{\sin^2\beta_1}{\cos\beta_1}k_1\right).$$

The total phase difference Δ is found to be

$$\begin{aligned}\Delta &= k_2\,A\,C - k_1\,A\,B - k\,B\,D\\ &= \left(\frac{k_2}{\cos\beta_2} - \frac{k_1}{\cos\beta_1} - \frac{k_2\sin^2\beta_2}{\cos\beta_2} + \frac{k_1\sin^2\beta_1}{\cos\beta_1}\right)d \qquad (3)\\ &= (k_2\cos\beta_2 - k_1\cos\beta_1)\,d.\end{aligned}$$

This result is to be specialized to small angles of incidence α, hence also to small β_1, β_2. A mean angle of refraction defined by

$$\sin\beta = \sqrt{\sin\beta_1\sin\beta_2}$$

is to be introduced. By multiplying the two laws of refraction (2) we obtain

$$\sin^2\alpha = \sin^2\beta\,\frac{k_1\,k_2}{k^2}$$

and

$$\frac{1}{\cos\beta} = \left(1 - \frac{k^2}{k_1\,k_2}\sin^2\alpha\right)^{-1/2} = 1 + \frac{1}{2}\frac{k^2}{k_1\,k_2}\sin^2\alpha + \dots \qquad (4)$$

On the other hand, we have for $i = 1, 2$

$$\cos\beta_i = 1 - \frac{1}{2}\sin^2\beta_i + \dots = 1 - \frac{1}{2}\sin^2\alpha\,\frac{k^2}{k_i^2} + \dots$$

$$\begin{aligned}k_2\cos\beta_2 - k_1\cos\beta_1 &= k_2 - k_1 - \frac{k^2}{2}\sin^2\alpha\left(\frac{1}{k_2} - \frac{1}{k_1}\right) + \dots\\ &= (k_2 - k_1)\left\{1 + \frac{1}{2}\frac{k^2}{k_1\,k_2}\sin^2\alpha + \dots\right\}.\end{aligned}$$

Substituting this in (3) and applying (4) one obtains

$$\Delta = \frac{(k_2 - k_1)\,d}{\cos\beta}, \qquad (5)$$

which agrees with eq. (31.9).

Index

C

D

S